P9-AQL-190

The General Topology of Dynamical Systems

Graduate Studies in Mathematics

Volume 1

The General Topology of Dynamical Systems

Ethan Akin

American Mathematical Society

1991 *Mathematics Subject Classification*. Primary 58Fxx;
Secondary 34Cxx, 34Dxx.

ABSTRACT. Recent work in smooth dynamical systems theory has highlighted certain topics from topological dynamics. This book organizes these ideas to provide the topological foundations for dynamical systems theory in general. The central theme is the importance of chain recurrence. The theory of attractors and different notions of recurrence and transitivity arise naturally as do various Lyapunov function constructions. The results are applied to the study of invariant measures and topological hyperbolicity.

Library of Congress Cataloging-in-Publication Data

Akin, Ethan, 1946–
The general topology of dynamical systems/Ethan Akin.
p. cm.—(Graduate studies in mathematics, ISSN 1065-7339; v. 1.)
Includes bibliographical references and index.
ISBN 0-8218-3800-8 hardcover (alk. paper)
1. Differentiable dynamical systems. 2. Topological dynamics. I. Title. II. Series.
QA614.8.A39 1993 92-41669
515′.35—dc20 CIP

Printed in the United States of America.
The paper used in this book is acid-free and falls within the guidelines
established to ensure permanence and durability. ♾

This publication was typeset using AMS-TEX,
the American Mathematical Society's TEX macro system.

10 9 8 7 6 5 4 3 2 1 98 97 96 95 94 93

For Paul Akin
May 30, 1908 – May 26, 1992

Contents

Preface

A large branch of modern dynamical systems theory has grown out of the work of Smale and his colleagues. The germinal technical concept, hyperbolicity, extended from a fixed point to more general invariant sets, consists of conditions imposed upon the tangent maps of the system. However, development of the subject revealed the fruitfulness of a number of purely topological concepts such as attractor, basic set, filtration, and chain recurrence. While some of these ideas were new, many were familiar objects of study in topological dynamics. The latter was a well established subject, flourishing, and somewhat separated from the differentiable theory. However, perusal of surveys like Bhatia and Szegö (1970) and Nemytskii and Stepanov (1960) reveals that topological dynamics drew much of its motivation, as well as many of its examples, from the still older qualitative theory of differential equations originating with Poincaré and exemplified in Andronov, Vitt, and Khaikin's great book (1937).

The recent global results associated with hyperbolicity have provided a new perspective on topological dynamics. For me this new view began with a look at Shub and Smale's 1972 paper, *Beyond hyperbolicity.* This book is the result of an often interrupted contemplation of the best way to organize the parts of topological dynamics which are most useful for the nonspecialist. John Kelley wrote in the preface to his justly famous book, *General topology*, that he was, with difficulty, prevented by his friends from using the title "What every young analyst should know". The reader will note that I have adapted his title. This is partly gratitude (and an attempt at sympathetic magic), but mostly because my intent is inspired by his. I hope to have described what every dynamicist should know, or at least be acquainted with, from topological dynamics.

While the book is thus intended as a service text and reference, its subject eventually organized itself into a unified story whose central theme is the role of chain recurrence in the study of dynamical systems on compact metric spaces. The assumption of metrizability is, for most of the results, just a convenience, but compactness is essential. We repeatedly use the preservation of compactness by continuous maps. Even more often we need the observation that for a decreasing sequence of nonempty compact sets, $\{A_n\}$, the intersection, A, is nonempty and if U is any neighborhood of A, then $A_n \subset U$ for sufficiently large n. On the other hand, we study the iterations not just of continuous maps but of more general closed relations on the space. At first glance, this appears to be one of those tedious and mechanical generalizations more honored in the omission than in the transcribing. Instead, even the homeomorphisms which are our primary interest are best studied by thickening them up to relations in various ways (Joseph Auslander's prolongations). Also, the relation results can be used to partly mitigate the unfortunate demand for compactness. Given a homeomorphism of a locally compact space we can restrict to a large compact subset, A. Of course, if A is an invariant set then the restriction is still a homeomorphism. But even if A is not even positive invariant, the restriction is a closed relation on A, though not a mapping. Furthermore, the relation results yield constructions on A, e.g., Lyapunov functions, more powerful than would be obtained by a further restriction to the largest invariant subset of A.

As for prerequisites, except for the measure theory in Chapter 8 and occasional forays into differentiable territory, what is needed is fluency in the topology of metric spaces. However, a reader whose background includes a modern treatment of differential equations like Hirsch and Smale (1974) or Arnold (1973) will have a better understanding of why we take up the topics that we do.

In Chapters 1–3 we develop the fundamentals of the dynamics of a closed relation. We introduce and apply various kinds of recurrence and invariant sets, the theory of attractors, and the construction of Lyapunov functions. With Chapters 4 and 5 we return to mappings to discuss topological transitivity, minimal subsets, decompositions and constructions converging upon the chain recurrent set. In Chapter 6 we derive the related results for flows and obtain special results for Lyapunov functions and chain recurrence in the vector field case. Chapter 7 concerns perturbation theory. Since our perturbations are topological rather than differentiable, the structural stability results associated with hyperbolicity do not apply, but we describe Takens' results on Zeeman's "tolerance stability conjecture". In Chapter 8 we describe invariant measures and compare topological notions of ergodicity and mixing with the measure theoretic versions. In Chapter 9 we apply the results to some important examples, e.g., shift maps on spaces of symbols and flows on the torus. Finally, in Chapters 10 and 11 we describe the hyperbolicity results for fixed points and for Axiom A homeomorphisms, respectively. This latter is the topological generalization of Smale's differential idea.

The results from the exercises in the text are used as lemmas and so should at least be read. The straightforward proofs are better performed by the reader (guided by the hints) or omitted entirely than laid out in detail on the printed page.

Chapter 0. Introduction: Gradient Systems

In a first course in differential equations you learn methods for solving the associated initial value problem. Given an initial position you follow the solution path forward or backward in time by using an explicit analytic formula. But these solution methods apply only in very special cases. The geometric study of a differential equation attempts instead to visualize the behavior of the entire system at once. The first picture which allows this heroic perspective is the gradient picture. Once glimpsed this description reappears in applications to all of the sciences and whenever it applies it displays its power by organizing the system with a single vision. When compelled by the complexities of more sophisticated systems to move beyond it, you leave with regret and never abandon it completely. Always you look first for those aspects of the system where the intuitions of that initial vision still endure.

Imagine a flat plane each point of which represents a state of the system. Over this plane appears a landscape, the graph of a smooth real-valued function. The associated gradient dynamic represents a tendency to move upward on this landscape in the direction of steepest ascent. Rest points, equilibria, occur at critical points, i.e., local maxima, minima or saddle points. In the usual case, illustrated in Figure 0.1, the critical points are isolated and each solution path approaches an equilibrium asymptotically as time tends to infinity. Figure 0.1A shows the motion on the landscape while Figure 0.1B shows its image on the state plane below, the *phase portrait* of the system. The local maxima, M_1 and M_2, are attracting equilibria (locally asymptotically stable). Each is contained in an open set called its *domain of attraction* consisting—as the name suggests—of points tending toward it in the limit. In applications these are the observable equilibria. The relationship between any real system and its dynamic model is inevitably noisy, that is, there are real effects ignored by the model. To assert that the model represents reality accurately is to presume that these disturbances are relatively small. Nonetheless, they suffice to perturb a point off a saddle point like S or the trajectories tending to S and into one of the neighboring domains of attraction. In contrast the dynamic tends to restore an equilibrium like M_1 or M_2 against such perturbations. Despite this phantom existence, the unstable equilibria are important because the—usually lower dimensional—set

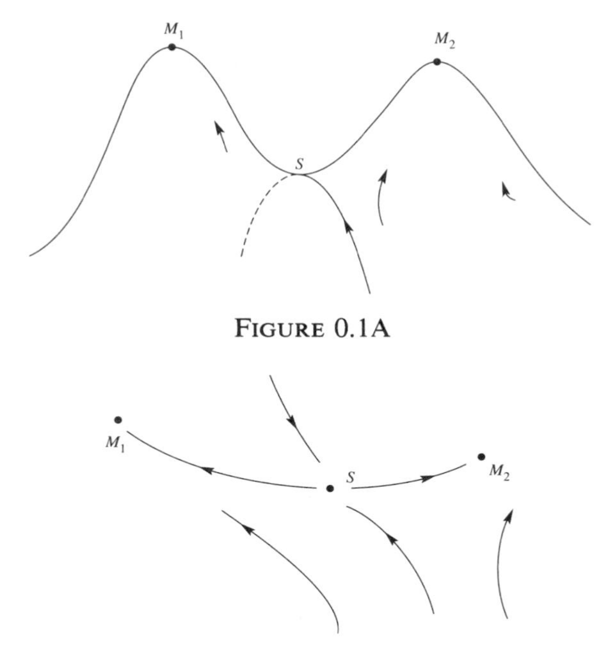

FIGURE 0.1A

FIGURE 0.1B

of trajectories tending toward them form the separatrices between domains of attraction, i.e., removing them leaves an open subset which is the union of the disjoint domains of attraction. Thus, the important points to locate and understand are the recurrent points, in this case the equilibria, while the remaining points, the transients, are best described by relating their destiny to the recurrents.

The same general analysis applies even if the system is not the gradient, provided that the height function is a strict Lyapunov function for the system. This means that, except at equilibrium, the motion is still always ascending. Instead of the steepest direction of ascent of a mountain goat, imagine the spiraling upward of a mountain road. Furthermore, even in the complex systems which follow we will retain the crucial distinction between transient and recurrent points and the associated emphasis on recurrence. In fact, our central theme will be to elaborate the meaning and consequences of different notions of recurrence.

The simplest recurrence phenomenon beyond equilibrium is periodicity and with it we leave the gradient picture behind. In the phase portrait of Figure 0.2, we illustrate a limit cycle. If the system has a periodic nonequilibrium solution then it cannot admit a strict Lyapunov function and so cannot be a gradient system. You cannot ascend continuously and yet return to your initial position if the height function is single-valued. Notice that in the Figure 0.2 example, angle θ is constantly increasing but θ is not a

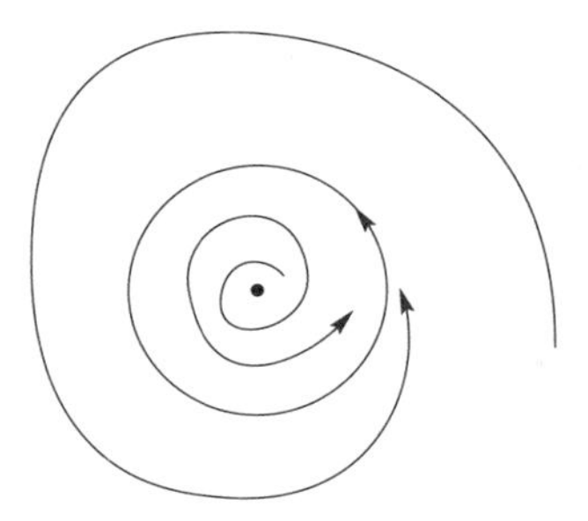

FIGURE 0.2. $d\theta/dt = 1$, $dr/dt = r(1-r)$.

single-valued function. In this example the periodic orbit and the central equilibrium are the recurrent points. The periodic orbit is attracting with domain of attraction the punctured plane, while the center is a repellor, i.e., an attracting equilibrium if time is run backward, with domain the interior of the unit circle.

Periodicity is the simplest example of nonequilibrium recurrence. Furthermore, the limit cycle example of Figure 0.2 is important because it is, to use an important but vague term, robust. A property of a dynamical system is robust if it is preserved under small perturbations, that is, if the property still holds for dynamical systems sufficiently close to the original. In any application the vagueness is eliminated by specifying the meaning of "sufficiently close", i.e., topologizing the dynamical systems, and by describing precisely in what way the property "still holds". For the example in Figure 0.2, given $\varepsilon > 0$ there exists $\delta > 0$ such that if a vector field δ small in the C^1 sense (small at every point and with continuous small derivatives at every point) is added to the original, then the new vector field has a unique limit cycle ε close to the original one.

The dynamical systems which we will analyze come in two closely related varieties: continuous vs. discrete time models, or, differential vs. difference equations, or again, flows vs. maps. Given a smooth vectorfield on a compact manifold X we can integrate it to obtain the corresponding flow $\varphi: X \times \mathbb{R} \to X$ where $\varphi(x, t)$ for fixed x is the solution path with initial condition x. In general, a continuous map $\varphi: X \times [0, \infty) \to X$ is called a *semiflow* if $f^0(x) = x$ and $f^t \circ f^s = f^{t+s}$, where $f^t: X \to X$ is the map defined by $f^t(x) = \varphi(x, t)$. If the maps f^t are homeomorphisms, then we can define $f^{-t} = (f^t)^{-1}$ and extend φ to the associated flow, mapping $X \times \mathbb{R}$ to X. The particular map f^1 is called the *time-one map* of the semiflow.

1. EXERCISE. *Prove: If f^1 is invertible, then f^t is invertible for all t. (Hint: For $0 < t < 1$; $f^1 = f^t \circ f^{1-t} = f^{1-t} \circ f^t$.) Prove continuity on $X \times \mathbb{R}$ of the associated flow. (Hint: From $X \times [-1, 1] \to X$ compose with the homeomorphism f). If X is a compact C^r $(1 \le r \le \infty)$ closed*

manifold and $\varphi: X \times [0, \infty) \to X$ *is a* C^r *semiflow, prove that* f^1 *is a* C^r *diffeomorphism and the associated flow is* C^r *if the tangent map* $T_x f^1$ *is a linear isomorphism at each point* x *of* X. (*Use the Implicit Function Theorem and the fact that* f^1 *is homotopic to the identity*). □

A discrete time dynamical system is defined by a continuous map $f: X \to X$ and letting $x_{t+1} = f(x_t)$ for $t = 0, 1, 2, \ldots$. The discrete time analogue of the semiflow is defined by $f^n(x)$ where f^n is the n-fold composition of f ($n = 0, 1, \ldots$, where f^0 is defined to be the identity). Again, if f is invertible, we can define $f^{-n} = (f^{-1})^n$ and get the discrete time analogue of the flow.

This emphasis on flows ignores the original vector field. In practical applications, the vector field provides some extra leverage so that the continuous time model of a phenomenon is usually easier to use than the discrete time version. But the theoretical apparatus which we develop here is best constructed for the dynamics of a map. We then translate the results back to vector fields by using the time-one map of the flow. On the other hand, a homeomorphism can always be embedded in a flow via the following "suspension construction".

2. EXERCISE. *Let* $f: X \to X$ *be a homeomorphism of a compact metric space. Construct the compact metric space* Y *from* $X \times [0, 1]$ *by identifying the pairs of points* $\{(x, 1), (f(x), 0)\}$ *for all* x *in* X. *Define a flow on* Y *so that the time-one map is the quotient of the function* $(x, t) \to (f(x), t)$. *If* f *is a* C^r, *orientation preserving diffeomorphism of a* C^r *compact manifold* X, *prove that* Y *is naturally a* C^r *orientable manifold and that the associated flow is* C^r. □

N.B. Unless otherwise mentioned, all spaces throughout the book are assumed to be compact metric spaces.

Chapter 1. Closed Relations and Their Dynamic Extensions

What does it mean to view a map $f: X \to X$ as a dynamical system? Think of a dynamical system as imposing a temporal ordering on the space. Given a point $x \in X$ we regard $f(x)$ as the next point, the point which immediately follows x in time. More generally, y follows x if $y = f^n(x)$ for some $n = 1, 2, \ldots$. If we identify f and its iterates with their graphs in $X \times X$, this amounts to including f in a larger relation

$$\mathscr{O}f = \bigcup_{n=1}^{\infty} f^n, \text{ i.e., } (x, y) \in \mathscr{O}f \Leftrightarrow y = f^n(x) \text{ for some } n = 1, 2, \ldots.$$

Notice that $\mathscr{O}f$ is a transitive relation. If $y = f^n(x)$ and $z = f^m(y)$, then $z = f^{n+m}(x)$. However, $\mathscr{O}f$ is rarely a closed subset of $X \times X$. In particular, the set of followers of x, $\mathscr{O}f(x) = \{f(x), f^2(x), \ldots\}$, is rarely closed. We would like to regard the limit points of this positive orbit of x as future related to x as well.

The desire to get closure conditions as well as transitivity leads us to include $\mathscr{O}f$ in still larger relations $F \subset X \times X$. Each extension leads to a larger set $F(x) = \{y: (x, y) \in F\}$ of future states associated with x.

Each extension has an associated concept of recurrence. With respect to F, x is recurrent if it lies in its own future: $x \in F(x)$ or $(x, x) \in F$, and so x returns to itself in the F sense. Thus, $x = f(x)$ means x is a rest point or equilibrium, while $x \in \mathscr{O}f(x)$ means x is a periodic point: $f^n(x) = x$ for some positive n. This is the discrete time analogue of the periodic orbits discussed in the introduction.

The original map f gets a bit lost in the tower of relations we construct. In fact, most of the early results remain unchanged if we begin with f itself just a relation on X. Because we will want to define concepts such as invariant set, recurrent point, etc., for the extensions as well as for f, it is useful to introduce these ideas and constructions for such a general relation. The special case where f is a map does have associated special properties and we will gradually shift our emphasis back to this case after laying a general foundation.

A relation $f: X_1 \to X_2$ can be thought of as a map from X_1 to the power set of X_2 associating to each $x \in X_1$ a subset $f(x)$ of X_2. We will adopt the alternative view, regarding f as a subset of $X_1 \times X_2$ so that

$$y \in f(x) \quad \text{means} \quad (x, y) \in f.$$

This allows us to apply subset operations such as union, intersection, and closure directly to relations. Since any subset of $X_1 \times X_2$ can be regarded as a relation, it is clear that $f(x)$ may be empty. We define the *domain* of f by $\mathrm{Dom}(f) = \{x : f(x) \neq \emptyset\}$.

Thus, we identify a function with its graph, relaxing into the common, if slightly abusive, habit of using the same symbol for a singleton set and for the point it contains, e.g. $f(x)$ for a map f. In particular, the identity map $1_X : X \to X$ is identified with the diagonal subset of $X \times X$. The ε neighborhoods of the diagonal are important examples of relations which are not functions.

$$V_\varepsilon = \{(x_1, x_2) \in X \times X : d(x_1, x_2) < \varepsilon\},$$
$$\overline{V}_\varepsilon = \{(x_1, x_2) \in X \times X : d(x_1, x_2) \leq \varepsilon\},$$

V_ε is open. $\overline{V}_\varepsilon$ is closed although it may be larger than the closure of V_ε (i.e., $\overline{V}_\varepsilon$ need not equal $\overline{V_\varepsilon}$).

For relations $f : X_1 \to X_2$ and $g : X_2 \to X_3$ we define the *inverse* $f^{-1} : X_2 \to X_1$ and the *composition* $g \circ f : X_1 \to X_3$ by

$$x \in f^{-1}(y) \Leftrightarrow y \in f(x),$$

i.e., $f^{-1} = \{(y, x) : (x, y) \in f\}$.

$$y \in g \circ f(x) \Leftrightarrow z \in f(x) \quad \text{and} \quad y \in g(z) \text{ for some } z \in X_2,$$

i.e., $g \circ f$ is the projection to $X_1 \times X_3$ of the subset $\{(x, z, y) \in X_1 \times X_2 \times X_3 : (x, z) \in f \text{ and } (z, y) \in g\}$.

The usual composition properties of associativity, identity, and inversion generalize to relations, e.g., $1_{X_2} \circ f = f = f \circ 1_{X_1}$ and $(g \circ f)^{-1} = f^{-1} \circ g^{-1}$. There are additional algebraic properties as well. For example, composition distributes over union:

$$\left(\bigcup_m \{g_m\}\right) \circ \left(\bigcup_n \{f_n\}\right) = \bigcup_{m, n} \{g_m \circ f_n\}.$$

For $f : X \to X$ we define f^n to be the n-fold composition of f ($n = 0, 1, 2, \ldots$ with $f^1 = f$ and $f^0 = 1_X$ by definition). f^{-n} is defined to be $(f^{-1})^n$ (which equals $(f^n)^{-1}$). From associativity follows $f^{m+n} = f^m \circ f^n$ if $m, n \geq 0$ or $m, n \leq 0$. But, for example, $1_X \subset f^{-n} \circ f^n$ $(n > 0)$ only when the domain of f is all of X and even then the inclusion is strict unless f is an injective mapping.

For a relation $f\colon X_1 \to X_2$ and a subset A of X_1 the *image* $f(A) \subset X_2$ is defined by

$$f(A) = \{y : (x, y) \in f \text{ for some } x \in A\} = \bigcup\{f(x) : x \in A\}.$$

A useful device, which we will call *regarding the subset as a relation*, exchanges for $A \subset X_1$ the relation $p \times A$ from a singleton space p to X_1. Then $p \times f(A) = f \circ (p \times A)$. For example, the equation $g \circ f(A) = g(f(A))$, easily checked directly, follows from associativity when A is regarded as a relation.

$f\colon X_1 \to X_2$ is called a *closed relation* if it is a closed subset of $X_1 \times X_2$. Notice that a subset A of X is closed exactly when the associated relation $p \times A$ is a closed relation. The following proposition lists some convenient properties of closed relations. Compactness is essential for several of the results and the reader should note where it is used in the proofs.

1. PROPOSITION. *Let $f\colon X_1 \to X_2$ and $g\colon X_2 \to X_3$ be closed relations.*

(a) *The domain* $\mathrm{Dom}(f)$ *is a closed subset of X_1.*
(b) *The inverse $f^{-1}\colon X_2 \to X_1$ is a closed relation.*
(c) *The composition $g \circ f\colon X_1 \to X_3$ is a closed relation.*
(d) *If A is a closed subset of X_1 then the image $f(A)$ is a closed subset of X_2.*
(e) *If B is a closed subset of X_2, then $\{x : f(x) \cap B \neq \emptyset\}$ is a closed subset of X_1.*
(f) *If U is an open subset of X_2, then $\{x : f(x) \subset U\}$ is an open subset of X_1.*

PROOF. (b) is obvious and (c) follows by using the projection π_{13} to $X_1 \times X_3$:

$$g \circ f = \pi_{13}[(f \times X_3) \cap (X_1 \times g)].$$

The proof of (d) is similar. In fact, (d) follows from (c) when we regard A as a relation. (e) follows from (b) and (d) because

$$\{x : f(x) \cap B \neq \emptyset\} = f^{-1}(B).$$

In particular, with $B = X_2$ we get the domain of f: $\mathrm{Dom}(f) = f^{-1}(X_2)$ and (a) follows from (e).

Finally, the set described in (f) is the complement of $f^{-1}(X_2 - U)$ and so (f) follows from (e) as well. □

The assumption of closure is thus a kind of continuity assumption. To see this we rephrase part (f):

2. COROLLARY. *Let $f\colon X_1 \to X_2$ be a closed relation. For every closed subset A of X_1 and every $\varepsilon > 0$ there exists a $\delta > 0$ such that*

$$f \circ \overline{V}_\delta(A) = f(\overline{V}_\delta(A)) \subset V_\varepsilon(f(A)) = V_\varepsilon \circ f(A).$$

PROOF. $\{x : f(x) \subset V_\varepsilon(f(A))\}$ is open in X_1 by (f) above and it contains A. Hence, it contains some δ neighborhood of A. □

In particular, a mapping $f\colon X_1 \to X_2$ is continuous if and only if it is closed regarded as a relation. However, it is important to note that δ depends on A as well as ε. The analogue of uniform continuity: $f \circ V_\delta \subset V_\varepsilon \circ f$ for some $\delta > 0$ is not true for all closed relations.

Of course for any relation we can apply the closure operator to get a closed relation:

3. LEMMA. *Let $f\colon X_1 \to X_2$ be a relation. For $\varepsilon > 0$ let $\overline{V}^1_\varepsilon, \overline{V}^2_\varepsilon$ be the ε neighborhoods of the diagonal for X_1 and X_2 respectively. The composition $\overline{V}^2_\varepsilon \circ f \circ \overline{V}^1_\varepsilon$ is a closed relation whose interior contains the closure $\overline{f}$ of f. Furthermore,*

$$\overline{f} = \bigcap\{\overline{V}^2_\varepsilon \circ f \circ \overline{V}^1_\varepsilon : \varepsilon > 0\}.$$

PROOF. On $X_1 \times X_2$ use the metric

$$d((x_1, y_1), (x_2, y_2)) = \max(d_1(x_1, x_2), d_2(y_1, y_2)).$$

Then $\overline{V}^2_\varepsilon \circ f \circ \overline{V}^1_\varepsilon$ is precisely the closed ε neighborhood of the set f in $X_1 \times X_2$ and the lemma follows. □

The utility of this lemma is illustrated by the following continuity result for composition:

4. PROPOSITION. *Let $\{F_n\}$ and $\{G_n\}$ be decreasing sequences of closed relations from X_1 to X_2 and from X_2 to X_3, respectively. If $\bigcap_n\{F_n\} = f$ and $\bigcap_n\{G_n\} = g$ then*

$$\bigcap_n\{G_n \circ F_n\} = g \circ f.$$

PROOF. Letting $\overline{V}^i_\varepsilon$ denote the ε neighborhood of the diagonal in X_i $(i = 1, 2, 3)$ we show first that

$$\bigcap\{g \circ \overline{V}^2_\varepsilon \circ f : \varepsilon > 0\} = g \circ f.$$

If (x, y) lies in the intersection then there exist $z_n, \tilde{z}_n$ in X_2 with $(x, z_n) \in f$, $(\tilde{z}_n, y) \in g$ and $d(z_n, \tilde{z}_n) \leq 1/n$. By passing to a subsequence we can assume that $\{z_n\}$ and $\{\tilde{z}_n\}$ converge. They clearly have the same limit point which we denote by z. Because f and g are closed, $(x, z) \in f$ and $(z, y) \in g$. So $(x, y) \in g \circ f$.

From this result and the lemma we see that given $\varepsilon > 0$, there exists a positive $\delta \leq \varepsilon$ such that $g \circ \overline{V}^2_\delta \circ f$ is contained in the neighborhood $V^3_{\varepsilon/2} \circ g \circ f \circ V^1_{\varepsilon/2}$. Furthermore, we can choose N such that for $n \geq N$:

$$G_n \subset V^3_{\delta/2} \circ g \circ V^2_{\delta/2} \quad \text{and} \quad F_n \subset V^2_{\delta/2} \circ f \circ V^1_{\delta/2}.$$

$V^2_{\delta/2} \circ V^2_{\delta/2} \subset V^2_\delta$ (the triangle inequality) then implies that

$$G_n \circ F_n \subset V^3_{\delta/2} \circ g \circ V^2_\delta \circ f \circ V^1_{\delta/2} \subset V^3_\varepsilon \circ g \circ f \circ V^1_\varepsilon .$$

So $\bigcap_n \{G_n \circ F_n\}$ lies in every ε neighborhood of the closed relation $g \circ f$. Hence, $\bigcap_n \{G_n \circ F_n\} \subset g \circ f$. The reverse inclusion is obvious because the sequences are decreasing.

REMARK. If $\{A_n\}$ is a decreasing sequence of closed subsets of X_1 with intersection A, it follows, by regarding the closed subsets as relations, that

$$\bigcap_n \{F_n(A_n)\} = f(A). \quad \square$$

We now begin our study of a closed relation $f: X \to X$, regarded as a dynamical system, by including f in larger relations. First is the positive orbit relation:

(1.1) $\quad \mathcal{O}f = \bigcup_{n=1}^\infty f^n$, i.e., $y \in \mathcal{O}f(x) \Leftrightarrow y \in f^n(x)$ for some $n = 1, 2, \dots$.

N.B. We start with $n = 1$ and f, not with $n = 0$ and 1_X.

A relation $F: X \to X$ is called *transitive* if $y \in F(x)$ and $z \in F(y)$ implies $z \in F(x)$, or equivalently, if $F \circ F \subset F$. Clearly, $\mathcal{O}f$ is transitive. It is, in fact, the smallest transitive relation containing f. But $\mathcal{O}f$ is usually not closed. $\mathcal{O}f$ is further enlarged by applying closure operators in various ways. We will need the lim sup operator whose properties are reviewed in the following.

5. EXERCISE. *Let $\{C_n\}$ be a sequence of closed subsets of X. The* lim sup *of the sequence is defined by*:

$$\limsup\{C_n\} = \bigcap_n \overline{\bigcup_{k \geq n} C_k}.$$

Letting $C = \limsup\{C_n\}$, prove:

(a) *$x \in C$ if and only if there exist a sequence of integers $\{n_i\}$ tending to infinity, and a sequence $\{x_i\}$ with $x_i \in C_{n_i}$ such that $x = \lim_{i\to\infty}\{x_i\}$.*

(b) $\overline{\bigcup_n C_n} = C \cup \bigcup_n C_n$.

(c) *C is the smallest closed subset of X such that for U any neighborhood of C, $C_n \subset U$ for n sufficiently large.*

(d) *If the sequence $\{C_n\}$ is decreasing then $\limsup\{C_n\} = \bigcap_n C_n$. If the sequence $\{C_n\}$ is increasing then the $\limsup\{C_n\} = \overline{\bigcup_n C_n}$.*

(e) *If $\{F_n\}$ and $\{G_n\}$ are sequences of closed relations in $X_1 \times X_2$ and $X_2 \times X_3$, respectively, with $f = \limsup\{F_n\}$ and $g = \limsup\{G_n\}$, then $\limsup\{G_n \circ F_n\} \subset g \circ f$. (Use Proposition 4 or its proof.) Show by example that the inclusion can be strict.*

(f) *If $f: X_1 \to X_2$ is a continuous map and $\{A_n\}$ is a sequence of closed subsets of X_1, then $\limsup\{f(A_n)\} = f(\limsup\{A_n\})$. (For one direction use*

(e) *for the other use* $f(\overline{B}) = \overline{f(B)})$. *If* $\{\widetilde{A}_n\}$ *is a sequence of closed subsets of* X_2, *then* $\limsup\{f^{-1}(\widetilde{A}_n)\} \subset f^{-1}(\limsup\{\widetilde{A}_n\})$. □

Now if A is any closed subset of X, we define

$$\omega f[A] = \limsup\{f^n(A)\}. \tag{1.2}$$

Thus, $\omega f[A]$ is the closed set upon which the successive images of A pile up. From (c) of the exercise, if U is any neighborhood of $\omega f[A]$ then $f^n(A) \subset U$ for all n sufficiently large.

By applying this operator to $A = x$ for all x in X we define the ω limit relation $\omega f \colon X \to X$ by

(1.3) $\omega f(x) = \limsup\{f^n(x)\}$, i.e., $y \in \omega f(x) \Leftrightarrow$ there exist sequences $\{y_i\}$, $\{n_i\}$ with $y_i \in f^{n_i}(x)$, $n_i \to \infty$, and $y_i \to y$.

We mentioned earlier that it is desirable that the set of states in the "future of x" be closed while $\mathscr{O}f(x)$ is usually not a closed set. So our first enlargement of $\mathscr{O}f$ consists of defining $\mathscr{R}f \colon X \to X$ by taking the closure of each such set:

$$\mathscr{R}f = \mathscr{O}f \cup \omega f, \text{ i.e., } \mathscr{R}f(x) = \mathscr{O}f(x) \cup \omega f(x) = \overline{\mathscr{O}f(x)}, \tag{1.4}$$

where the latter equation follows from (b) of the exercise.

Although each $\mathscr{R}f(x)$ and $\omega f(x)$ is closed, the resulting assemblies of slices, $\mathscr{R}f$ and ωf need not be closed relations. Instead of taking the lim sup of $\{f^n(x)\}$ for each x, we get a closed relation by defining:

(1.5) $\Omega f = \limsup\{f^n\}$, i.e., $y \in \Omega f(x) \Leftrightarrow$ there exist sequences $\{x_i\}$, $\{y_i\}$, $\{n_i\}$ with $y_i \in f^{n_i}(x_i)$, $n_i \to \infty$, $x_i \to x$, and $y_i \to y$.

Thus, $y \in \omega f(x)$ if y is near $f^n(x)$ for some large n while $y \in \Omega f(x)$ if y is near $f^n(\tilde{x})$ for some $\tilde{x}$ near x and some large n.

We obtain the closure of $\mathscr{O}f$ by using Ωf instead of ωf:

$$\mathscr{N}f = \mathscr{O}f \cup \Omega f = \overline{\mathscr{O}f}. \tag{1.6}$$

By proving the following useful formulae the reader may develop a feel for the difference between the two sorts of closure:

6. EXERCISE. *For a closed relation* $f \colon X \to X$,

$$\mathscr{R}f = \bigcap\{\overline{V}_\varepsilon \circ \mathscr{O}f : \varepsilon > 0\}, \qquad \mathscr{N}f = \bigcap\{\overline{V}_\varepsilon \circ \mathscr{O}f \circ \overline{V}_\varepsilon : \varepsilon > 0\}. \tag{1.7}$$

(*Use Lemma* 3. *Notice that* $\overline{V}_\varepsilon \circ \mathscr{O}f \circ \overline{V}_\varepsilon$ *is closed but* $\overline{V}_\varepsilon \circ \mathscr{O}f$ *need not be.*)

If A *is a closed subset of* X *then the image* $\Omega f(A) = \bigcup\{\Omega f(x) : x \in A\}$ *is closed and satisfies*:

$$\bigcap\{\Omega f(\overline{V}_\varepsilon(A)) : \varepsilon > 0\} = \Omega f(A) = \bigcap\{\omega f[\overline{V}_\varepsilon(A)] : \varepsilon > 0\}. \tag{1.8}$$

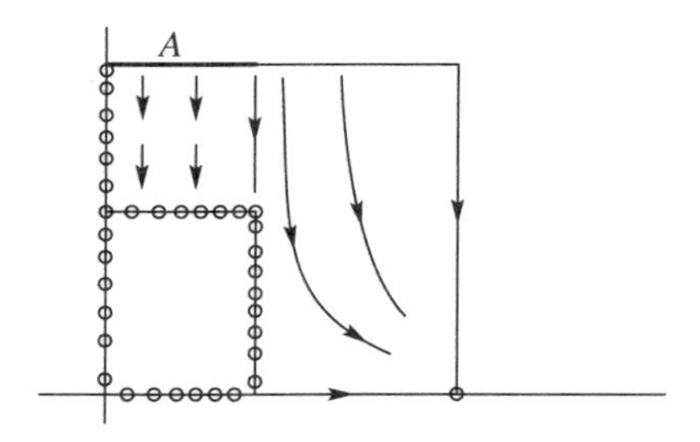

FIGURE 1.1

(*Use Lemma* 4 *for the first equation.*) *In particular,* $\Omega f(x) = \bigcap\{\omega f[\overline{V}_\varepsilon(x)] : \varepsilon > 0\}$. □

Notice the square brackets in the definition (1.2). These are intended to emphasize that $\omega f[A]$ is usually *not* the image of A under a relation. Clearly, $\omega f[A]$ contains the image $\omega f(A) \equiv \bigcap\{\omega f(x) : x \in A\}$ and by (e) of Exercise 5, $\omega f[A]$ is contained in $\Omega f(A)$. Thus we have a sequence of inclusions

$$(1.9) \qquad \Omega f(A) \supset \omega f[A] \supset \overline{\omega f(A)} \supset \omega f(A),$$

but any or all of these inclusions may be strict.

7. EXERCISE. *Interpret Figure* 1.1 *as the phase portrait of a flow, with the points marked by* $\circ$ *all fixed. Let* f *be the time one map of the flow. For the closed interval* A *in Figure* 1.1 *show that all of the inclusions are strict.* ((1.8) *is helpful for computing* $\Omega f(A)$). *Furthermore,* $\bigcap\{\omega f(\overline{V}_\varepsilon(A)) : \varepsilon > 0\}$ *is also distinct from these sets.* □

With $\mathscr{N} f$ we have returned to the family of closed relations. However, in taking the closure of $\mathscr{O} f$ we generally lose transitivity. Since we want both closure and transitivity we define

$$(1.10) \qquad \mathscr{G} f = \text{ the smallest closed, transitive relation containing } f.$$

The intersection of a family of transitive relations is transitive. So we can define $\mathscr{G} f$, from above as it were, to be the intersection of all closed, transitive relations containing f. This family is nonempty since it contains $X \times X$. It is useful, at least conceptually, to generate $\mathscr{G} f$ from below by some constructive procedure starting with f. A relation F on X is closed and transitive if and only if $F = \mathscr{N} F$. So the obvious procedure is to keep iterating the operator $\mathscr{N}$. This will work but, as we will see in the supplementary exercises, it requires transfinite induction. As a result the construction is not too helpful for applications.

There is an alternate procedure due to Conley which constructs a closed transitive relation, generally larger than $\mathscr{G} f$, in a simple and direct fashion.

Given a closed relation f on X, a *chain* or 0-*chain* is a finite or infinite sequence $\{x_n\}$ such that $x_{n+1} \in f(x_n)$ along the way. Given $\varepsilon \geq 0$ an

ε-chain is a finite or infinite sequence $\{x_n\}$ such that each x_{n+1} is within ε of the set $f(x_n)$, i.e., $x_{n+1} \in \overline{V}_\varepsilon(f(x_n))$. (This requires that $f(x_n)$ be nonempty.) If the chain is finite, i.e., the sequence consists of finitely many—but at least two—terms, then the first term is called the beginning and the last term the end of the chain. Now we define the relation $\mathscr{C}f$ on X by:

(1.11) $\mathscr{C}f = \bigcap\{\mathscr{O}(\overline{V}_\varepsilon \circ f) : \varepsilon > 0\}$, i.e., $y \in \mathscr{C}f(x) \Leftrightarrow$ for every $\varepsilon > 0$ there is an ε-chain beginning at x and ending at y.

The relation $\mathscr{C}f$ and its associated concept of recurrence are of fundamental theoretical importance. The patient reader will discover its central role in the chapters to come. Let us pause to consider why this idea of Conley is so important in applied studies as well.

For a continuous map $f: X \to X$ a 0 chain beginning with x is just an initial segment of the positive orbit: $x, f(x), f^2(x), \dots$. If for no other reason than round-off error it is usually impossible to calculate this sequence exactly. However, given $\varepsilon > 0$ it is usually possible with sufficient expenditure to get within ε of $f(x)$ for any x. These errors occur with each calculation. So in attempting to calculate the orbit of x, we are likely to get an ε-chain instead: $\{x_n\}$ with $d(x_{n+1}, f(x_n)) \le \varepsilon$. Plotting the resulting points we thus observe a thickening of $\mathscr{C}f(x)$ rather than of $\mathscr{R}f(x)$, the closure of the positive orbit. Thus arises the need to understand the relationship between $\mathscr{C}f$ and the original system f, a need which our theoretical apparatus is intended to meet.

Contrast $\mathscr{C}f$ with the chain interpretation of $\mathscr{N}f$. By (1.7) $y \in \mathscr{N}f(x)$ if for every $\varepsilon > 0$ there is 0-chain beginning ε near x and ending ε near y. The errors are made only at the beginning and the end.

$\mathscr{C}f$ is defined in (1.11) as the intersection of transitive relations. So $\mathscr{C}f$ is clearly transitive. $\mathscr{C}f$ is closed as well.

8. PROPOSITION. *Let $f: X \to X$ be a closed relation.*

$$\begin{aligned}\mathscr{C}f &= \bigcap\{\mathscr{O}(V_\varepsilon \circ f \circ V_\varepsilon) : \varepsilon > 0\} \\ &= \bigcap\{\mathscr{N}(\overline{V}_\varepsilon \circ f \circ \overline{V}_\varepsilon) : \varepsilon > 0\}.\end{aligned}$$

In particular, $\mathscr{C}f$ is a closed, transitive relation.

PROOF. Given $x \in X$ and $\varepsilon > 0$, apply Corollary 2 to choose $\delta \le \varepsilon/3$ such that $f \circ \overline{V}_\delta(x) \subset V_{\varepsilon/3} \circ f(x)$. For $n \ge 1$:

$$\begin{aligned}(\overline{V}_\delta \circ f \circ \overline{V}_\delta)^n(x) &\subset (\overline{V}_{2\delta} \circ f)^{n-1} \circ \overline{V}_{2\delta} \circ f \circ \overline{V}_\delta(x) \\ &\subset (\overline{V}_{2\delta} \circ f)^{n-1} \circ \overline{V}_{2\delta} \circ \overline{V}_{\varepsilon/3} \circ f(x) \subset (\overline{V}_\varepsilon \circ f)^n(x).\end{aligned}$$

Hence, $\bigcap\{\mathscr{O}(\overline{V}_\delta \circ f \circ \overline{V}_\delta)(x) : \delta > 0\} \subset \mathscr{O}(\overline{V}_\varepsilon \circ f)(x)$. Now intersect over all $\varepsilon > 0$ and note that if $\{F_n\}$ is any sequence of relations then

$\bigcap\{F_n(x)\} = \bigcap\{F_n\}(x)$. Thus,

$$\bigcap\{\mathscr{O}(\overline{V}_\delta \circ f \circ \overline{V}_\delta)\}(x) \subset \mathscr{C}f(x).$$

The reverse inclusion is clear and so we have proved the first equation.

Notice now that $\varepsilon_1 + \varepsilon_2 < \varepsilon$ implies

$$\overline{V}_{\varepsilon_2} \circ \mathscr{O}(\overline{V}_{\varepsilon_1} \circ f \circ \overline{V}_{\varepsilon_1}) \circ \overline{V}_{\varepsilon_2} \subset \mathscr{O}(V_\varepsilon \circ f \circ V_\varepsilon). \tag{1.12}$$

Intersect over all $\varepsilon_2 > 0$ sufficiently small and apply Lemma 3 to get for $\varepsilon_1 < \varepsilon$:

$$\mathscr{N}(\overline{V}_{\varepsilon_1} \circ f \circ \overline{V}_{\varepsilon_1}) \subset \mathscr{O}(V_\varepsilon \circ f \circ V_\varepsilon). \tag{1.13}$$

Intersect over ε_1 and then over ε to complete the proof. □

9. EXERCISE. (a) *Despite the use of the metric on X in the definition* (1.11), *$\mathscr{C}f$ is a topological invariant, i.e., prove that we can replace d with an equivalent metric $\tilde{d}$ on X (yielding the same topology on X) without changing the set $\mathscr{C}f$. (Hint: d and $\tilde{d}$ are uniformly equivalent. For every $\varepsilon > 0$ there exists a $\delta > 0$ such that $d < \delta$ implies $\tilde{d} < \varepsilon$ and $\tilde{d} < \delta$ implies $d < \varepsilon$.)*

(b) *If 1_X is the identity map on X prove that $\mathscr{G}1_X = 1_X$. If X is connected prove that $\mathscr{C}1_X = X \times X$. More generally prove that*

$$\mathscr{C}1_X = \bigcup\{C \times C : C \text{ is a component of } X\}.$$

(Hint: If C_1 and C_2 are distinct components there exists an open-and-closed subset A of X with $C_1 \subset A$ and $A \cap C_2 = \varnothing$. There exists $\varepsilon > 0$ such that $d(x_1, x_2) > \varepsilon$ if $x_1 \in A$ and $x_2 \in X - A$.) □

Thus for any closed relation $f\colon X \to X$ we have defined a tower of relations:

$$f \subset \mathscr{O}f \subset \mathscr{R}f \subset \mathscr{N}f \subset \mathscr{G}f \subset \mathscr{C}f. \tag{1.14}$$

For any relation F on X there is an associated concept of recurrence. The *cyclic set of* F, denoted by $|F|$, is defined by:

$$|F| = \mathrm{Dom}(F \cap 1_X) = \{x : (x, x) \in F\} = \{x : x \in F(x)\}. \tag{1.15}$$

10. EXERCISE. (a) *If F is a closed relation then $|F|$ is a closed subset of X.*

(b) *Assume F is a transitive relation. Prove that $|F| = \mathrm{Dom}(F \cap F^{-1})$ and that $F \cap F^{-1}$ restricts to an equivalence relation (reflexive, symmetric, and transitive) on $|F|$. Notice that this is the true portion of the classical false result that symmetry and transitivity imply reflexivity.* □

For the relations of (1.14) the points of the associated cyclic sets have names as follows:

$x \in |f| : x$ is a *fixed point* of f.

$x \in |\mathscr{O} f| : x$ is a *periodic point* of f.

$x \in |\mathscr{R} f| : x$ is a *positive recurrent point* of f.

$x \in |\mathscr{N} f| : x$ is a *nonwandering point* of f.

$x \in |\mathscr{G} f| : x$ is a *generalized nonwandering point* of f.

$x \in |\mathscr{C} f| : x$ is a *chain recurrent point* of f.

The chain recurrent set, $|\mathscr{C} f|$, is the most important of these, but the sharper forms of recurrence associated with the finer relations are usually easier to detect.

The remainder of the chapter consists of technical properties and identities. The results are useful, but the details are tedious. They are best looked over lightly at the first reading.

11. PROPOSITION. (a) *monotonicity: For* $\mathscr{A} = \mathscr{O}, \omega, \mathscr{R}, \Omega, \mathscr{N}, \mathscr{G}$ *or* $\mathscr{C}$: $f_1 \subset f_2$ *implies* $\mathscr{A} f_1 \subset \mathscr{A} f_2$.

(b) *For* $\mathscr{A} = \mathscr{O}, \Omega, \mathscr{N}, \mathscr{G}$ *or* $\mathscr{C}$: $\mathscr{A}(f^{-1}) = (\mathscr{A} f)^{-1}$.

(c) *For* $\mathscr{A} = \mathscr{O}, \mathscr{G}, \mathscr{C}$: $\mathscr{A}(\mathscr{A} f) = \mathscr{A} f$.

(d) *For any relation* f *on* X:

$$f \cup (f \circ \mathscr{O} f) = \mathscr{O} f = f \cup (\mathscr{O} f \circ f),$$

and if f *is closed*

$$f \cup (f \circ \mathscr{G} f) = \mathscr{G} f = f \cup (\mathscr{G} f \circ f), \qquad f \cup (f \circ \mathscr{C} f) = \mathscr{C} f = f \cup (\mathscr{C} f \circ f).$$

(e) *For a closed relation* f *on* X:

$$\begin{aligned} \omega f &\subset f \circ \omega f, & \mathscr{R} f &\subset f \cup (f \circ \mathscr{R} f), \\ \Omega f &\subset f \circ \Omega f, & \mathscr{N} f &\subset f \cup (f \circ \mathscr{N} f), \\ \Omega f &\subset \Omega f \circ f, & \mathscr{N} f &\subset f \cup (\mathscr{N} f \circ f). \end{aligned}$$

(f) *For a closed relation* f *on* X: $\Omega f \subset \Omega f \circ \Omega f$.

PROOF. (a) is obvious and (b) is an easy exercise (for $\mathscr{C}$ use Proposition 8). Note that F^{-1} is transitive if F is.

(c) If F is transitive then $\mathscr{O} F = F$ and if F is also closed $\mathscr{G} F = F$. In particular, $\mathscr{O}(\mathscr{O} f) = \mathscr{O} f$ and $\mathscr{G}(\mathscr{G} f) = \mathscr{G} f$. For $\mathscr{C}$ note first that $\overline{V}_{\varepsilon_1} \circ \mathscr{O}(\overline{V}_{\varepsilon_2} \circ f) \subset \mathscr{O}(\overline{V}_{\varepsilon} \circ f)$ when $\varepsilon_1 + \varepsilon_2 \le \varepsilon$ (compare with (1.12)). By applying $\mathscr{O}$ we obtain $\mathscr{O}(\overline{V}_{\varepsilon_1} \circ \mathscr{O}(\overline{V}_{\varepsilon_2} \circ f)) \subset \mathscr{O}(\mathscr{O}(\overline{V}_{\varepsilon} \circ f)) = \mathscr{O}(\overline{V}_{\varepsilon} \circ f)$. Hence, from the definition of $\mathscr{C}$:

$$\mathscr{C}(\mathscr{C} f) \subset \mathscr{O}(\overline{V}_{\varepsilon_1} \circ \mathscr{C} f) \subset \mathscr{O}(\overline{V}_{\varepsilon_1} \circ \mathscr{O}(\overline{V}_{\varepsilon_1} \circ f)) \subset \mathscr{O}(\overline{V}_{\varepsilon} \circ f).$$

Now intersect over $\varepsilon > 0$ to get $\mathscr{C}(\mathscr{C}f) \subset \mathscr{C}f$.

(d) The result for $\mathscr{O}f$ is obvious. In general, if F is transitive and contains f then $f \subset F_1 \subset F$ with $F_1 = f \cup (f \circ F)$ or $f \cup (F \circ f)$. We prove that $\mathscr{C}f \subset f \cup (f \circ \mathscr{C}f)$ and $\mathscr{G}f \subset f \cup (f \circ \mathscr{G}f)$ when f is closed. The dual results follow from these applied to f^{-1}.

For $\mathscr{C}f$, let $\varepsilon > 0$:

$$\begin{aligned}\mathscr{C}f \subset \mathscr{O}(\overline{V}_\varepsilon \circ f) &= (\overline{V}_\varepsilon \circ f) \cup (\overline{V}_\varepsilon \circ f) \circ \mathscr{O}(\overline{V}_\varepsilon \circ f)\\ &\subset (\overline{V}_\varepsilon \circ f) \cup (\overline{V}_\varepsilon \circ f) \circ \mathscr{N}(\overline{V}_\varepsilon \circ f).\end{aligned}$$

By Propositions 4 and 8 the intersection over ε of the last expression is $f \cup (f \circ \mathscr{C}f)$.

When f is closed and $F = \mathscr{G}f$, then $F_1 = f \cup (f \circ F)$ is closed and contains f. When we show F_1 is transitive it follows that $\mathscr{G}f \subset F_1$.

$$F_1 \circ F_1 = f \circ (F_1 \cup F \circ F_1) \subset f \circ F \subset F_1.$$

(e) For $x \in X$:

$$\{f^{n+1}(x)\} = \{f \circ f^n(x)\}, \quad \{f^{n+1}\} = \{f \circ f^n\}, \quad \{f^{n+1}\} = \{f^n \circ f\}.$$

Apply the lim sup operator to these sequences and use (e) of Exercise 5 to get the three inclusions in the left column. The corresponding inclusions on the right then follow from (d) with $\mathscr{O}f$. (f) follows similarly from $f^{m+n} = f^m \circ f^n$. Note that as the pair (m, n) tends to ∞ the lim sup of $\{f^{m+n}\}$, $\{f^m\}$ and $\{f^n\}$ are all Ωf.

REMARK. The reader should take special note of the equation $f \cup (f \circ F) = F = f \cup (F \circ f)$ for $F = \mathscr{G}f$ and $\mathscr{C}f$ as these will be applied repeatedly. □

When f is a continuous mapping some of these results can be strengthened. Notice that a relation $f: X_1 \to X_2$ is a mapping exactly when

$$1_{X_1} \subset f^{-1} \circ f \quad \text{and} \quad f \circ f^{-1} \subset 1_{X_2}, \tag{1.16}$$

because the first is equivalent to $\mathrm{Dom}(f) = X_1$ and the second is equivalent to $y_1, y_2 \in f(x)$ implies $y_1 = y_2$. Recall from Corollary 2 that a map is a closed relation exactly when it is a continuous map.

Attending to this special case we have an opportunity to understand the dynamic meaning of the relation operators, to observe how the tableau of relations (1.14) portrays the motion of the system.

For a map f, $\omega f(x)$ is the set of limit points of a particular sequence of points in X: the positive orbit $f(x), f^2(x), \ldots$. Thus, $y \in \omega f(x)$ when for every neighborhood U of y the sequence $\{f^n(x)\}$ enters U infinitely often. By contrast, $y \in \Omega f(x)$ when for every neighborhood U of y and V of x the sequence of images $\{f^n(V)\}$ meets U infinitely often.

12. **Proposition.** *Let $f\colon X \to X$ be a continuous map.*

(a)

$$f \circ \omega f = \omega f = \omega f \circ f, \qquad f \cup (f \circ \mathscr{R} f) = \mathscr{R} f = f \cup (\mathscr{R} f \circ f),$$
$$f \circ \Omega f = \Omega f \subset \Omega f \circ f, \qquad f \cup (f \circ \mathscr{N} f) = \mathscr{N} f \subset f \cup (\mathscr{N} f \circ f).$$

(b) *For $\mathscr{A} = \mathscr{O}, \omega, \mathscr{R}, \Omega, \mathscr{N}, \mathscr{G}$ or $\mathscr{C}$: $\mathscr{A} f \subset f^{-1} \circ \mathscr{A} f \circ f$. In other words, if $y \in \mathscr{A} f(x)$ then $f(y) \in \mathscr{A} f(f(x))$.*

(c) $\Omega f \subset \Omega f \circ \omega f$.

(d) $\omega f \circ (\omega f)^{-1} \subset \Omega f$.

In other words, $y_1, y_2 \in \omega f(x)$ implies $y_2 \in \Omega f(y_1)$.

Proof. (a) The sequences $\{f^{n+1}(x) = f^n(f(x))\}$ and $\{f^n(x)\}$ have the same limit points proving $\omega f(f(x)) = \omega f(x)$. If the subsequence $f^{n_i}(x)$ converges to y then $f^{n_i+1}(x)$ converges to $f(y)$. So $f(\omega f(x)) \subset \omega f(x)$ while the reverse inclusion is Proposition 11(e).

$f \circ \Omega f(x) \subset \Omega f(x)$ says that $y \in \Omega f(x)$ implies $f(y) \in \Omega f(x)$. To prove this start with $\varepsilon > 0$ and choose $\delta < \varepsilon$ an ε modulus of uniform continuity, so that $d(y_1, y_2) \le \delta$ implies $d(f(y_1), f(y_2)) < \varepsilon$. There exist x_1 and large n such that $d(x, x_1) < \delta$ and $d(y, f^n(x_1)) < \delta$. Hence, $d(x, x_1) < \varepsilon$ and $d(f(y), f^{n+1}(x_1)) < \varepsilon$. The reverse inequality and $\Omega f \subset \Omega f \circ f$ are in Proposition 11(e).

The remaining identities follow from $\mathscr{R} f = \mathscr{O} f \cup \omega f$ and $\mathscr{N} f = \mathscr{O} f \cup \Omega f$.

(b) For $\mathscr{O}$ the result is trivial. For ω and Ω the result follows formally from (1.16) and (a), e.g.,

$$\Omega f \subset f^{-1} \circ f \circ \Omega f \subset f^{-1} \circ \Omega f \circ f.$$

As usual, the results for $\mathscr{R}$ and $\mathscr{N}$ follow from those for $\mathscr{O}$, ω, and Ω.

For $\mathscr{G}$ the result will follow when we show that the closed relation $f^{-1} \circ \mathscr{G} f \circ f$ contains f and is transitive. These properties follow from (1.16):

$$f \subset f^{-1} \circ f \circ f \subset f^{-1} \circ \mathscr{G} f \circ f,$$
$$(f^{-1} \circ \mathscr{G} f \circ f) \circ (f^{-1} \circ \mathscr{G} f \circ f) \subset f^{-1} \circ \mathscr{G} f \circ \mathscr{G} f \circ f \subset f^{-1} \circ \mathscr{G} f \circ f.$$

For $\mathscr{C}$ the result follows by choosing δ an ε modulus of uniform continuity and observing that if $\{x_n\}$ is a δ-chain from x to y then $\{f(x_n)\}$ is an ε-chain from $f(x)$ to $f(y)$.

(c) To prove $y \in \Omega f(\omega f(x))$ it is enough to find, given $\varepsilon > 0$, large n and $z \in V_\varepsilon(\omega f(x))$ such that $d(f^n(z), y) < \varepsilon$ (see (1.8) of Exercise 6). Now suppose $y \in \Omega f(x)$. There exists N such that $f^n(x) \in V_\varepsilon(\omega f(x))$ for $n \ge N$ and there exists $\varepsilon_1 > 0$ such that $V_{\varepsilon_1}(f^N(x)) \subset V_\varepsilon(\omega f(x))$. Now choose $\delta \le \varepsilon$ an ε_1 modulus of continuity for f^N. Because $y \in \Omega f(x)$, we can find large $n > N$ and z_0 such that $d(x, z_0) < \delta$ and $d(y, f^n(z_0)) < \delta$. Let $z = f^N(z_0)$. $z \in V_\varepsilon(\omega f(x))$ because $d(f^N(x), z) < \varepsilon_1$. Finally,

$$d(y, f^{n-N}(z)) = d(y, f^n(z_0)) < \delta \leq \varepsilon.$$

(d) Given $\varepsilon > 0$ we can choose N such that $d(f^N(x), y_7) < \varepsilon$ and large n such that $d(f^{n+N}(x), y_2) < \varepsilon$. So $z = f^N(x)$ is ε near y_7, and $f^n(z)$ is ε near y_2.

REMARK. In (d) letting $y_1 = y_2$ we see that all of the points of $\omega f(x)$ are nonwandering, i.e., $\omega f(x) \subset |\Omega f| = |\mathcal{N} f|$ for all $x \in X$. □

13. EXERCISE. (a) *Show by example that the inclusion* $\Omega f \subset \Omega f \circ f$ *can be strict for continuous maps. On the other hand prove that* $\Omega f = \Omega f \circ f$ *and* $\mathcal{N} f = f \cup (\mathcal{N} f \circ f)$ *if* $f: X \to X$ *is a homeomorphism. Although this follows from* $f^{-1} \circ \Omega f^{-1} = \Omega f^{-1}$, *it is useful to prove it directly.*

(b) *For a homeomorphism* f *prove*:

$$\mathcal{N} f \cup (\mathcal{N} f)^2 \cup \cdots \cup (\mathcal{N} f)^n = \mathcal{O} f \cup (\Omega f)^n. \tag{1.17}$$

(*Hint*: $\{(\Omega f)^n\}$ *is an increasing sequence by Proposition* 11(f).) *In particular, if* $(\Omega f)^n = (\Omega f)^{n+1}$ *show that* $\mathcal{O}(\mathcal{N} f) = \mathcal{O} f \cup (\Omega f)^n = \mathcal{G} f$. □

We conclude this section by describing mappings of relations. In category terminology, relations are the objects of our study and we want to know what is a morphism between them. This leads to the appropriate notion of isomorphism, namely conjugacy.

14. EXERCISE. (a) *For a relation* $h: X_1 \to X_2$ *check* (1.16) *in detail proving*: (i) $h \circ h^{-1} \subset 1_{X_2}$ *if and only if for any* $x \in X_1$ *there is at most one* $y \in X_2$ *with* $y \in h(x)$; (ii) $1_{X_1} \subset h^{-1} \circ h$ *if and only if for any* $x \in X_1$ *there is at least one* $y \in X_2$ *with* $y \in h(x)$.

Assume now that h *is a map so that* (i) *and* (ii) *hold. Prove*: (i′) $h^{-1} \circ h = 1_{X_1}$ *if and only if* h *is injective, i.e., a one-to-one map;* (ii′) $h \circ h^{-1} = 1_{X_2}$ *if and only if* h *is surjective, i.e., an onto map.* (*Apply* (i) *and* (ii), *respectively, to* h^{-1}.)

For any map h *prove that* $E_h \equiv h^{-1} \circ h$ *is an equivalence relation on* X_1 (*use* (i) *and* (ii) *to check formally that* $E_h \circ E_h \subset E_h$, $E_h^{-1} = E_h$, $1_{X_1} \subset E_h$). E_h *is closed if* h *is continuous. Note that* $(x, \tilde{x}) \in E_h$ *if and only if* $h(x) = h(\tilde{x})$.

(b) *Let* $A_1 \subset X_1$, $A_2 \subset X_2$, *and* $h: X_1 \to X_2$. *If* h *is a map check that*

$$A_1 \subset h^{-1}(A_2) \Leftrightarrow h(A_1) \subset A_2. \tag{1.18}$$

Give examples to show that neither implication need hold if h *is a general relation.*

(c) *If* $h: X_1 \to X_2$ *and* $\tilde{h}: \tilde{X}_1 \to \tilde{X}_2$ *are relations we define the product relation* $h \times \tilde{h}: X_1 \times \tilde{X}_1 \to X_2 \times \tilde{X}_2$ *by* $(y, \tilde{y}) \in h \times \tilde{h}(x, \tilde{x})$ *if and only if* $y \in h(x)$ *and* $\tilde{y} \in \tilde{h}(\tilde{x})$. *So as a subset of* $(X_1 \times \tilde{X}_1) \times (X_2 \times \tilde{X}_2)$, $h \times \tilde{h}$ *is just the image of the product subset of* $(X_1 \times X_2) \times (\tilde{X}_1 \times \tilde{X}_2)$ *under the obvious permutation of factors. When* h *and* $\tilde{h}$ *are maps,* $h \times \tilde{h}$ *is the product mapping. Check that* $(h \times \tilde{h})^{-1} = h^{-1} \times \tilde{h}^{-1}$ *and that the obvious composition rules apply. In particular,*

$$(1.19) \qquad (1_{X_1} \times \tilde{h}) \circ (h \times 1_{X_2}) = h \times \tilde{h} = (h \times 1_{X_2}) \circ (1_{X_1} \times \tilde{h}).$$

Now let $f_1: X_1 \to \tilde{X}_1$ *and* $f_2: X_2 \to \tilde{X}_2$. *Prove:*

$$(1.20) \qquad h \times \tilde{h}(f_1) = \tilde{h} \circ f_1 \circ h^{-1}, \qquad (h \times \tilde{h})^{-1}(f_2) = \tilde{h}^{-1} \circ f_2 \circ h,$$

where, for example, $h \times \tilde{h}(f_1)$ *is the image under the product relation of* f_1 *regarded as a subset of* $X_1 \times \tilde{X}_1$, *while* $\tilde{h} \circ f_1 \circ h^{-1}$ *is the composed relation regarded as a subset of* $X_2 \times \tilde{X}_2$. *(Observe that the second equation follows from the first applied to* h^{-1} *and* $\tilde{h}^{-1}$.*)* □

15. DEFINITION. *Let* f_1 *and* f_2 *be relations on* X_1 *and* X_2. *A continuous map* $h: X_1 \to X_2$ *is said to map* f_1 *to* f_2, *written* $h: f_1 \to f_2$ *if* $(x, \tilde{x}) \in f_1$ *implies* $(h(x), h(\tilde{x})) \in f_2$. *This condition is equivalent to each of the following inclusions:*

(a) $h \circ f_1 \circ h^{-1} \subset f_2$,
(b) $h \circ f_1 \subset f_2 \circ h$,
(c) $f_1 \subset h^{-1} \circ f_2 \circ h$,
(d) $f_1 \circ h^{-1} \subset h^{-1} \circ f_2$.

Such a continuous map h *is called a semiconjugacy from* f_1 *to* f_2. *A conjugacy is a homeomorphism* h *such that* h *maps* f_1 *to* f_2 *and* h^{-1} *maps* f_2 *to* f_1, *or equivalently a homeomorphism such that any (and hence all) of* (a)–(d) *hold with the inclusions replaced by equalities.*

PROOF OF EQUIVALENCE. Clearly, the original definition says $h \times h(f_1) \subset f_2$ which is equivalent to (a) by (1.20). We can prove the equivalence either by using set inclusions or composition identities. For example, for (a) ⇒ (b), use (1.19) and (1.18):

$$(h \times 1) \circ (1 \times h)(f_1) = (h \times h)(f_1) \subset f_2 \quad \text{implies} \quad (1 \times h)(f_1) \subset (h^{-1} \times 1)(f_2),$$

and apply (1.20) again.

Alternatively, compose with h and use (1.16):

$$h \circ f_1 \circ h^{-1} \subset f_2 \quad \text{implies} \quad h \circ f_1 \subset h \circ f_1 \circ h^{-1} \circ h \subset f_2 \circ h.$$

Similar arguments yield (b) ⇒ (c) ⇒ (d) ⇒ (a).

For the conjugacy results observe that for a homeomorphism h, condition (a) describing $h^{-1}: f_2 \to f_1$ is condition (c) with the inequality reversed.

REMARK. For mappings, relation inclusion is the same as equality. Hence, if $f_1: X_1 \to X_2$ and $f_2: X_2 \to X_2$ are maps then the continuous map $h: X_1 \to X_2$ is a semiconjugacy, mapping f_1 to f_2 if and only if

$$h \circ f_1 = f_2 \circ h, \tag{1.21}$$

which says that the diagram:

$$\begin{array}{ccc} X_1 & \xrightarrow{h} & X_2 \\ \downarrow{\scriptstyle f_1} & & \downarrow{\scriptstyle f_2} \\ X_1 & \xrightarrow[h]{} & X_2 \end{array}$$

commutes. In particular, a homeomorphism h which is a semiconjugacy between mappings f_1 and f_2 is automatically a conjugacy. □

Clearly, if h maps f_1 to f_2 and $\tilde{f}_1 \subset f_1$, $f_2 \subset \tilde{f}_2$ then h maps $\tilde{f}_1$ to $\tilde{f}_2$. Thus, if we define $h^* f_2$ and $h_* f_1$:

$$\begin{aligned} h^* f_2 &\equiv h^{-1} \circ f_2 \circ h = (h \times h)^{-1}(f_2), \\ h_* f_1 &\equiv h \circ f_1 \circ h^{-1} = (h \times h)(f_1), \end{aligned} \tag{1.22}$$

then condition (a) says $f_1 \subset h^* f_2$ while condition (c) says $h_* f_1 \subset f_2$. Notice that their equivalence immediately follows from (1.18).

Thus, the relation $h^* f_2$ is the largest relation on X_1 which is mapped by h to f_2, while $h_* f_1$ is the smallest relation on X_2 into which f_1 is mapped by h.

We first collect some elementary results about these operators.

16. EXERCISE. (a) *Prove that the operators h^* and h_* are monotone, i.e., $\tilde{f}_1 \subset f_1$ and $f_2 \subset \tilde{f}_2$ imply $h_* \tilde{f}_1 \subset h_* f_1$ and $h^* f_2 \subset h^* \tilde{f}_2$. If h is surjective then the operator h^* is injective, i.e., $h^* f_2 = h^* \tilde{f}_2$ implies $f_2 = \tilde{f}_2$. Similarly, if h is injective the operator h_* is injective.*

(b) *Prove: $h^*(\overline{f}_2) = \overline{h^*(f_2)}$ and $h_*(\overline{f}_1) = \overline{h_*(f_1)}$. In particular, if f_1 and f_2 are closed relations then $h^* f_2$ and $h_* f_1$ are closed as well.*

(c) *Prove that $h^*(1_{X_2})$ is the equivalence relation E_h and so $1_{X_1} \subset h^*(1_{X_2})$. Thus, any continuous map $h: X_1 \to X_2$ maps 1_{X_1} to 1_{X_2}.*

(d) *Prove that the operations are functorial: $(1_X)^* f = f$ and $(1_X)_* f = f$ for any relation f on X. If $h: X_1 \to X_2$ and $g: X_2 \to X_3$ then $(g \circ h)^* = h^* \circ g^*$ and $(g \circ h)_* = g_* \circ h_*$ as operators on relations. Thus, if h maps f_1 to f_2 and g maps f_2 to f_3 then $g \circ h$ maps f_1 to f_3. In particular, conjugacy is an equivalence relation.*

(e) *Prove that $h^*(f_2^{-1}) = (h^* f_2)^{-1}$ and $h_*(f_1^{-1}) = (h_* f_1)^{-1}$.*

(f) *For relations* f_2, $\tilde{f}_2$ *on* X_2 *and* f_1, $\tilde{f}_1$ *on* X_1:

$$h^*(f_2) \circ h^*(\tilde{f}_2) \subset h^*(f_2 \circ \tilde{f}_2) \tag{1.23}$$

with equality if h *is a surjective map.*

$$h_*(f_1 \circ \tilde{f}_1) \subset h_*(f_1) \circ h_*(\tilde{f}_1) \tag{1.24}$$

with equality if h *is an injective map.* □

Now for the main result which says that our constructions are preserved by a semiconjugacy.

17. **Proposition.** *Assume that the continuous map* $h: X_1 \to X_2$ *maps a relation* f_1 *on* X_1 *to a relation* f_2 *on* X_2. *Then* h *maps*:

$$\begin{array}{ll} \overline{f}_1 & \text{to } \overline{f}_2, \\ f_1^{-1} & \text{to } f_2^{-1}, \\ f_1^n & \text{to } f_2^n \quad \text{for } n - 1, 2, \dots, \\ \mathscr{A} f_1 & \text{to } \mathscr{A} f_2 \quad \text{for } \mathscr{A} = \mathscr{O}, \omega, \Omega, \mathscr{R}, \mathscr{N}, \mathscr{G}, \mathscr{C}. \end{array}$$

In particular, a conjugacy between f_1 *and* f_2 *is a conjugacy between* $\mathscr{A} f_1$ *and* $\mathscr{A} f_2$ *for* $\mathscr{A} = \mathscr{O}$, ω, *etc.*

Proof. We are given $f_1 \subset h^* f_2$. That $\overline{f}_1 \subset h^*(\overline{f}_2)$ and $f_1^{-1} \subset h^*(f_2^{-1})$ then follow respectively, from (b) and (e) of the exercise. Induction using (1.23) then implies $f_1^n \subset h^*(f_2^n)$ for all whole numbers n. Taking the union over n we get $\mathscr{O} f_1 \subset h^*(\mathscr{O} f_2)$. For $x \in X_1$:

$$\begin{aligned} \omega f_1(x) &= \limsup\{f_1^n(x)\} \subset \limsup\{h^*(f_2^n)(x)\} \\ &= \limsup\{h^{-1}(f_2^n(h(x)))\} \\ &\subset h^{-1} \limsup\{f_2^n(h(x))\} = h^*(\omega f_2)(x) \end{aligned}$$

by part (f) of Exercise 5. Similarly,

$$\begin{aligned} \Omega f_1 &= \limsup\{f_1^n\} \subset \limsup\{h^*(f_2^n)\} \\ &= \limsup\{(h \times h)^{-1}(f_2^n)\} \\ &\subset (h \times h)^{-1}(\limsup\{f_2^n\}) = h^*(\Omega f_2). \end{aligned}$$

So h maps $\mathscr{A} f_1$ to $\mathscr{A} f_2$ for $\mathscr{A} = \mathscr{O}$, ω, and Ω. Taking unions we get the result for $\mathscr{A} = \mathscr{R} = \mathscr{O} \cup \omega$ and $\mathscr{N} = \mathscr{O} \cup \Omega$.

By monotonicity,

$$f_1 \subset h^* f_2 \subset h^*(\mathscr{G} f_2).$$

On the other hand, by (1.23) and the transitivity of $\mathscr{G} f_2$:

$$h^*(\mathscr{G} f_2) \circ h^*(\mathscr{G} f_2) \subset h^*(\mathscr{G} f_2 \circ \mathscr{G} f_2) \subset h^*(\mathscr{G} f_2).$$

Thus, $h^*(\mathscr{G} f_2)$ is a closed, transitive relation containing f_1 and so it contains $\mathscr{G} f_1$.

Finally, for any $\varepsilon > 0$ we can choose an ε modulus of uniform continuity $\delta > 0$ for h. This says $V_\delta^1 \subset (h \times h)^{-1}(V_\varepsilon^2)$, i.e., $V_\delta^1 \subset h^*(V_\varepsilon^2)$. Consequently,

$$\begin{aligned}\mathscr{C} f_1 &\subset \mathscr{O}(V_\delta^1 \circ f_1) \subset \mathscr{O}(h^*(V_\varepsilon^2) \circ h^* f_2) \\ &\subset h^*(\mathscr{O}(V_\varepsilon^2 \circ f_2)) = (h \times h)^{-1}(\mathscr{O}(V_\varepsilon^2 \circ f_2)).\end{aligned}$$

Intersecting over all $\varepsilon > 0$ we obtain that $\mathscr{C} f_1$ is contained in $(h \times h)^{-1}(\mathscr{C} f_2) = h^*(\mathscr{C} f_2)$. □

Supplementary exercises

18. (a) Let $\mathscr{F}$ be a monotone family of closed subsets of X, i.e., $F_1, F_2 \in \mathscr{F}$ implies $F_1 \subset F_2$ or $F_2 \subset F_1$. Prove that there exists a countable subfamily $\mathscr{F}_0$ of $\mathscr{F}$ such that $\bigcup \mathscr{F}_0 = \bigcup \mathscr{F}$ and $\bigcap \mathscr{F}_0 = \bigcap \mathscr{F}$. (*Hint*: For the intersection result let $\{U_n\}$ be a countable base for the family of open neighborhoods of $\bigcap \mathscr{F}$. There exists $F_n \in \mathscr{F}$ with $F_n \subset U_n$. Let $\mathscr{F}_0 = \{F_n\}$. For the union result let $\{x_n\}$ be a sequence dense in $\bigcup \mathscr{F}$ and let $F_n \in \mathscr{F}$ with $x_n \in F_n$. Either there exists $F \in \mathscr{F}$ such that $F_n \subset F$ for all n—let $\mathscr{F}_0 = \{F\}$—or for all $F \in \mathscr{F}$, $F \subset F_n$ for some n—let $\mathscr{F}_0 = \{F_n\}$.)

(b) Given a closed relation $f \colon X \to X$ define $F_0 = \mathscr{N} f$. Define F_α for each ordinal α by transfinite induction: $F_{\alpha+1} = \mathscr{N} F_\alpha$ and for a limit ordinal α: $F_\alpha = \overline{\bigcup_{\beta<\alpha} F_\beta}$. Thus, $\{F_\alpha\}$ is an increasing family of closed relations containing f. The family stabilizes at α if $F_{\alpha+1} = F_\alpha$ in which case $F_\beta = F_\alpha$ for all $\beta > \alpha$ and $F_\alpha = \mathscr{G} f$. Prove that the family stabilizes at some countable ordinal α (use (a)).

(c) Let Γ be the first uncountable ordinal (= the set of all countable ordinals). The *long line* L is constructed by ordering $\Gamma \times [0, 1)$ lexographically, i.e., $(\alpha_1, t_1) < (\alpha_2, t_2)$ if $\alpha_1 < \alpha_2$ or $\alpha_1 = \alpha_2$ and $t_1 < t_2$, and imposing the order topology. The one point compactification L^* is obtained by attaching the point Γ with $(\alpha, t) < \Gamma$ for all $\alpha < \Gamma$. L^* is a compact Hausdorff space but is not metrizable as L is not even paracompact. Define $f \colon L^* \to L^*$ by $f(\alpha, t) = (\alpha, 2t/(1+t))$ and $f(\Gamma) = \Gamma$. Prove that f is a homeomorphism with fixed points the ordinals, identifying the ordinal α with $(\alpha, 0) \in L$. From f construct the family of closed relations $\{F_\alpha\}$ as in (b). Prove, by induction on α, that for α, β countable ordinals there exists a countable ordinal γ such that $\gamma_1 < \gamma$ for all $\gamma_1 \in F_\alpha(\beta)$ and hence $F_\alpha(\beta)$ is a compact subset of L. (Recall that if $\{\alpha_n\}$ is any sequence of countable ordinals the union α is a countable ordinal with $\alpha_n \leq \alpha$ for all n). On the other hand, prove that $\Gamma \in F_\Gamma(\beta)$ for all countable ordinals β. Thus, the family $\{F_\alpha\}$ does not stabilize at any countable ordinal. Explain the discrepancy with the result in (b). Recall, or prove by induction on α, that there exists

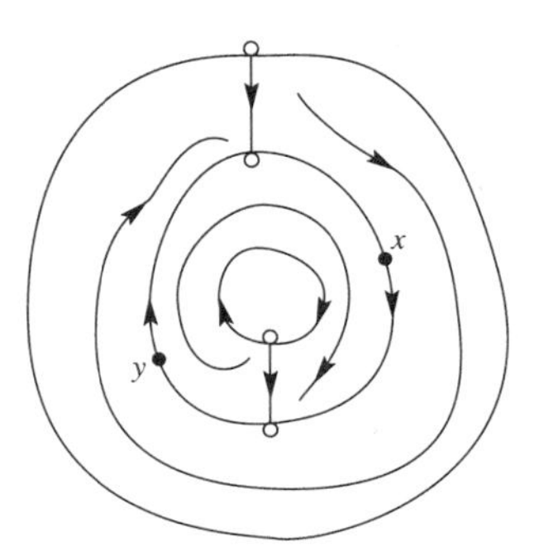

FIGURE 1.2

an order preserving homeomorphism of $L_\beta = \{(\alpha, t) : (\alpha, t) \leq (\beta, 0)\}$ onto the unit interval. Conclude that for any countable ordinal there exists a continuous homeomorphism f of the unit interval such that the associated tower of relations $\{F_\alpha\}$ does not stabilize before β.

19. Interpret Figure 1.2 as the phase portrait of a flow and let f be the associated time-one map. Show that the points labelled x and y satisfy $x \in \Omega f(y)$ and $y \in \Omega f(x)$ but neither x nor y lies in $|\Omega f|$. (Compare with Exercise 10(b).)

20. (a) Prove that for f_2 a closed relation on X_2

$$h_* h^* f_2 \subset f_2, \qquad (h^* f_2)^n \subset h^*(f_2^n),$$
$$\mathscr{A}(h^* f_2) \subset h^*(\mathscr{A} f_2) \quad \text{for } \mathscr{A} = \mathscr{O}, \omega, \Omega, \mathscr{R}, \mathscr{N}, \mathscr{G}, \mathscr{C},$$

with equalities in all cases when h is surjective. (The inclusions follow from Proposition 17 applied to $f_1 = h^* f_2$. The equalities in the surjective case follow from the equality in (1.23) and an adjustment of the previous proof, at least for $\mathscr{A} = \mathscr{O}, \omega, \Omega, \mathscr{R}, \mathscr{N}$. For $\mathscr{C}$, first prove

$$h^* f_2 = \bigcap \{h^* \overline{V}_\delta^2 \circ h^* f_2 : \delta > 0\},$$

and so for every $\varepsilon > 0$ there exists $\delta > 0$ such that $V_\varepsilon^1 \circ h^* f_2 \circ V_\varepsilon^1 \supset h^* \overline{V}_\delta^2 \circ h^* f_2$ and apply Proposition 8 to show $\mathscr{C}(h^* f_2) \supset h^*(\mathscr{C} f_2)$ when h is surjective. For $\mathscr{G}$ the inductive construction of Exercise 18 is needed.)

(b) Prove that for f_1 a closed relation on X_1

$$f_1 \subset h^* h_* f_1, \qquad h_*(f_1^n) \subset (h_* f_1)^n,$$
$$h_*(\mathscr{A} f_1) \subset \mathscr{A}(h_* f_1) \quad \text{for } \mathscr{A} = \mathscr{O}, \omega, \Omega, \mathscr{R}, \mathscr{N}, \mathscr{G}, \mathscr{C}.$$

(Apply Proposition 17 with $f_2 = h_* f_1$. The inclusions are equations when h is injective, a result which is much less useful than its dual in (a).)

(c) With E_h the equivalence relation $h^{-1} \circ h$ check that $h^* h_* f_1 = E_h \circ f_1 \circ E_h$. Assume that h is surjective and then prove

$$h_* \mathscr{A} f_1 = \mathscr{A}(h_* f_1) \quad (\text{for } \mathscr{A} = \mathscr{O}, \omega, \Omega, \mathscr{R}, \mathscr{N}, \mathscr{G}, \mathscr{C})$$

if and only if

$$E_h \circ (\mathscr{A} f_1) \circ E_h = \mathscr{A}(E_h \circ f_1 \circ E_h).$$

(*Hint*: Because h is surjective, the top equation is equivalent to

$$h^* h_* \mathscr{A}(h^* f_1) = h^* A(h_* f_1),$$

and because h is surjective, $h^* \mathscr{A}(h_* f_1) = \mathscr{A}(h^* h_* f_1))$.

(d) Assume that h is surjective and that f_1 is a continuous map. Prove that $h_* f_1$ is a continuous map on X_2 if and only if $E_h \circ f_1 = f_1 \circ E_h$, i.e., $x E_h \tilde{x}$ implies $f_1(x) E_h f_1(\tilde{x})$. Notice that this says that f_1 maps E_h to itself.

(e) If X_1 is a closed subset of X_2 and h is the inclusion map show that $h^* f_2 = f_2 \cap (X_1 \times X_1)$. This is called the *restriction* of f_2 to X_1.

Let $h: X_1 \to X_2$ and $k: X_2 \to X_1$ be continuous maps (we do not assume they are inverses). Define $f_1 = k \circ h$ and $f_2 = h \circ k$. Show that h maps f_1 to f_2 and k maps f_2 to f_1.

Chapter 2. Invariant Sets and Lyapunov Functions

We illustrate the constructions of the previous chapter by applying them to a useful family of homeomorphisms of the unit interval. These *interval examples* are all gradient systems but usually of a rather degenerate type.

Let K be a closed subset of $I = [0, 1]$ with $0, 1 \in K$. Recall that the complement of K in I is the union of a finite or countably infinite family of disjoint open intervals, the components of $I - K$. It is easy to construct a C^1 real valued function u on I such that $u(x) = 0$ exactly when $x \in K$. Furthermore the sign of u on each component of $I - K$ can be preassigned arbitrarily. We call the interval example *positive* if $u > 0$ on $I - K$. The differential equation

$$\frac{dx}{dt} = u(x), \qquad x \in I$$

determines a flow on I with time-one map $f: I \to I$. Clearly, $f(x) = x$ if $x \in K$. If x lies in J, a component of $I - K$, the bi-infinite sequence $\dots, f^{-2}(x), f^{-1}(x), x, f(x), f^2(x), \dots$ remains in J and is monotone increasing or decreasing according to whether u is positive or negative on J. Moving left or right along the sequence we approach the two endpoints of J. Since 0 and 1 are fixed points, f induces a homeomorphism $\tilde{f}$ of the circle S regarded as the quotient of I obtained by identifying 0 and 1.

1. Exercise. *The reader might find two dimensional pictures easier to visualize. Consider polar coordinates on the unit disc and define:*

$$\frac{dr}{dt} = u(r), \qquad \frac{d\theta}{dt} = 2\pi.$$

Prove that to each $x \in K$ there corresponds a periodic solution on the circle of radius $r = x$, while the remaining solutions spiral counter clockwise inward or outward. Show that the time-one map of this system restricts to f on the unit interval (in fact on each ray). Also compare this with the suspension construction of Exercise 0.2. □

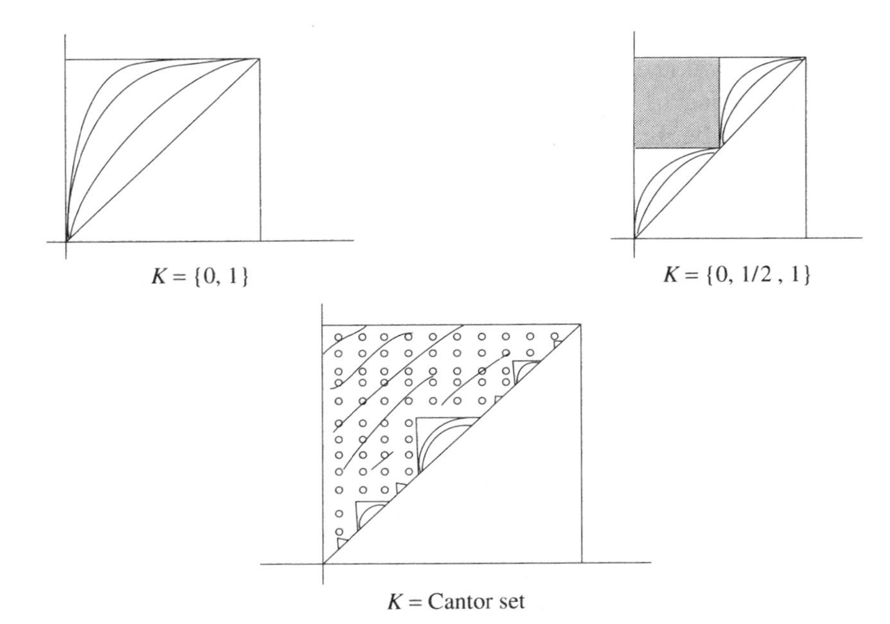

FIGURE 2.1

Figure 2.1 shows positive interval examples with K equal $\{0, 1\}$, $\{0, 1/2, 1\}$, and the classical Cantor set obtained by removing successive "middle thirds".

For $K = \{0, 1\}$ we observe a sequence of curves, the graphs of f, f^2, f^3, ..., approaching $0 \times I \cup I \times 1$. The union of these curves above the diagonal is $\mathscr{O}f$. $\mathscr{R}f = \mathscr{O}f \cup (0, 1] \times 1$ and $\mathscr{N}f = \mathscr{O}f \cup 0 \times I \cup I \times 1$. $\mathscr{N}f$ is transitive and so $\mathscr{N}f = \mathscr{G}f$. When $K = \{0, 1/2, 1\}$, $\mathscr{N}f$ is the union of two of these triangle pictures meeting at $(1/2, 1/2)$. This time $\mathscr{N}f$ is not transitive because $1/2 \in \mathscr{N}f(x_1)$ and $x_2 \in \mathscr{N}f(1/2)$ when $0 \le x_1 \le 1/2 \le x_2 \le 1$. So $\mathscr{N}f \cup (\mathscr{N}f)^2 = \mathscr{N}f \cup [0, 1/2] \times [1/2, 1]$, i.e., the shaded square is adjoined. This figure is transitive and so equals $\mathscr{G}f$. In each of these cases $\mathscr{C}f = \mathscr{G}f$.

Now look at the Cantor set. $\mathscr{N}f$ consists of countably many triangle pictures getting smaller and smaller, together with 1_K, the diagonal points over K. Because no two triangles meet, $\mathscr{N}f$ is again transitive and so $\mathscr{G}f = \mathscr{N}f$. But now if $x_2 > \omega f(x_1)$ for any x_1 it is easy to see that x_1 can be connected to x_2 by an ε-chain for any $\varepsilon > 0$. Just jump past endpoints, skipping intervals of length $< \varepsilon$ entirely. So $\mathscr{C}f$ consists of $\mathscr{G}f$ together with the dotted region above all the triangles.

When we identify 0 with 1 to obtain the associated homeomorphisms of the circle we get $\mathscr{C}\tilde{f} = S \times S$ in all three cases. Also, $\mathscr{G}\tilde{f} = \mathscr{C}\tilde{f}$ in the first two cases. But in the Cantor set example when we identify the points $(0, 0)$ and $(0, 1)$ in $\mathscr{G}f$ the result is a transitive relation on S. So $\mathscr{G}\tilde{f} = \mathscr{G}f/(0, 0) = (1, 1)$ which is a proper subset of $\mathscr{C}\tilde{f} = S \times S$.

2. Exercise. (a) *For any positive interval example with associated circle homeomorphism* $\tilde{f}: S \to S$ *prove that* $\mathscr{C}\tilde{f} = S \times S$.

(b) *Let* K *consist of* $n+1$ *points so that the associated circle homeomorphism* $\tilde{f}$ *has* n *fixed points. When the example is positive,* $(\Omega\tilde{f})^{n+1} = S \times S$ *and so* $\mathscr{G}\tilde{f} = S \times S$. □

Returning to general relations we look now at those subsets which are closed under applications of the relation. A subset A of X is called *positive invariant* with respect to a relation f on X (written "A is f +invariant") if $f(A) \subset A$, i.e., $x \in A$ implies $f(x) \subset A$. A is called f *invariant* if $f(A) = A$. This means that A is f +invariant and, in addition, $f^{-1}(x) \cap A \neq \varnothing$ for all x in A. The latter condition requires $A \subset \mathrm{Dom}(f^{-1}) = f(X)$ but does not imply $f^{-1}(x) \subset A$ for x in A, i.e., f^{-1} +invariance. Thus, A f +invariant, f^{-1} +invariant and $A \subset f(X)$ together imply that A is f invariant but the converse is true only when f is a one-to-one mapping.

3. Exercise. (a) *If* $f_1 \subset f_2$, *then* A f_2 *+invariant implies* A *is* f_1 *+invariant*.

(b) *The intersection and union of a family of* f *+invariant sets are* f *+invariant sets.*

(c) *If* f *is a closed relation and* $\{A_n\}$ *is a decreasing sequence of closed* f *invariant sets, then* $\bigcap\{A_n\}$ *is* f *invariant.*

(d) *If* A_1 *is* f *+invariant and* f^{-1} *+invariant and* A_2 *is* f *invariant, then* $A_1 \cap A_2$ *is* f *invariant. (Hint: For* $x \in A_1 \cap A_2$ *there exists* $y \in A_2$ *such that* $x \in f(y)$. $y \in A_1$ *as well by* f^{-1} *+invariance.)*

(e) *If* f *is a closed relation and* $F = \mathscr{G}f$ *or* $\mathscr{C}f$ *then* A F *+invariant implies* $F(A) = f(A)$. *(Hint: Use* $F = f \cup f \circ F$.*) So an* F *+invariant set is* F *invariant if and only if it is* f *invariant. If* f *is a continuous map, then the same results hold for* $F = \mathscr{R}f$ *and* $\mathscr{N}f$. □

Recall that for A a closed subset of X and f a closed relation on X, $\omega f[A] = \limsup\{f^n(A)\}$ is the subset of X upon which the images of A accumulate. When A is f +invariant, $\omega f[A]$ is a subset of A with special properties.

4. Proposition. (a) *Let* f *be a closed relation on* X *and let* A *be a closed* f *+invariant subset. Then*

$$\omega f[A] = \bigcap_{n \geq 0}\{f^n(A)\}.$$

$\omega f[A]$ *is a closed* f *invariant subset of* A *and it contains any other invariant subset of* A.

(b) *Let* F *be a closed, transitive relation on* X. *Then*

$$\Omega F = \omega F = \bigcap_{n \geq 0}\{F^n\}.$$

ΩF *is a closed, transitive relation contained in* F. *In fact,*

$$F \circ \Omega F = \Omega F = \Omega F \circ F.$$

(c) *If* f *is a closed relation on* X, *then for* $F = \mathscr{G}f$ *or* $\mathscr{C}f$:

$$F = \mathscr{O}f \cup \Omega F, \qquad f \circ \Omega F = \Omega F = \Omega F \circ f, \qquad \Omega F = \Omega F \circ \omega f.$$

In particular, a subset A *is* F $+$*invariant if and only if it is* f $+$*invariant and* ΩF $+$*invariant.*

PROOF. (a) If A is f $+$invariant then the sequence of images $\{f^n(A)\}$ is decreasing. If A and f are closed then all the $f^n(A)$ are closed and so $\omega f[A] = \limsup\{f^n(A)\} = \bigcap\{f^n(A)\}$. $\omega f[A]$ is the intersection of f $+$invariant sets and so it is f $+$invariant. If $x \in \omega f[A]$ then $f^{-1}(x) \cap f^n(A) \neq 0$ because $x \in f^{n+1}(A)$. So $f^{-1}(x) \cap \omega f[A] = \bigcap\{f^{-1}(x) \cap f^n(A)\}$ is nonempty as the decreasing intersection of closed nonempty sets. Thus $\omega f[A]$ is f invariant.

Finally, if $B \subset A$ and $f(B) = B$ then $B \subset f^n(A)$ for all n and so $B \subset \omega f[A]$.

(b) By transitivity $\{F^n\}$ is a decreasing sequence and so $\Omega F = \limsup\{F^n\} = \bigcap\{F^n\}$. Also $\omega F(x) = \limsup\{F^n(x)\} = \bigcap\{F^n(x)\} = \bigcap\{F^n\}(x) = \Omega F(x)$. $F^{n+1} = F \circ F^n = F^n \circ F$ so $\Omega F = F \circ \Omega F = \Omega F \circ F$ by Proposition 1.4.

(c) For $F = \mathscr{G}f$ or $\mathscr{C}f$ we have, by Proposition 1.10(d): $F = f \cup f \circ F = f \cup f \circ (f \cup f \circ F) = f \cup f^2 \cup f^2 \circ F$. So by induction

$$F = f \cup \cdots \cup f^n \cup f^n \circ F \subset \mathscr{O}f \cup F^{n+1} \subset F,$$

where the latter inclusion follows from the transitivity of F and $f \subset F$. Thus, for all n, $F = \mathscr{O}f \cup F^{n+1}$ and intersecting on n we have $F = \mathscr{O}f \cup \Omega F$.

Also, $F = f \cup f \circ F$ implies $F^{n+1} = f \circ F^n \cup f \circ F^{n+1} = f \circ F^n$ since $F^{n+1} \subset F^n$. Intersecting on n, we get $\Omega F = f \circ \Omega F$ by Proposition 1.4. Similarly, $F = f \cup F \circ f$ implies $\Omega F = \Omega F \circ f$. For each x, $\Omega F = \Omega F \circ f^n$ implies $\Omega F(x) = \Omega F(f^n(x))$. By Exercise 1.5(e) $\Omega F(x) \subset \Omega F \circ \omega f(x) \subset \Omega F \circ \Omega f(x) \subset \Omega F(x)$.

Finally, A is f $+$invariant if and only if it is $\mathscr{O}f$ $+$invariant. It is $\mathscr{O}f \cup \Omega F$ $+$invariant if and only if it is $\mathscr{O}f$ $+$invariant and ΩF $+$ invariant. □

Part (c) shows that for $F = \mathscr{G}f$ or $\mathscr{C}f$ the limit relation ΩF plays the same role for F that Ωf does for $\mathscr{N}f$ and ωf does for $\mathscr{R}f$. Thus, if f is a continuous map, for example, $y \in F(x)$ means either $y = f^n(x)$ for some $n = 1, 2, \ldots$ or $y \in \Omega F(x)$. The limit relation ΩF is important because the corresponding recurrent points appear naturally here. Clearly, if $x \in F(x)$, i.e., $x \in |F|$, then $x \in F^n(x)$ for all n and $x \in \Omega F(x)$. So if there exists $z \in |F|$ with $z \in F(x)$ and $y \in F(z)$ then $y \in \Omega F(x)$ because

$\Omega F = F \circ \Omega F \circ F$. As we will now show the converse is true. This means that for $y \in F(x)$ either $y \in f^n(x)$ for some n or the association from x to y "factors through" $|F|$.

5. LEMMA. *Let F be a closed relation on X and B be a closed set disjoint from $|F|$. There exists $\varepsilon > 0$ such that $y \in F(x)$ for x, y in B implies $d(x, y) > \varepsilon$. If in addition, F is transitive there exists N such that no F chain in B has length greater than N, i.e., $x_0, x_1, \dots, x_k \in B$ with $x_{i+1} \in F(x_i)$ implies $k \leq N$.*

PROOF. B disjoint from $|F|$ says that 1_B, the set of diagonal points of B, is disjoint from F. So there exists $\varepsilon > 0$ such that $\overline{V}_\varepsilon \cap B \times B$ is disjoint from F. Equivalently, $F \cap B \times B$ is disjoint from $\overline{V}_\varepsilon$.

Now suppose F is transitive. Choose a finite cover of B by sets of diameter $\leq \varepsilon$. Suppose that the cover has N members. If $\{x_n\}$ is a chain in B then by transitivity of F, $x_i \in F(x_j)$ whenever $i > j$. So $d(x_i, x_j) > \varepsilon$ and the points of the chain lie in distinct elements of the cover. So the chain contains at most N points. □

The following list of useful relationships between the limit relation ΩF and the recurrent set $|F|$ includes the factorization result we want (part (c)).

6. PROPOSITION. (a) *If A is a closed f +invariant set for a closed relation f, then*

$$A \cap |f| = \omega f[A] \cap |f|.$$

(b) *If A is a closed F +invariant set for a closed, transitive relation F, then*

$$\omega F(A) = \omega F[A] = \Omega F(A) = F(A \cap |F|), \qquad A \cap |F| = \Omega F(A) \cap |F|.$$

(c) *For a closed, transitive relation F,*

$$\Omega F(x) = F(F(x) \cap |F|),$$

i.e., $y \in \Omega F(x)$ if and only if there exists $z \in F(x) \cap |F|$ with $y \in F(z)$.

(d) $|F| = |\Omega F|$ *for a closed, transitive relation F.*

PROOF. (a) $x \in A$ with $x \in f(x)$ implies $x \in f^n(A)$ for $n = 1, 2, \dots$ and so $x \in \bigcap\{f^n(A)\} = \omega f[A] \subset A$.

(d) $\Omega F \subset F$ and so $|\Omega F| \subset |F|$. Conversely, $x \in F(x)$ implies $x \in F^n(x)$ for all n. So $x \in |\Omega F|$.

(c) As remarked before Lemma 5, $F(F(x) \cap |F|) \subset \Omega F(x)$ because $z \in F(x) \cap |F|$ and $y \in F(z)$ implies $y \in F \circ \Omega F \circ F(x)$. For the converse suppose $y \in \Omega F(x)$ and consider the closed set

$$B = F^{-1}(y) \cap F(x) = \{z : y \in F(z) \text{ and } z \in F(x)\}.$$

This is a closed set and $y \in F^n(x)$ for all n implies that B contains chains of length n for all n. So by Lemma 5 $B \cap |F| \neq \varnothing$.

(b) Because A is F +invariant $\omega F[A] = \bigcap\{F^n(A)\}$. By Proposition 4(b) $\omega F(A) = \Omega F(A) = \bigcap\{F^n\}(A)$. These intersections are equal by Proposition 1.4. By (a) $A \cap |F| \subset \omega F[A]$ and so by the F +invariance of $\omega F[A]$: $F(A \cap |F|) \subset \omega F[A]$. Conversely, if $y \in \omega F(x)$ for some x in A then by (c) $y = F(z)$ for some $z \in F(x) \cap |F| \subset A \cap |F|$. Finally, by (a), $A \cap |F| = \omega F[A] \cap |F|$.

REMARKS. (a) If F is transitive and $A \subset X$ then $F(A)$ and $A \cup F(A)$ are F +invariant. In fact, $A \cup F(A)$ is the smallest F +invariant set containing A.

(b) An important example of a set which is usually *not* F +invariant is $|F|$. Notice that for $F = \mathscr{N} f$, $\mathscr{G} f$, or $\mathscr{C} f$ the set $|F| = K$ is not F +invariant for any of the interval examples of Figure 2.1.

(c) Part (d) is a special case of a more general result: For any closed relation f: $|\mathscr{N} f| = |\Omega f|$. This follows from $|\mathscr{N} f| = |\mathscr{O} f| \cup |\Omega f|$ and $|\mathscr{O} f| \subset |\Omega f|$. □

Hidden beneath the formalism there are some useful ideas here. For A a closed f +invariant subset, $B = \omega f[A]$ is the maximum f invariant subset of A. From (a) we have $B \cap |f| \subset A \cap |f| \subset B$ and so $B \cap |f| = A \cap |f|$. If, in addition, $f = F$ is transitive then by (b) $B = F(B \cap |F|)$. This means that the F invariant set B can be "recovered" from the intersection with the cyclic set $|F|$, $B \cap |F|$, which we call the *trace of* B *on* $|F|$.

We now contrast the different concepts of F +invariance for the various extensions F of f.

Notice first that a subset A is f +invariant if and only if it is $\mathscr{O} f$ +invariant. If A and f are closed A is then $\mathscr{R} f$ +invariant as well because $x \in A$ implies $f^n(x) \subset A$ for all n and so $\mathscr{R} f(x) = \overline{\bigcup\{f^n(x)\}} \subset A$. However, $\mathscr{N} f$ +invariance is usually a stronger condition and $\mathscr{G} f$ +invariance is stronger still (except, of course, when f is transitive in which case $f = \mathscr{N} f = \mathscr{G} f$).

7. PROPOSITION. (a) *Let* f *be a closed relation on* X *and* A *be a closed subset. The following conditions on* A *are equivalent*:

(1) A *is* $\mathscr{N} f$ *+invariant.*

(2) *For every neighborhood* G *of* A *there exists a neighborhood* U *of* A *contained in* G *such that* $x \in U$ *implies* $\mathscr{O} f(x) \subset G$. (*This is usually called* stability *of* A *with respect to* f.)

(3) *For every neighborhood* G *of* A *there exists a neighborhood* U *of* A *contained in* G *with* U f *+invariant, i.e.,* $x \in U$ *implies* $\mathscr{O} f(x) \subset U$. (*This means the* f *+invariant sets include a base for the neighborhood system of* A.)

(b) *Let* F *be a closed transitive relation on* X *and* A *be a closed* F *+invariant subset. For every neighborhood* G *of* A *there exist* U *and* U_0, F *+invariant subsets, with* U_0 *open,* U *closed, and* $A \subset U_0 \subset U \subset G$.

PROOF. (a) $(3) \Rightarrow (2)$. This is obvious.

$(2) \Rightarrow (1)$. Given $\varepsilon > 0$ there exists a positive $\delta < \varepsilon$ such that $x \in V_\delta(A)$ implies $\mathscr{O}f(x) \subset V_\varepsilon(A)$. Now suppose $x \in A$ and $y \in \mathscr{N}f(x)$ Then there exists a chain beginning δ close to x and ending δ close to y. As the chain remains in $V_\varepsilon(A)$ it follows that $y \in V_{2\varepsilon}(A)$. Intersecting on ε we get $y \in A$.

$(1) \Rightarrow (3)$. By Corollary 1.2 the $\mathscr{N}f$ +invariance of A implies that for some $\delta > 0$ $\mathscr{N}f(\overline{V}_\delta(A))$ is contained in the neighborhood G. We can choose δ so that $\overline{V}_\delta(A) \subset G$ as well. Hence

$$G \supset U_1 \equiv \overline{V}_\delta(A) \cup \mathscr{N}f(\overline{V}_\delta(A)) \supset U_2 \equiv V_\delta(A) \cup \mathscr{O}f(V_\delta(A)) \supset V_\delta(A).$$

U_2 is clearly an f +invariant neighborhood of A.

(b) If $f = F$ is also transitive, then $\mathscr{N}f = \mathscr{O}f = F$ as well. So in the above proof that $(1) \Rightarrow (3)$ $U_1 = \overline{V}_\delta(A) \cup F(\overline{V}_\delta(A))$ is closed and is F +invariant. Also by transitivity

$$U_0 \equiv \{x : F(x) \subset \operatorname{Int} U_1\}$$

is F +invariant. It is open by Proposition 1.1 and contains A because A is F +invariant and $V_\delta(A) \subset U_1$.

REMARK. If f is a continuous map then $f \circ \mathscr{N}f \subset \mathscr{N}f$ (Proposition 1.12(a)) implies that U_1 is f +invariant. So U in (a) can be chosen to be closed. Similarly, the open set $\{x : \mathscr{N}f(x) \subset \operatorname{Int} U_1\}$ is f +invariant and contains A. So, alternatively, U in (a) can be chosen to be open. □

8. EXERCISE. *Let* $K = \{n^{-1} : n = 1, 2, \ldots\} \cup \{0\}$. *Illustrate a positive interval example for* K. *Show that* $\{0\}$ *is* $\mathscr{N}f$ *invariant with* f *+invariant neighborhoods* $[0, n^{-1}]$ *for* $n = 1, 2, \ldots$. *Show that the only* $\mathscr{N}f$ *+invariant set containing* 0 *in its interior is* I. *Thus, the neighborhoods of* A *in Proposition* 7(a) *cannot necessarily be chosen* $\mathscr{N}f$ *invariant. Show that* $\{0\}$ *is not* $\mathscr{G}f$ *+invariant*. □

We now introduce a useful way of partially measuring the direction of motion by using a real parameter. A continuous real valued function L on X is called a *Lyapunov function* for a closed relation f if $y \in f(x)$ implies that $L(y) \geq L(x)$, i.e.,

$$f \subseteq \leq_L \equiv \{(x, y) : L(x) \leq L(y)\}.$$

Because $\leq_L$ is a closed, transitive relation on X $f \subseteq \leq_L$ implies $\mathscr{G}f \subseteq \leq_L$. Thus, a Lyapunov function for f is automatically a Lyapunov function for $\mathscr{G}f$ and a fortiori for $\mathscr{N}f$. However, the condition that L be a Lyapunov function for $\mathscr{C}f$ is usually a stronger demand.

L is a Lyapunov function precisely when $y_1 \in f^{-1}(x)$ and $y_2 \in f(x)$ imply $L(y_1) \leq L(x) \leq L(y_2)$. x is called a *regular point* if these inequalities are strict for all such y_1 and y_2. Because the infimum and supremum are attained by L on closed sets we can rephrase this as: x is a regular point if

$$\sup L|f^{-1}(x) < L(x) < \inf L|f(x). \tag{2.1}$$

If $\{x_n\}$ is any chain in X then the sequence $\{L(x_n)\}$ is nondecreasing. If the points of the chain are regular then the sequence is increasing.

A point which is not regular is called a *critical point*. We denote the set of critical points of L by $|L|$ (despite the dependence of the set upon f as well as L). Clearly, $|f| \subset |L|$ since $x \in f(x) \cap f^{-1}(x)$ for x in $|f|$.

Notice that $-L$ is a Lyapunov function for f^{-1} if L is a Lyapunov function for f and the symmetry of (2.1) implies $|-L| = |L|$.

9. PROPOSITION. *Let f be a closed relation on X.*

(a) *The set of Lyapunov functions for f includes constant functions and is closed under addition and under multiplication by positive scalars. If $\{L_n\}$ is a sequence of Lyapunov functions converging pointwise to L, then L is a Lyapunov function if it is continuous. In particular, if $L = \sum_n L_n$ is either a finite or uniformly convergent series of Lyapunov functions, then L is a Lyapunov function, and, in addition,*

$$|L| \subset \bigcap_n \{|L_n|\}. \tag{2.2}$$

(b) *If x is a regular point for a Lyapunov function L, then x is regular with respect to $\mathscr{G}f$, i.e.,* (2.1) *is equivalent to the apparently stronger condition*:

$$\sup L|(\mathscr{G}f)^{-1}(x) < L(x) < \inf L|(\mathscr{G}f)(x). \tag{2.3}$$

(c) *The set of critical points of a Lyapunov function is closed and contains* $|\mathscr{G}f|$.

PROOF. (a) Each inequality $L(x) \leq L(y)$ for $(x, y) \in f$ is preserved under addition, multiplication by positive scalars, and pointwise limit. Furthermore, the inequality is strict for $L = \sum_n L_n$ if it is strict for any summand. Hence, (2.2).

(b) Recall that $\mathscr{G}f = f \cup \mathscr{G}f \circ f$ by Proposition 1.11(d). So if $y \in \mathscr{G}f(x)$ either $y \in f(x)$ or $y \in \mathscr{G}f(z)$ for some $z \in f(x)$. Because an f Lyapunov function is automatically a $\mathscr{G}f$ Lyapunov function $L(y) \geq L(z)$ in the latter case. Thus, $\inf L|\mathscr{G}f(x) = \inf L|f(x)$. $\sup L|(\mathscr{G}f)^{-1}(x) = \sup L|f^{-1}(x)$ is the same equation applied to $-L$ for f^{-1}. So (2.3) follows from (2.1).

(c) Since (2.3) is false for points in $|\mathscr{G}f|$ such points are critical. Finally, $|L|$ is closed because it is the domain of the closed relation:

$$(f \cup f^{-1}) \cap (L \times L)^{-1} 1_{\mathbb{R}} = \{(x, y) \in f \cup f^{-1} : L(x) = L(y)\}.$$

REMARK. If L is a $\mathscr{C}f$ Lyapunov function (recall that this is not automatic) then the proof of (b) shows that

$$\inf L|\mathscr{C}f(x) = \inf L|f(x) \quad \text{and} \quad \sup L|(\mathscr{C}f)^{-1}(x) = \sup L|f^{-1}(x).$$

So the critical points concepts relative to f and relative to $\mathscr{C}f$ then agree as well. Of course, in that case, $|\mathscr{C}f| \subset |L|$. □

If $f: X \to X$ is a homeomorphism then we call L a *strict Lyapunov function* if $L(x) \leq L(f(x))$ with strict inequality unless $f(x) = x$. This is the discrete time analogue of the strict Lyapunov functions described in the introduction. To say that a Lyapunov function L is strict for a homeomorphism f is precisely to say that the only critical points are fixed points, i.e., $|L| = |f|$. By part (c) above this requires that $|\mathscr{G}f| = |f|$, i.e., the only generalized nonwandering points are fixed points. This condition is sufficient for the existence of a strict Lyapunov function. In fact, we will now show that for any closed relation f there exist Lyapunov functions with critical set as small as part (c) allows namely $|L| = |\mathscr{G}f|$.

Notice that if L is a Lyapunov function then for any real c: $L^{-1}[c, \infty) = \{x : L(x) \geq c\}$ is a closed f +invariant set. In fact, because L is a $\mathscr{G}f$ Lyapunov function this set is $\mathscr{G}f$ +invariant. The key idea in constructing Lyapunov functions is the following converse:

10. LEMMA. *Let A be a closed, $\mathscr{G}f$ +invariant subset. Then there exists $L: X \to [0, 1]$ a Lyapunov function for f with $A = L^{-1}(1)$.*

PROOF. We use the standard Urysohn Lemma argument together with Proposition 7(b). Let $\{0, 1, r_2, r_3, \dots\}$ be a counting of the rationals in $[0, 1]$. Define $U_0 = X$ and $U_1 = A$. Proceeding inductively using Proposition 7(b) we can choose U_n a closed, $\mathscr{G}f$ +invariant neighborhood of A such that $r_n < r_m$ implies $\operatorname{Int} U_n \supset U_m$. We can also make sure that $\bigcap\{U_n : n \geq 2\} = A$. Define $L(x)$ to be the value of the associated Dedekind cut:

$$L(x) = \sup\{r_n : x \in U_n\} = \inf\{r_n : x \notin U_n\}.$$

As usual, L is continuous because $\{L < t\} = \bigcup_{r_n > t}\{X - U_n\}$ and $\{L > t\} = \bigcap_{r_n > t}\{\operatorname{Int} U_n\}$. Because U_n is +invariant $x \in U_n$ implies $f(x) \subset U_n$ and so for $y \in f(x)$ we have

$$\{r_n : y \in U_n\} \supset \{r_n : x \in U_n\}.$$

Thus, $L(y)$, the sup of the larger set, is at least $L(x)$.

Since $x \in A$ if and only if $x \in U_n$ for all $r_n < 1$, $L^{-1}(1) = A$. □

11. EXERCISE. *That L is a Lyapunov function follows because the U_n are f +invariant. $\mathscr{G}f$ +invariance was not required. Where did we use the stronger demand that A be $\mathscr{G}f$ +invariant? Why would $\mathscr{N}f$ +invariance not suffice by using Proposition* 7(a) *instead of* (b)? □

12. THEOREM. *Let F be a closed, transitive relation on X. There exists a continuous real-valued function L on X such that $y \in F(x)$ implies $L(x) \leq L(y)$ with equality only when, in addition, $x \in F(y)$. In particular, L is a Lyapunov function for F with $|L| = |F|$.*

Proof. Notice first that if $y \in F(x)$ and $x \in F(y)$ (i.e., $(x, y) \in F \cap F^{-1}$) then $L(x) = L(y)$ for any Lyapunov function. Our proof begins by constructing for each pair (x, y) in $F - F^{-1}$ (i.e., $y \in F(x)$ but $x \notin F(y)$) a Lyapunov function $L_{(x,y)}$ such that $L_{(x,y)}(x) < L_{(x,y)}(y)$.

Define the closed sets: $A_+ = \{y\} \cup F(y)$ and $A_- = \{x\} \cup F^{-1}(x)$. A_+ is F +invariant and A_- is F^{-1} +invariant. Furthermore, $A_+ \cap A_- = \emptyset$ because $z \in (1_X \cup F)(y)$ and $x \in (1_X \cup F)(z)$ would imply $x \in (1_X \cup F)(y)$ by transitivity of F. This contradicts the assumptions $y \in F(x)$ and $x \notin F(y)$.

Because F and F^{-1} are closed and transitive $F = \mathscr{G}F$ and $F^{-1} = \mathscr{G}F^{-1}$. So we can apply Lemma 10 to construct $[0, 1]$ valued Lyapunov functions L_+ and L_- for F and F^{-1} respectively such that $L_+^{-1}(1) = A_+$ and $L_-^{-1}(1) = A_-$. Since $A_+ \cap A_- = \emptyset$, we have $L_+ < 1$ on A_- and $L_- < 1$ on A_+. Define the Lyapunov function $L_{(x,y)}$ for F to be $L_+ - L_-$. Notice that

$$L_{(x,y)}(x) < 1 - 1 < L_{(x,y)}(y).$$

Now define the open subset $G_{(x,y)}$ of $X \times X$ by

$$G_{(x,y)} = \{(x_1, y_1) : L_{(x,y)}(x_1) < L_{(x,y)}(y_1)\}.$$

By construction $(x, y) \in G_{(x,y)}$. The family $\{G_{(x,y)} : (x, y) \in F - F^{-1}\}$ covers the open subset $F - F^{-1}$ of F. $F - F^{-1}$ is a subset of a compact metric space. So, while it is not compact, its topology does have a countable base. In particular, $F - F^{-1}$ satisfies the Lindelöf property: every open cover has a countable subcover. So there is a countable sequence $\{(x_n, y_n)\}$ such that

$$F - F^{-1} \subset \bigcup \{G_{(x_n, y_n)}\}.$$

Because $-1 \leq L_{(x_n, y_n)} \leq 1$ for all n, the series $\sum 2^{-n} L_{(x_n, y_n)}$ converges uniformly and the limit L is a Lyapunov function by Proposition 9(a). Thus, $L(x) \leq L(y)$ for $(x, y) \in F$. The inequality is strict for some summand at each point of $F - F^{-1}$ because this set is covered by the $G_{(x_n, y_n)}$'s. So $L(x) < L(y)$ for $(x, y) \in F - F^{-1}$.

In particular, $y \in F(x)$ and $L(x) = L(y)$ imply $x \in F(y)$ and so by transitivity $x, y \in |F|$. Thus, the only critical points lie in $|F|$. □

13. **Corollary.** *Let f be a closed relation on X. Then there exists a Lyapunov function L for f with $|L| = |\mathscr{G}f|$.*

Proof. Use $F = \mathscr{G}f$. □

Of course by using $F = \mathscr{C}f$ we can construct a $\mathscr{C}f$ Lyapunov function with $|L| = |\mathscr{C}f|$ as well. In the next chapter we will construct $\mathscr{C}f$ Lyapunov functions by a different method. To illustrate the difference between the two types consider $f = 1_X$. Any real valued L is a $1_X = \mathscr{G}1_X$ Lyapunov function but by Exercise 1.9 L is a $\mathscr{C}1_X$ Lyapunov function if and only if it is constant on the components of X.

The positive interval examples and their associated homeomorphisms of the circle are more interesting.

For a positive interval example f with fixed point set $K \subset I$ the identity $1: I \to [0, 1]$ is a $\mathscr{C}f$ Lyapunov function with $|1| = K$ because $f^{-1}(x) < x < f(x)$ unless $x \in K$. But the results for $\tilde{f}$ on the circle depend on the nature of K. For $K = \{0, 1/2, 1\}$ (or any finite set) $\mathscr{G}\tilde{f} = \mathscr{C}\tilde{f} = S \times S$ by Exercise 2 and so the only Lyapunov functions are constants. For any positive interval example $\mathscr{C}\tilde{f} = S \times S$ and so the only $\mathscr{C}\tilde{f}$ Lyapunov functions are constant. But consider $K = \{0, 1\} \cup [1/3, 2/3]$ and define:

$$L(x) = \begin{cases} x, & 0 \le x \le 1/3, \\ 1 - 2x, & 1/3 \le x \le 2/3, \\ x - 1, & 2/3 \le x \le 1. \end{cases}$$

Since L is increasing on $(0, 1/3)$ and $(2/3, 1)$ it is a Lyapunov function for f. Since $L(0) = L(1)$ it induces a Lyapunov function for $\tilde{f}$. $|L|$ on the circle is $|\mathscr{G}\tilde{f}| = K$ (or more precisely the image of K in the quotient circle).

But the real puzzle is the Cantor set example. In that case as well $\mathscr{G}\tilde{f} = K$. So there must exist a Lyapunov function L on the circle with $|L| = K$. Pulled back to I, L can be chosen strictly increasing on each middle third interval *but* $L(0) = L(1)$. What is going on here? For the solution see the next chapter.

We conclude this chapter with an explicit description of $\Omega\mathscr{C}f$ and some consequences of it which will be useful later. Again, beware of tedium.

14. LEMMA. *Let f be a closed relation on X. For x, $y \in X$ the following are equivalent.*

(1) $y \in \Omega\mathscr{C}f(x)$.

(2) *There exists ε_k-chains of length n_k connecting x with y, where $\{\varepsilon_k\}$ is a sequence of positive numbers approaching* 0 *and $\{n_k\}$ is a sequence of whole numbers approaching ∞.*

(3) *There exists $\{x_i : i = 0, 1, \ldots\}$ an infinite* 0*-chain beginning with x, i.e., $x_0 = x$ and $x_{i+1} \in f(x_i)$, such that $y \in \mathscr{C}f(x_i)$ for all i.*

PROOF. (1) $\Rightarrow$ (2). If $y \in (\mathscr{C}f)^n(x)$ then for every $\varepsilon > 0$ there exists an ε chain connecting x with y of length at least n.

(3) $\Rightarrow$ (1). $x_n \in f^n(x) \subset (\mathscr{C}f)^n(x)$ and so $y \in (\mathscr{C}f)^{n+1}(x)$ for all n.

(2) $\Rightarrow$ (3). We can restate (2) by saying there exists an infinite subset N_0 of the whole numbers and for each $n \in N_0$ there exists $\varepsilon_n > 0$ with $\lim_n \varepsilon_n = 0$ and an ε_n chain of length n $\{x_{in} : i = 0, 1, \ldots, n\}$ with

$x_{0n} = x$, $x_{nn} = y$. By induction we can refine $N_0 \supset N_1 \supset N_2 \supset \cdots \supset N_p$ with N_p an infinite subset such that for $i = 0, 1, \dots, p$ the subsequence $\{x_{in}\}$ converges to a point x_i as $n \to \infty$ in N_p. Define N_∞ by the diagonal process so that the kth element of N_∞ is the kth element of N_k. N_∞ is an infinite subset of N_0 and as $n \to \infty$ in N_∞ $\lim\{x_{in}\} = x_i$ for $i = 0, 1, 2, \dots$. Clearly $x_0 = x$. As $(x_{in}, x_{i+1n}) \in \overline{V}_{\varepsilon_n} \circ f$ for all $n > i$ in N_0, we let n tend to ∞ in N_∞ to get $(x_i, x_{i+1}) \in \bigcap_n\{\overline{V}_{\varepsilon_n} \circ f\} = f$, i.e., $x_{i+1} \in f(x_i)$. Now given $\varepsilon > 0$ and i, choose $n \in N_\infty$ with $n > i$, $\varepsilon_n < \varepsilon$ and $d(x_{in}, x_i) < \varepsilon$. The sequence $\{x_i, x_{i+1\,n}, \dots, x_{nn} = y\}$ shows that $y \in \mathcal{O}(V_\varepsilon \circ f \circ V_\varepsilon)(x_i)$. So by Proposition 1.8, $y \in \mathcal{C}f(x_i)$. □

For the next result define

$$\mathcal{O}_n f = \bigcup_{k \geq n} f^k. \tag{2.4}$$

So by definition of the lim sup,

$$\Omega f = \bigcap_n \overline{\mathcal{O}_n f}. \tag{2.5}$$

15. **Proposition.** *For f a closed relation on X*

$$\begin{aligned}\Omega\mathcal{C} f &= \bigcap_{n,\varepsilon} \mathcal{O}_n(V_\varepsilon \circ f) = \bigcap_{n,\varepsilon} \mathcal{O}_n(V_\varepsilon \circ f \circ V_\varepsilon) \qquad (\varepsilon > 0)\\ &= \bigcap_\varepsilon \Omega(\overline{V}_\varepsilon \circ f) = \bigcap_\varepsilon \Omega(\overline{V}_\varepsilon \circ f \circ \overline{V}_\varepsilon).\end{aligned}$$

Furthermore, we have the identities

$$\Omega\mathcal{C} f = \mathcal{C} f \circ \Omega f = \mathcal{C} f \circ \omega f = \mathcal{C}\Omega f.$$

Proof. That $\Omega\mathcal{C} f = \bigcap_{n,\varepsilon} \mathcal{O}_n(V_\varepsilon \circ f)$ follows from the equivalence of (1) and (2) of the lemma. Clearly,

$$\begin{array}{ccc}\bigcap_{n,\varepsilon} \mathcal{O}_n(V_\varepsilon \circ f) & \subset & \bigcap_{n,\varepsilon} \mathcal{O}_n(V_\varepsilon \circ f \circ V_\varepsilon)\\ \cap & & \cap \\ \bigcap_{n,\varepsilon} \overline{\mathcal{O}_n(\overline{V}_\varepsilon \circ f)} & \subset & \bigcap_{n,\varepsilon} \overline{\mathcal{O}_n(\overline{V}_\varepsilon \circ f \circ \overline{V}_\varepsilon)}.\end{array}$$

The proof that all of these intersections agree is identical to the proof of the analogous result for $\mathcal{C} f$, Proposition 1.8. Just replace $\mathcal{O}$ by $\mathcal{O}_n$ throughout. The replacement of $\bigcap_n \overline{\mathcal{O}_n}$ by Ω follows from (2.5).

Now $\mathcal{C} f \circ \omega f \subset \mathcal{C} f \circ \Omega f \subset \mathcal{C} f \circ \Omega\mathcal{C} f = \Omega\mathcal{C} f$ by Proposition 4(b). On the other hand, by Proposition 4(c) $\Omega\mathcal{C} = \Omega\mathcal{C} f \circ \omega f \subset \mathcal{C} f \circ \omega f$. Thus, all of the inclusions are equalities.

Next we follow the proof that $\mathcal{C}\mathcal{C} f = \mathcal{C} f$ (Proposition 1.11(c)): if $\varepsilon_1 + \varepsilon_2 < \varepsilon$ then

$$\mathcal{O}(V_{\varepsilon_1} \circ \mathcal{O}_n(V_{\varepsilon_2} \circ f \circ V_{\varepsilon_2}) \circ V_{\varepsilon_1}) \subset \mathcal{O}_n(V_\varepsilon \circ f \circ V_\varepsilon),$$

and this implies $\mathscr{C}\Omega\mathscr{C}f \subset \Omega\mathscr{C}f$. Clearly, then $\mathscr{C}\Omega f \subset \mathscr{C}\Omega\mathscr{C}f \subset \Omega\mathscr{C}f$.

For the reverse inclusion it suffices to show that for each $\varepsilon > 0$ $\Omega\mathscr{C}f \subset \mathscr{O}(V_\varepsilon \circ \Omega f \circ V_\varepsilon)$ since by Proposition 1.8 the intersection of the latter family is $\mathscr{C}\Omega f$. Choose n and $\varepsilon_1 > 0$ so that $V_{\varepsilon_1} \circ \mathscr{O}_n f \circ V_{\varepsilon_1} \subset V_\varepsilon \circ \Omega f \circ V_\varepsilon$ (cf. (2.5) and Lemma 1.3). By Lemma 1.3 and Proposition 1.4 we can choose $\delta > 0$ so that $(\overline{V}_\delta \circ f \circ \overline{V}_\delta)^k \subset V_{\varepsilon_1} \circ f^k \circ V_{\varepsilon_1}$ for all k between n and $2n$. Hence,

$$\begin{aligned}\mathscr{O}(V_\varepsilon \circ \Omega f \circ V_\varepsilon) &\supset \mathscr{O}\left(V_{\varepsilon_1} \circ \bigcup_{k=n}^{2n} f^k \circ V_{\varepsilon_1}\right) \\ &\supset \mathscr{O}\left(\bigcup_{k=n}^{2n} (\overline{V}_\delta \circ f \circ \overline{V}_\delta)^k\right).\end{aligned}$$

But for any relation g: $\mathscr{O}(\bigcup_{k=n}^{2n} g^k) = \mathscr{O}_n(g)$ because any $m \geq n$ can be written in the form $m = k_1 + k_2 + \cdots + k_j$ with $n \leq k_i \leq 2n$. Thus,

$$\mathscr{O}(V_\varepsilon \circ \Omega f \circ V_\varepsilon) \supset \mathscr{O}_n(V_\delta \circ f \circ V_\delta) \supset \Omega\mathscr{C}f$$

by the earlier results. □

Supplementary exercises

16. If $f: X \to X$ is a continuous map and A is a nonempty closed subset then $\omega f[A]$ is nonempty. In particular $X = \operatorname{Dom} \omega f$. However, for general closed relations this need not be true because $\operatorname{Dom}(f)$ might be a proper subset of X. Prove that $\omega f[A] = \varnothing$ if and only if $f^n(A) = \varnothing$ for some whole number n and that this is equivalent to $A \cap \operatorname{Dom}(\omega f) = \varnothing$, i.e., $\omega f(x) = \varnothing$ for all x in A.

    In general, we will call a closed relation f on X *nilpotent* if it satisfies the following equivalent conditions:

    (a) $\omega f[X] = \varnothing$, (a′) $\omega(f^{-1})[X] = \varnothing$,
    (b) $f^n = \varnothing$ for some n, (b′) $f^{-n} = \varnothing$ for some n,
    (c) $\omega f = \varnothing$, (c′) $\omega(f^{-1}) = \varnothing$,
    (d) $\Omega f = \varnothing$, (d′) $(\overline{V}_\varepsilon \circ f)^n = \varnothing$ for some $\varepsilon > 0$ and some n,
    (e) $\Omega\mathscr{G}f = \varnothing$, (e′) $\Omega\mathscr{C}f = \varnothing$,
    (f) $|\mathscr{G}f| = \varnothing$, (f′) $|\mathscr{C}f| = \varnothing$,
    (g) $\varnothing$ is the only invariant subset of X.

    Prove the equivalence of these conditions (use Lemma 5 for (f) and (f′)). If f is nilpotent show that $\operatorname{Dom}(f) \cup f(X)$ can be written as a disjoint union of closed sets $X_1, X_2, \ldots, X_n$ such that $f(X_i \cup X_{i+1} \cup \cdots \cup X_n) \subset X_{i+1} \cup \cdots \cup X_n$ $(1 \leq i \leq n$ with $X_{n+1} = \varnothing)$. (Use induction on n such that $f^n = \varnothing$ and begin with $X_1 = \operatorname{Dom} f^{n-1}$ which is disjoint from $f(X)$.)

The difficulties described in the first paragraph can be eliminated by restriction. Prove:

$$\text{(2.6)} \qquad \begin{aligned} \mathrm{Dom}(\omega f) &= \bigcap_{n\geq 0} \mathrm{Dom}(f^n) = \omega(f^{-1})[X] \\ &= \bigcap_{n\geq 0} f^{-n}(X) = \Omega\mathscr{C} f^{-1}(X) = \mathrm{Dom}(\Omega\mathscr{C} f), \end{aligned}$$

and show that this set, which we denote by X_- contains $|\mathscr{C} f| = |\Omega\mathscr{C} f|$. (*Hint*: X is $+$ invariant with respect to any relation on X.) Prove that $x \in X_-$ if and only if there is an infinite chain beginning at x, i.e., there is an infinite sequence $\{x_0, x_1, x_2, \dots\}$ with $x_{i+1} \in f(x_i)$ for all $i \geq 0$ and $x_0 = x$. When we restrict f to X_- we preserve the interesting part of the dynamics. In particular, show that (with f_{X_-} denoting the restriction of f to X_-, $f \cap X_- \times X_-$)

$$\text{(2.7)} \qquad \begin{aligned} \omega f[X] \cap X_- &= \omega f[X] \cap \omega(f^{-1})[X] \\ &= \omega(f_{X_-})[X_-] \supset |\mathscr{C} f| \end{aligned}$$

and that $x \in \omega f[X] \cap \omega(f^{-1})[X]$ if and only if there is a bi-infinite sequence $\{\dots, x_{-2}, x_{-1}, x_0, x_1, x_2, \dots\}$ with $x_{i+1} \in f(x_i)$ for all i and $x_0 = x$.

17. For a closed relation f on X define $X_f = \{\{x_i\} \in \prod_{-\infty}^{+\infty} X : x_{i+1} \in f(x_i)\}$ with maps

$$\begin{aligned} &\sigma_f : X_f \to X_f & &\pi_f : X_f \to X, \\ &\sigma_f(\{x_i\})_i = x_{i+1}, & &\pi_f(\{x_i\}) = x_0. \end{aligned}$$

Let X_f^+ be the projection of X_f to the product $\prod_0^\infty X$ with $\sigma_f^+ : X_f^+ \to X_f^+$ and $\pi_f^+ : X_f^+ \to X$ defined analogously.

(a) Prove that X_f and X_f^+ are closed subsets of the products and so are compact metric spaces. σ_f, π_f, σ_f^+, π_f^+ are all continuous maps and σ_f is a homeomorphism.

(b) Prove that the image of π_f^+ is $\omega f[X]$ and the image of π_f is $\omega f[X] \cap \omega(f^{-1})[X]$. Consequently, π_f^+ is onto if and only if $\mathrm{Dom}(f) = X$ and π_f is onto if and only if $\mathrm{Dom}(f) = \mathrm{Dom}(f^{-1}) = X$. π_f^+ is a homeomorphism if and only if f is a mapping. π_f is a homeomorphism if and only if f is a homeomorphism.

(c) Prove that π_f maps σ_f on X_f to f on X and π_f^+ maps σ_f^+ on X_f^+ to f on X. If $\mathrm{Dom}(f) = X$ then $\pi_{f*}^+(\sigma_f^+) = f$, and $\pi_{f*}^+(\mathscr{O}\sigma_f^+) = \mathscr{O} f$. If $\mathrm{Dom}(f) = \mathrm{Dom}(f^{-1}) = X$ then $\pi_{f*}(\sigma_f) = f$ and $\pi_{f*}(\mathscr{O}\sigma_f) = \mathscr{O} f$.

(d) This construction is functorial, i.e., if h maps f_1 on X_1 to f_2 on X_2, define $\sigma(h) : X_{f_1} \to X_{f_2}$ and $\sigma^+(h) : X_{f_1}^+ \to X_{f_2}^+$ mapping σ_{f_1} to σ_{f_2} and $\sigma_{f_1}^+$ to $\sigma_{f_2}^+$. Show that all the obvious diagrams commute.

These constructions allow us to "replace a relation by a map" (on a different space) or even by a homeomorphism. Notice that by Proposition 1.17 $\pi_f^{(+)}$ maps $\mathscr{A}\sigma_f^{(+)}$ to $\mathscr{A}f$ but the inclusions $\pi_{f*}^{(+)}\mathscr{A}\sigma_f^{(+)}$ in $\mathscr{A}f$ might be strict.

18. With $f_1: X_1 \to Y_1$ and $f_2: X_2 \to Y_2$ recall the product relation $f_1 \times f_2: X_1 \times X_2 \to Y_1 \times Y_2$ (cf. Exercise 1.14).

(a) If f is a closed relation on X then $f \times f$ is a closed relation on $X \times X$. Prove that the diagonal $1_X \subset X \times X$ is an $f \times f$ +invariant subset if and only if f is a partially defined map i.e., if and only if $f \circ f^{-1} \subset 1_X$. Prove that 1_X is $f \times f$ invariant and $(f \times f)^{-1} = f^{-1} \times f^{-1}$ invariant if and only if f is a homeomorphism on X.

(b) Let f be a continuous map on X. Prove that the diagonal 1_X is $\mathscr{N}(f \times f)$ +invariant if and only if there exists a metric d on X with respect to which the Lipschitz constant of f is at most 1, i.e.,

$$d(f(x_1), f(x_2)) \leq d(x_1, x_2) \quad (\text{for } x_1, x_2 \in X).$$

(*Hint*: Just as Lemma 10 is the f invariant version of the Urysohn Lemma, the construction of d is the f invariant version of the metrization theorem for uniform spaces. Use Lemma 7 to inductively construct a sequence $\{U_n\}$ of open $f \times f$ +invariant neighborhoods of the diagonal with $U_0 = X \times X$ and such that $U_n^{-1} = U_n$ $(U_{n+1})^3 \subset U_n$, and $\bigcap_n \{U_n\} = 1_X$. Then follow the proof of the metrization theorem from Kelley (1955) [Theorem 6.12, p. 185].)

19. A relation F on X is called a *total order* if it is transitive $(F \circ F \subset F)$ and any two elements are comparable $(F \cup F^{-1} = X \times X)$. Observe that F is then reflexive $(F \cap F^{-1} \supset 1_X)$ and so $F \circ F = F$. Prove that a total order is closed if and only if $F(x)$ and $F^{-1}(x)$ are closed subsets of X for all x in X. (For any total order F: $F \subset F^{-1}(x) \times X \cup X \times F(y)$ whenever $X = F^{-1}(x) \cup F(y)$, e.g., when $x = y$. Prove that the intersection is F.)

A closed total order is called a *preference*. Prove that associated with any preference on X is a utility function $u: X \to \mathbb{R}$, a continuous map such that $(x, y) \in F$ if and only if $u(x) \leq u(y)$ (apply Theorem 12). Furthermore, if $u_1, u_2: X \to \mathbb{R}$ are utility functions for F there exists an increasing homeomorphism $\varphi: \mathbb{R} \to \mathbb{R}$ such that $u_2 = \varphi \circ u_1$ (u_1 and u_2 induce order preserving injections of the space of equivalence classes $X/F \cap F^{-1}$ onto closed subsets of $\mathbb{R}$. Extend $u_2 \circ u_1^{-1}$ linearly across the complementary subintervals).

20. Let L be a Lyapunov function for a closed relation f on X with $\text{Dom}(f) = X$. Suppose that A is a closed, nonempty + invariant subset. Prove that $\sup L|A$ is a critical value. (*Hint*: Let $A_1 = \{x \in A : L(x) = \sup L|A\}$. Prove $\varnothing \neq |\mathscr{G} f_{A_1}| \subset |\mathscr{G} f| \subset |L|$ where f_{A_1} is the restriction $f \cap A_1 \times A_1)$. In particular if f is a map on X then $\sup L$ is a critical value and if f is surjective $\inf L$ is a critical value as well.

Chapter 3. Attractors and Basic Sets

This work began with the hope that we could use our intuition about gradient systems to understand more complex patterns. Over the past two sections the desired analogies have been appearing. The chain recurrent set $|\mathscr{C}f|$ is the analogue, for a closed relation f, of the set of fixed points of a gradient system. On $|\mathscr{C}f|$ the relation $\mathscr{C}f$ is reflexive as well as transitive. So $\mathscr{C}f \cap \mathscr{C}f^{-1}$ is a closed equivalence relation on $|\mathscr{C}f|$ and the associated equivalence classes form a partition of $|\mathscr{C}f|$ into disjoint closed sets which we call the *basic sets* for f. These will prove to be the analogues of the individual fixed points of a gradient dynamic. A $\mathscr{C}f$ Lyapunov function L with $|L| = |\mathscr{C}f|$ is the analogue of the height function or a strict Lyapunov function. Such an L is strictly increasing on any chain disjoint from $|\mathscr{C}f|$.

Certain fixed points of a gradient system, the isolated local maxima of the height function, are of special importance because they are locally asymptotically stable. The positive orbits of points near such an equilibrium remain nearby and approach it in the limit. Generalizing this phenomenon is the very important concept of attractor. To introduce this new idea we review some earlier constructions.

For a closed relation f on X and a closed subset A of X $\omega f[A] \equiv \limsup\{f^n(A)\}$ is the subset approached by the positive images of A. If U is any neighborhood of $\omega f[A]$ then $f^n(A) \subset U$ for large enough n. In particular, if A is f +invariant $(f(A) \subset A)$ then the sequence of images is decreasing and $\omega f[A] = \bigcap\{f^n(A)\}$. So in that case $\omega f[A]$ is a subset of A. In fact, it is the largest f invariant subset of A, where B f invariant means $f(B) = B$. Now suppose that $B = \omega f[A]$ is contained in the interior of A. Then for U any neighborhood of B contained in $\operatorname{Int} A$ $\omega f[U] = B$ and so the sequence $\{f^n(U)\}$ eventually enters any smaller neighborhood of B. In fact, we will see that we can construct a sequence $\{U_n\}$ of closed neighborhoods of B such that $f(U_n) \subset U_{n+1} \subset \operatorname{Int} U_n$ and $\bigcap\{U_n\} = B$. This monotonic squeezing of neighborhoods is the characteristic pattern of an attractor. This pattern suggests a strengthening of the concept of +invariance.

For f a closed relation we call a closed subset U *inward* if $f(U) \subset \operatorname{Int} U$. This is obviously stronger than f +invariance. In fact, U is inward precisely when it is $\overline{V}_\varepsilon \circ f$ +invariant for some $\varepsilon > 0$. U is f +invariant if any 0-chain beginning in U remains in U, while v is inward if for some $\varepsilon > 0$ any ε chain beginning in U remains in U. In particular, if U is inward then $\mathscr{C}f(U) \subset \operatorname{Int} U$. (Recall that $\mathscr{C}f$ is contained in $\mathscr{O}(V_\varepsilon \circ f)$ for any $\varepsilon > 0$.)

In particular, inwardness with respect to f is equivalent to the apparently stronger condition of inwardness with respect to $\mathscr{C}f$ (and a fortiori with respect to $\mathscr{N}f$ or $\mathscr{G}f$). Contrast this with f +invariance. Proposition 2.7 characterizes the strictly stronger condition of $\mathscr{N}f$ +invariance. While $\mathscr{C}f$ and $\mathscr{G}f$ +invariance are stronger conditions still.

Notice also, that if A is F +invariant for $F = \mathscr{G}f$ or $\mathscr{C}f$ then the largest f invariant subset of A equals the largest F invariant subset because

$$(3.1) \qquad f(A) = F(A), \qquad \omega f[A] = \omega F[A] = \Omega F(A) = F(A \cap |F|),$$

where $\bigcap\{F^n\} = \limsup\{F^n\} = \Omega F$ (cf. Proposition 2.4). The first equation is Exercise 2.3(e) and the latter equations follow by induction and Proposition 2.6.

1. Exercise. *By using a positive interval example with* $K = \{0, 1/2, 1\}$, *show that even for a* $\mathscr{C}f$ *invariant set* A, $\omega f(A) = \{\omega f(x) : x \in A\}$ *may be a proper subset of* $A = \omega f[A]$. (*Use* $A = [1/2, 1]$ *or all of* I.) □

Natural examples of inward sets arise from Lyapunov functions. Recall that if L is a Lyapunov function for f then for any c the closed set $U = \{x : L(x) \geq c\}$ and the open set $G = \{x : L(x) > c\}$ are f +invariant. We call c a *regular value* of the Lyapunov function if all the points in $L^{-1}(c)$ are regular points. Otherwise we call c a *critical value.* Thus, the critical values are the image $L(|L|)$. If c is a regular value then $L(y) > c$ when $y \in f(x)$ and $L(x) = c$. This means that $f(U) \subset G \subset \operatorname{Int} U$. It follows that $\{x : L(x) \geq c\}$ is an inward set when c is a regular value.

In the following we characterize in various ways three successively stronger concepts: $\mathscr{C}f$ +invariant set, preattractor, and attractor. The power of the attractor concept comes from the equivalence between descriptions of different apparent strength. In practice the "weak" characterizations, being easier to check, are used to recognize an attractor. The "strong" characterizations can then be applied.

We begin with a continuity result sharpening Corollary 1.2.

2. Lemma. *Let* f *be a closed relation on* X. *For every closed subset* A, *positive integer* n *and* $\varepsilon > 0$ *there exists* $\delta > 0$ *such that*

$$(\overline{V}_\delta \circ f \circ \overline{V}_\delta)^n(A) \subset V_\varepsilon \circ f^n(A) = V_\varepsilon(f^n(A)).$$

Proof. By an inductive application of Proposition 1.4 the family of closed sets $\{(\overline{V}_\delta \circ f \circ \overline{V}_\delta)^n(A)\}$ decreases as δ decreases, has intersection $f^n(A)$. So for δ small enough the elements of the family are contained in the ε neighborhood of $f^n(A)$. □

3. THEOREM. *Let f be a closed relation on X and A be a closed subset of X.*

(a) *The following conditions are equivalent and define* A is a preattractor.

(1) *A is f +invariant and there exists a closed neighborhood U of A such that $\omega f[U] \subset A$.*

(2) *A is f +invariant and there exists a closed f +invariant neighborhood U of A such that $\bigcap_{n\geq 0}\{f^n(U)\} \subset A$.*

(3) *A is f +invariant and there exists U an inward neighborhood of A such that $\bigcap_{n\geq 0}\{f^n(U)\} \subset A$.*

(4) *A is f +invariant and there exists an inward set U containing A with $\bigcap_{n\geq 0}\{f^n(U)\} \subset A$.*

(5) *A is $\mathscr{G}f$ +invariant and $A \cap |\mathscr{G}f|$ is open (as well as closed) in $|\mathscr{G}f|$.*

(6) *A is $\mathscr{C}f$ +invariant and $A \cap |\mathscr{C}f|$ is open (as well as closed) in $|\mathscr{C}f|$.*

(7) *A is f +invariant and $\{x : \Omega\mathscr{G}f(x) \subset A\}$ is a neighborhood of A.*

(8) *A is f +invariant and $\{x : \Omega\mathscr{C}f(x) \subset A\}$ is a neighborhood of A.*

A preattractor is the intersection of its inward neighborhoods. In fact, the set of inward neighborhoods U of A with $\bigcap_{n\geq 0}\{f^n(U)\} = \omega f[U] \subset A$ form a base for the neighborhood system of a preattractor A.

The class of preattractors is closed under finite union and intersection and contains all inward sets.

(b) *The following conditions are equivalent and define* A is an attractor.

(1) *A is f invariant and there exists a closed neighborhood U of A such that $\bigcap_{n\geq 0}\{f^n(U)\}$ is contained in A (and so equals A).*

(2) *A is an f invariant preattractor.*

(3) *A is a $\mathscr{C}f$ invariant preattractor.*

(4) *$A = \omega f[B]$ $(= \Omega\mathscr{G}f(B) = \Omega\mathscr{C}f(B))$ for some preattractor B.*

(5) *$A = \omega f[U]$ $(= \Omega\mathscr{G}f(U) = \Omega\mathscr{C}f(U))$ for some inward set U.*

(c) *The following conditions are equivalent:*

(1) *A is $\mathscr{C}f$ +invariant.*

(2) *A is the intersection of a (possibly infinite) family of preattractors.*

(3) *The inward neighborhoods of A form a base for the neighborhood system of A.*

These conditions imply that $A \cap |\mathscr{C}f|$ is a closed union of components of $|\mathscr{C}f|$.

PROOF. (a) We prove the equivalence by showing

$$\begin{array}{ccccccccc} (2) & \Rightarrow & (1) & \Rightarrow & (3) & \Rightarrow & (8) & \Rightarrow & (7) \\ & & & & \Downarrow & & & & \Downarrow \\ & & & & (4) & \Rightarrow & (6) & \Rightarrow & (5) & \Rightarrow & (2). \end{array}$$

$(2) \Rightarrow (1)$. If U is f +invariant then $\omega f[U] = \bigcap_{n\geq 0}\{f^n(U)\}$.

(1) $\Rightarrow$ (3). Because $\limsup\{f^n(U)\} \subset A \subset \operatorname{Int} U$ there exists $n > 0$ such that $k \geq n$ implies $f^k(U) \subset \operatorname{Int} U$. Now choose $\varepsilon > 0$ such that $V_\varepsilon \circ f^k(U) \subset U$ for $n \leq k \leq 2n-1$. Next apply Lemma 2 to obtain a $\delta > 0$ so that $(\overline{V}_\delta \circ f)^k(U) \subset V_\varepsilon \circ f^k(U) \subset U$ for $n \leq k \leq 2n-1$. Define

$$U_1 = U \cup (\overline{V}_\delta \circ f)(U) \cup \cdots \cup (\overline{V}_\delta \circ f)^{n-1}(U).$$

Since U_1 is closed and contains U it is a closed neighborhood of A. It is inward because

$$(\overline{V}_\delta \circ f)(U_1) = (\overline{V}_\delta \circ f)(U) \cup \cdots \cup (\overline{V}_\delta \circ f)^n(U),$$

and this is a subset of U_1 because $(\overline{V}_\delta \circ f)^n(U) \subset U$.

Finally, $f^n(U_1) \subset (\overline{V}_\delta \circ f)^n(U_1) \subset U$ because $(\overline{V}_\delta \circ f)^k(U) \subset U$ for $n \leq k \leq 2n-1$. Hence, $\bigcap_{i\geq 0}\{f^i(U_1)\} = \bigcap_{i \geq 0}\{f^i(U)\} \subset A$.

(3) $\Rightarrow$ (4). Obvious.

(4) $\Rightarrow$ (6) and (3) $\Rightarrow$ (8). Since U is inward it is $\mathscr{C} f$ +invariant and so by (3.1) $\Omega\mathscr{C} f(U) = \omega f[U] = \bigcap_{n\geq 0}\{f^n(U)\} \subset A$. Thus

$$\Omega\mathscr{C} f(U) \subset A \subset U.$$

$\Omega\mathscr{C} f(U)$ is $\mathscr{C} f$ invariant and so

$$\Omega\mathscr{C} f(U) = \Omega\mathscr{C} f \circ \Omega\mathscr{C} f(U) \subset \Omega\mathscr{C} f(A) \subset \Omega\mathscr{C} f(U),$$

i.e.,

$$\Omega\mathscr{C} f(A) = \Omega\mathscr{C} f(U) \subset A.$$

So A is $\Omega\mathscr{C} f$ as well as f +invariant. By Proposition 2.4(c) A is $\mathscr{C} f$ +invariant. By Proposition 2.6(b) applied to A and U:

$$A \cap |\mathscr{C} f| = \Omega\mathscr{C} f(A) \cap |\mathscr{C} f| = \Omega\mathscr{C} f(U) \cap |\mathscr{C} f| = U \cap |\mathscr{C} f|.$$

But U is inward so that $\Omega\mathscr{C} f(U) \subset \mathscr{C} f(U) \subset \operatorname{Int} U$ and thus

$$A \cap |\mathscr{C} f| = (\operatorname{Int} U) \cap |\mathscr{C} f|.$$

So $A \cap |\mathscr{C} f|$ is open in $|\mathscr{C} f|$ as well as closed. This proves (6) given (4). Given (3), i.e., $A \subset \operatorname{Int} U$, (8) follows because $\{x : \Omega\mathscr{C} f(x) \subset A\}$ includes U.

(8) $\Rightarrow$ (7) and (6) $\Rightarrow$ (5). $\mathscr{G} f \subset \mathscr{C} f$ implies $|\mathscr{G} f| \subset |\mathscr{C} f|$ and $\Omega\mathscr{G} f \subset \Omega\mathscr{C} f$.

(7) $\Rightarrow$ (5). $A \subset \operatorname{Int}\{x : \Omega\mathscr{G} f(x) \subset A\}$, implies, in particular, that A is $\Omega\mathscr{G} f$ +invariant, i.e., $x \in A \Rightarrow \Omega\mathscr{G} f(x) \subset A$. As A is f +invariant, too, it is $\mathscr{G} f$ +invariant by Proposition 2.4(c). Now choose U closed such that

$$A \subset \operatorname{Int} U \subset U \subset \{x : \Omega\mathscr{G} f(x) \subset A\},$$
$$A \supset \Omega\mathscr{G} f(U) \supset \operatorname{Int} U \cap |\mathscr{G} f| \supset A \cap |\mathscr{G} f|.$$

Thus, $A \cap |\mathscr{G} f| = (\operatorname{Int} U) \cap |\mathscr{G} f|$ is open in $|\mathscr{G} f|$.

(5) $\Rightarrow$ (2). Because $A \cap |\mathscr{G}f|$ is open in $|\mathscr{G}f|$, $|\mathscr{G}f| - A$ is closed. By Proposition 2.7 applied to the $\mathscr{G}f$ +invariant set A we can find a $\mathscr{G}f$ +invariant closed neighborhood U of A disjoint from $|\mathscr{G}f| - A$, i.e.,

$$A \subset \operatorname{Int} U \quad \text{and} \quad A \cap |\mathscr{G}f| = U \cap |\mathscr{G}f|.$$

Because U is $\mathscr{G}f$ +invariant it is f +invariant and by (3.1) and Proposition 2.6(b) we have

$$\omega f[U] = \Omega\mathscr{G}f(U) = \mathscr{G}f(U \cap |\mathscr{G}f|) = \mathscr{G}f(A \cap |\mathscr{G}f|) \subset \mathscr{G}f(A).$$

Finally, $\mathscr{G}f(A) \subset A$ because A is $\mathscr{G}f$ +invariant.

This completes the chain of equivalences. Any inward set A satisfies (4) (let $U = A$). Closure under finite union and intersection follow from (6).

Now suppose A is a preattractor and $A \subset G$ with G open. We complete the proof of (a) by finding an inward set U_1 such that

$$\omega f[U_1] \subset A \subset \operatorname{Int} U_1 \subset U_1 \subset G.$$

Because A is $\mathscr{G}f$ +invariant we can choose U a closed $\mathscr{G}f$ +invariant neighborhood of A with $U \subset G$ (apply Proposition 2.7(b)). By (1) we can choose U small enough so that $\omega f[U] \subset A$. Since $f^k(U) \subset U \subset G$ for all k we can choose δ in the proof that (1) $\Rightarrow$ (3) small enough so that $(V_\delta \circ f)^k(U) \subset G$ for $k = 1, \dots, n-1$. Then the inward set U_1 constructed by the proof is contained in G and $\omega f[U_1] = \omega f[U] \subset A$.

(b). We prove

$$(5) \Rightarrow (4) \Rightarrow (3) \Rightarrow (2) \Rightarrow (1) \Rightarrow (5).$$

(5) $\Rightarrow$ (4). An inward set is a preattractor.

(4) $\Rightarrow$ (3). The preattractor B is $\mathscr{C}f$ +invariant and so $A = \omega f[B] = \Omega\mathscr{C}f(B)$ is $\mathscr{C}f$ +invariant. Also, $A \cap |\mathscr{C}f| = B \cap |\mathscr{C}f|$ by Proposition 2.6(b). So A satisfies (6) of (a) because B does.

(3) $\Rightarrow$ (2). For a $\mathscr{C}f$ +invariant set, $\mathscr{C}f$ invariance is equivalent to f invariance by (3.1).

(2) $\Rightarrow$ (1). A satisfies (2) of (a).

(1) $\Rightarrow$ (5). $\{x : f(x) \subset \operatorname{Int} U\}$ is open and contains A because A is f +invariant and U is a neighborhood of A. So we can find in U a closed neighborhood U_1 of A with $f(U_1) \subset U$ and so $\bigcap_{n\geq 1}\{f^n(U_1)\} \subset \bigcap_{n\geq 0}\{f^n(U)\}$. Let $A_n = \bigcap_{k=1}^n\{f^k(U_1)\}$. $\{A_n\}$ is a decreasing sequence of closed sets and $\bigcap\{A_n\} \subset A \subset \operatorname{Int} U_1$. So there exists n such that $A_n \subset \operatorname{Int} U_1$, i.e.,

$$f(U_1) \cap f^2(U_1) \cap \cdots \cap f^n(U_1) \subset \operatorname{Int} U_1.$$

So by Proposition 1.4 we can choose $\delta > 0$ such that

$$\overline{V}_\delta \circ f(U_1) \cap (\overline{V}_\delta \circ f)^2(U_1) \cap \cdots \cap (\overline{V}_\delta \circ f)^n(U_1) \subset \operatorname{Int} U_1$$

(compare the proof of Lemma 2).

Now define

$$U_2 = U_1 \cap (\overline{V}_\delta \circ f)(U_1) \cap \cdots \cap (\overline{V}_\delta \circ f)^{n-1}(U_1).$$

So

$$\overline{V}_\delta \circ f(U_2) \subset \overline{V}_\delta \circ f(U_1) \cap \cdots \cap (\overline{V}_\delta \circ f)^n(U_1).$$

Because the latter set is contained in U_1 it follows that $\overline{V}_\delta \circ f(U_2) \subset U_2$, i.e., U_2 is inward.

On the other hand, A is an f invariant subset of U_1. So $A = f^k(A) \subset f^k(U_1)$ for $k = 0, \dots, n-1$ and thus $A \subset U_2 \subset U_1$. Hence,

$$A = \omega f[A] \subset \omega f[U_2] = \bigcap_{n \geq 1} \{f^n(U_2)\} \subset \bigcap_{n \geq 1} \{f^n(U_1)\} \subset A.$$

Finally, $\omega f[U_2] = \Omega \mathscr{C} f(U_2)$ because U_2 is inward and consequently $\mathscr{C} f$ +invariant.

(c) $(3) \Rightarrow (2)$. Inward sets are preattractors.

$(2) \Rightarrow (1)$. Preattractors are $\mathscr{C} f$ +invariant and $\mathscr{C} f$ + invariance is preserved by arbitrary intersection.

$(1) \Rightarrow (3)$. Suppose G is an open set containing the $\mathscr{C} f$ +invariant set A. The family of closed sets $\{\overline{V}_\varepsilon(A) \cup \mathscr{N}(\overline{V}_\varepsilon \circ f \circ \overline{V}_\varepsilon)(A)\}$ decreases with $\varepsilon > 0$ and by Propositions 1.8 and 1.4 the intersection is $A \cup \mathscr{C} f(A) = A$. So there exists $\varepsilon > 0$ such that the closed set $\overline{V}_\varepsilon(A) \cup \mathscr{N}(\overline{V}_\varepsilon \circ f \circ \overline{V}_\varepsilon)(A)$ is contained in G. A fortiori

$$U_0 = V_\varepsilon(A) \cup \mathscr{O}(V_\varepsilon \circ f \circ V_\varepsilon)(A) \subset G,$$

and clearly,

$$V_\varepsilon \circ f(U_0) \subset U_0.$$

U_0 is open rather than closed. To get an inward set we shrink U_0 slightly:

$$\begin{aligned} U &= X - V_{\varepsilon/2}(X - U_0) \\ &= \{x : V_{\varepsilon/2}(x) \subset U_0\}. \end{aligned}$$

By the first equation U is closed. From the second it is clear that

$$V_{\varepsilon/2}(V_{\varepsilon/2} \circ f(U)) \subset V_\varepsilon \circ f(U_0) \subset U_0$$

implies $V_{\varepsilon/2} \circ f(U) \subset U$ and so U is inward. Also

$$V_{\varepsilon/2}(V_{\varepsilon/2}(A)) \subset V_\varepsilon(A) \subset U_0$$

implies $V_{\varepsilon/2}(A) \subset U$. So U is a neighborhood of A. Finally, $U \subset U_0 \subset G$.

This completes the proof of the equivalences. Condition (2) implies that $A \cap |\mathscr{C} f|$ is the intersection of open-and-closed subsets of $|\mathscr{C} f|$ by (6) of (a). □

From the third part of the theorem we can describe $\mathscr{N} f$ and $\mathscr{C} f$ + invariance of A as contrasting stability conditions. That A is a preattractor then corresponds to asymptotic stability.

4. EXERCISE. *Let f be a closed relation on X and let A be a closed subset of X.*

(a) *Prove that A is $\mathcal{N}f$ +invariant if and only if for every $\varepsilon > 0$ there exists $\delta > 0$ such that any δ chain beginning δ close to A remains ε close to A (i.e., $\mathcal{O}f(V_\delta(A)) \subset V_\varepsilon(A)$; use Proposition 2.7(a)).*

(b) *Prove that A is $\mathcal{C}f$ +invariant if and only if for every $\varepsilon > 0$ there exists $\delta > 0$ such that any δ chain beginning δ close to A remains ε close to A. (To prove necessity find inward neighborhoods of A by part (c). To prove sufficiency use $\mathcal{C}f \subset \mathcal{O}(V_\delta \circ f)$.)*

(c) *Prove that A is an attractor if and only if there is a sequence of closed neighborhoods $\{U_n\}$ of A with $f(U_n) \subset U_{n+1} \subset \operatorname{Int} U_n$ and $\bigcap\{U_n\} = A$.* □

When the closed relation is a continuous map there are additional characterizations of an attractor.

As a preliminary we list some properties of +invariance satisfied by continuous maps and their inverses. While the inverse of a closed relation is a closed relation, the inverse of a continuous map is a continuous map only for a homeomorphism. Thus we obtain separate and different results for a map and its inverse.

5. EXERCISE. *Let $f: X \to X$ be a continuous map. Prove:*

(a) *$A \subset X$ is f +invariant if and only if $X - A$ is f^{-1} +invariant. (Hint: (1.18).)*

(b) *A is f^{-1} invariant means $x \in A \Leftrightarrow f(x) \in A$ and so implies A is f +invariant. A is then f invariant if and only if $A \subset f(X)$.*

(c) *$A \subset X$ f +invariant implies the closure $\overline{A}$ is f +invariant, while $B \subset X$ f^{-1} +invariant implies $\operatorname{Int} B$ is f^{-1} +invariant. In particular, for any closed subset B of X $B \cup \overline{\mathcal{O}f(B)} = \bigcup_{n\geq 0} f^n(B) \cup \omega f[B]$ is closed and f +invariant and this is the smallest closed, f +invariant set containing B.*

(d) *A f +invariant implies $f^{-1}(A)$ is f +invariant.*

(e) *U inward with respect to f implies $f^{-1}U$ is inward with respect to f. (Hint: $f^{-1}U \supset \operatorname{Int} f^{-1}U \supset f^{-1} \operatorname{Int} U \supset f^{-1}f(U) \supset U \supset ff^{-1}(U)$.)* □

Recall that for any closed relation $\mathcal{N}(f^{-1}) = (\mathcal{N}f)^{-1}$ as well as $\mathcal{G}(f^{-1}) = (\mathcal{G}f)^{-1}$ and $\mathcal{C}(f^{-1}) = (\mathcal{C}f)^{-1}$, and we omit the parentheses in these cases. However, not even for a homeomorphism is it usually true that $\omega(f^{-1}) = (\omega f)^{-1}$. In fact, for general relations ωf is not a very useful tool. $\omega f(x) = \limsup\{f^n(x)\}$ includes the limit points of all infinite chains beginning at x but it may include more. However, for a map $\omega f(x)$ is the set of limit points of a particular sequence, namely the positive orbit of x: $f(x), f^2(x), \ldots$. Furthermore, in that case Proposition 1.11(d) says that for $y_1, y_2 \in \omega f(x)$ $y_1 \in \mathcal{N}f(y_2)$. In particular, letting $y_1 = y_2$ we obtain

$\omega f(x) \subset |\mathcal{N} f|$, the set of nonwandering points. So if we define

$$l_+[f] = \overline{\omega f(X)} = \overline{\bigcup\{\omega f(x) : x \in X\}}, \tag{3.2}$$

then for a continuous map f,

$$l_+[f] \subset |\mathcal{N} f| = |\mathcal{N} f^{-1}|. \tag{3.3}$$

6. THEOREM. *Let $f\colon X \to X$ be a continuous map.*

(a) *Let A be a closed f +invariant subset. Each of the following is equivalent to the condition that A is a preattractor.*

(1) *There exists a closed neighborhood U of A such that $\bigcap_{n\geq 0}\{f^n U\} \subset A$.*

(2) *$\{x : \Omega f(x) \subset A\}$ is a neighborhood of A.*

(3) *A is $\mathcal{N} f$ +invariant and $\{x\colon \omega f(x) \subset A\}$ is a neighborhood of A.*

(4) *A is $\mathcal{N} f$ +invariant and $A \cap l_+[f]$ is open (as well as closed) in $l_+[f]$.*

(b) *Let A be a closed f^{-1} invariant set, i.e., $f^{-1}(A) = A$. Each of the following is equivalent to the condition that A be a repellor for f, i.e., an attractor for f^{-1}.*

(1) *A is $\mathcal{N} f^{-1}$ +invariant and $A \cap l_+[f]$ is open (as well as closed) in $l_+[f]$.*

(2) *$A \cap [l_+[f]$ is open (as well as closed) in $l_+[f]$ and $(\omega f)^{-1}(A) \subset A$, i.e., $x \notin A$ implies $\omega f(x) \cap A = \varnothing$.*

(3) *$\omega f(x) \subset A$ implies $x \in A$ and A is* isolated *as an f invariant set, i.e., there exists U a closed neighborhood of A such that any f invariant subset of U is contained in A.*

PROOF. (a) If A is a preattractor then it is $\mathcal{G} f$ +invariant and $\{x : \Omega\mathcal{G} f(x) \subset A\}$ is a neighborhood of A by (5) and (7) of Theorem 3(a). These conditions clearly imply (3). We will now prove: (3) $\Rightarrow$ (4) $\Rightarrow$ (2) $\Rightarrow$ (1) $\Rightarrow$ Theorem 3(a)(4).

(3) $\Rightarrow$ (4). Let U be an open set containing A and contained in $\{x : \omega f(x) \subset A\}$. We prove that $A \cap l_+[f] = U \cap l_+[f]$ and so this set is open-and-closed in $l_+[f]$. It is sufficient to prove $U \cap \omega f(X) \subset A$, because $U \cap \overline{K} \subset \overline{U \cap K}$ for U open and any set K.

If $z \in U$ with $z \in \omega f(y)$ then z is a limit point of the sequence $\{f^n(y)\}$ and so $f^k(y) \in U$ for some $k > 0$. $\omega f(y) = \omega f(f^k(y)) \subset A$ because $f^k(y) \in U \subset \{x : \omega f(x) \subset A\}$.

(4) $\Rightarrow$ (2). Because $l_+[f] - A$ is closed and A is $\mathcal{N} f$ +invariant, Proposition 2.7(a) and the associated remark imply that an open f +invariant set U exists with $A \subset U$ and $\overline{U} \cap l_+[f] = A \cap l_+[f]$. We prove that $U \subset \{x : \Omega f(x) \subset A\}$. Fix $x \in U$ and choose $\varepsilon > 0$ such that $V_\varepsilon(x) \subset U$. For every $y \in \Omega f(x)$ and every $\varepsilon_1 > 0$ with $\varepsilon_1 < \varepsilon$ there exist x_1 and n_1

such that $d(x, x_1) < \varepsilon_1$ and $d(y, f^{n_1}(x_1)) < \varepsilon_1$. Since $\varepsilon_1 < \varepsilon$, $x_1 \in U$. Because U is f +invariant $f^{n_1}(x_1) \in U$. So for every ε_1, $y \in V_{\varepsilon_1}(U)$, i..e, $y \in \overline{U}$. Thus $\Omega f(x) \subset \overline{U}$ for every $x \in U$.

By Proposition 1.12(c) $\Omega f \subset \Omega f \circ \omega f$, i.e.,

$$\Omega f(x) \subset \Omega f(\omega f(x)) \subset \mathscr{N} f(\omega f(x)).$$

But $\omega f(x) \subset \Omega f(x) \subset \overline{U}$ and so $\omega f(x) \subset \overline{U} \cap l_+[f] \subset A$. Because A is $\mathscr{N} f$ +invariant, $\mathscr{N} f(\omega f(x)) \subset A$ and so, finally, $\Omega f(x) \subset A$.

(2) $\Rightarrow$ (1). Let U be a closed neighborhood of A contained in $\{x : \Omega f(x) \subset A\}$. We claim that $\bigcap_{n \geq 0}\{f^n(U)\} \subset A$. Suppose $y \in f^n(U)$ for all $n \geq 0$. There exists $x_n \in U$ such that $f^n(x_n) = y$. Choose a convergent subsequence $\{x_{n_i}\}$ with limit $x \in U$. Then $x_{n_i} \to x$ and $f^{n_i}(x_{n_i}) = y$, i.e., $y \in \Omega f(x)$. But $\Omega f(x) \subset A$ for x in U and so $y \in A$.

(1) $\Rightarrow$ (4) of Theorem 3(a). The construction from the proof that (1) $\Rightarrow$ (5) for Theorem 3(b) yields an inward set $U_2 \subset U_1$. This time, however, we do not have f invariance of A and so A need not be contained in U_2. All we have is that $A \cap f(A) \cap \cdots \cap f^{n-1}(A) \subset U_2$. Because A is f +invariant, the sequence $\{f^k(A)\}$ is decreasing and so we have $f^{n-1}(A) \subset U_2$. But Exercise 5(e) then implies that $U_3 \equiv (f^{n-1})^{-1}(U_2) = (f^{-1})^{n-1}(U_2)$ is inward and contains A. Furthermore, $f^{n-1}(U_3) \subset U_2$ and so $\bigcap_{k \geq 0}\{f^k(U_3)\} \subset \bigcap_{k \geq 0}\{f^k(U_2)\} \subset \bigcap_{k \geq 0}\{f^k(U_1)\} \subset A$ as before.

(b) If A is an f^{-1} attractor then A is f^{-1} +invariant and $A \cap |\mathscr{G} f^{-1}| = A \cap |\mathscr{G} f|$ is open-and-closed in $|\mathscr{G} f^{-1}| = |\mathscr{G} f|$. This implies (1) because $\mathscr{N} f^{-1} \subset \mathscr{G} f^{-1}$ and $l_+[f] \subset |\mathscr{N} f| \subset |\mathscr{G} f|$.

(1) $\Rightarrow$ (2). Obvious because $(\omega f)^{-1} \subset \mathscr{N} f^{-1}$. So $\mathscr{N} f^{-1}(A) \subset A$ implies $(\omega f)^{-1}(A) \subset A$.

(2) $\Rightarrow$ (1) of Theorem 3(b) for f^{-1}: As usual we can choose U a closed neighborhood of A with $U \cap l_+[f] = A \cap l_+[f]$. We prove that $\bigcap_{n \geq 0}\{f^{-n}(U)\} \subset A$.

Suppose $x \notin A$. We will show x is not in the intersection. Because $x \notin A$, $\omega f(x) \cap A = \varnothing$ and so $\omega f(x) \cap U = \omega f(x) \cap l_+[f] \cap U \subset \omega f(x) \cap A = \varnothing$. So the sequence $\{f^k(x)\}$ eventually enters the open set $X - U$, i.e., $f^n(x) \notin U$ for large n. This means $x \notin f^{-n}(U)$.

(3) $\Rightarrow$ (1) of Theorem 3(b) for f^{-1}: If U is an isolating neighborhood for A, we prove that $\bigcap_{n \geq 0} f^{-n} U = A$. If not, then there exists $x \in U - A$ such that $f^n(x) \in U$ for all $n \geq 0$ and so $\omega f(x) \subset U$. By Proposition 1.12(a) $f \circ \omega f = \omega f$ and so $\omega f(x)$ is an invariant subset of U. Because U isolates A, $\omega f(x) \subset A$.

Conversely, if $\bigcap_{n \geq 0} f^{-n} U = A$ then U is an isolating neighborhood for A and the rest of (3) follows from (2) for a repellor A. $\square$

7. EXERCISE. (a) *Consider a positive interval example with* $K = \{0, 1\}$. *For the associated homeomorphism* $\tilde{f}$ *on the circle, let* A *be the singleton* $0 = 1$ *and prove:* A *is* $\tilde{f}$ *invariant and* $\tilde{f}^{-1}$ *invariant.* $A = l_+[\tilde{f}] = l_+[\tilde{f}^{-1}]$. *For all* x *in the circle* $\omega\tilde{f}(x) = A$ *and* $\omega(\tilde{f}^{-1})(x) = A$. *In particular,* $A \cap l_+[\tilde{f}]$ *is open-and-closed in* $l_+[\tilde{f}]$ *and* $\{x : \omega\tilde{f}(x) \subset A\}$ *is a neighborhood of* A. *Also* A *is an isolated invariant set. Prove that* A *is not* $\mathscr{N}\tilde{f}$ *+invariant or* $\mathscr{N}\tilde{f}^{-1}$ *+invariant and so* A *is not an attractor for* $\tilde{f}$ *or* $\tilde{f}^{-1}$.

(b) *Consider a positive interval example with* $K = \{0, 1/2, 1\}$. *For the homeomorphism* f *on the interval show that* $K = |\mathscr{C}f|$. *Let* $A = 1/2$. *Prove that* $A \cap |\mathscr{C}f|$ *is open-and-closed in* $|\mathscr{C}f|$ *and* A *is* f *invariant and* f^{-1} *invariant. Prove that* A *is not* $\mathscr{N}f$ *+invariant or* $\mathscr{N}f^{-1}$ *invariant and so is not an attractor for* f *or* f^{-1}.

(c) *Consider an interval example with* $K = \{0, 1/2\} \cup \{1/2 + 1/n : n \geq 2\}$ *and with the real valued function* u *on* I *such that* $u(x) = 0$ *for* $x \in K$, > 0 *for* $x \in (0, 1/2)$ *and* < 0 *otherwise. Let* f *be the associated homeomorphism of the interval and* $A = 1/2$. *Show that* A *is* $\mathscr{C}f$ *invariant and* $A = \omega f(x)$ *for* $x \in (0, 1/2)$, *but* A *is not an attractor.* □

Returning to the case of a general closed relation, we will refer to the intersection $B \cap |\mathscr{C}f|$ for any preattractor B as the *trace of* B *on* $|\mathscr{C}f|$.

For every inward set, or more generally for any preattractor B, there is a unique attractor A with the same trace, namely, $A = \omega f[B] = \Omega\mathscr{C}f(B) = \bigcap_{n\geq 0}\{f^n(B)\} = \mathscr{C}f(B \cap |\mathscr{C}f|)$ (cf. (3.1) and Theorem 3(b)(4)). In particular, if A is an attractor then $A = \omega f[A]$ and so A is determined by its trace on $|\mathscr{C}f|$. This proves the first part of

8. PROPOSITION. *Let* f *be a closed relation on* X. *Any attractor* A *is determined by its trace on* $|\mathscr{C}f|$ *via*

$$A = \mathscr{C}f(A \cap |\mathscr{C}f|) = \Omega\mathscr{C}f(A \cap |\mathscr{C}f|). \tag{3.4}$$

The set of attractors for f *is countable and has a natural lattice structure.*

PROOF. As the trace of an attractor A determines A it suffices to show that there are countably many traces of attractors. The trace of A is open-and-closed in the compact metric space $|\mathscr{C}f|$ by Theorem 3(a)(6). A compact metric space has only countably many open-and-closed sets: Begin with a countable base for the topology. Each open-and-closed set is a finite union of sets in the base.

If A_1 and A_2 are attractors, then $A_1 \cup A_2$ is f invariant and so is an attractor. $(A_1 \cup A_2) \cap |\mathscr{C}f| = (A_1 \cap |\mathscr{C}f|) \cup (A_2 \cap |\mathscr{C}f|)$. $A_1 \cap A_2$ is a preattractor but need not be f invariant when f is not a homeomorphic mapping. However, we can define $A_1 \wedge A_2 = \omega f[A_1 \cap A_2]$. It is easy to check that $(A_1 \wedge A_2) \cap |\mathscr{C}f| = (A_1 \cap |\mathscr{C}f|) \cap (A_2 \cap |\mathscr{C}f|)$.

Finally, for $x \in |\mathscr{C}f|$, $\mathscr{C}f(x) = \Omega\mathscr{C}f(x)$ and so the second equation in (3.4) follows. □

9. PROPOSITION. *Let* f *be a closed relation on* X. *An attractor for* f^{-1} *is called a* repellor *for* f. *An* attractor-repellor pair *consists of an attractor* A_+ *and a repellor* A_- *such that* $A_+ \cap A_- = \varnothing$ *and* $|\mathscr{C}f| \subset A_+ \cup A_-$. *For each attractor* A_+ *there is a unique dual repellor* A_- *such that* A_+, A_- *is an attractor-repellor pair.* A_- *is defined by*

$$A_- = \mathscr{C}f^{-1}(|\mathscr{C}f| - (A_+ \cap |\mathscr{C}f|)). \tag{3.5}$$

If A_+, A_- *is an attractor repellor pair, then* $x \in X - (A_+ \cup A_-)$ *implies* $\Omega\mathscr{C}f(x) \subset A_+$ *and* $(\Omega\mathscr{C}f)^{-1}(x) \subset A_-$.

PROOF. With A_- defined by (3.5) suppose $x \in A_-$. There exists $y \in |\mathscr{C}f| - (A_+ \cap |\mathscr{C}f|)$ with $x \in \mathscr{C}f^{-1}(y)$, or equivalently $y \in \mathscr{C}f(x)$. If x were also in A_+ then $y \in \mathscr{C}f(A_+) = A_+$ and so $y \in A_+ \cap |\mathscr{C}f|$ which is not true. Thus, $A_+ \cap A_- = \varnothing$. Clearly, $A_- \cap |\mathscr{C}f| \supset |\mathscr{C}f| - (A_+ \cap |\mathscr{C}f|)$ and equality follows from $A_+ \cap A_- = \varnothing$. So $A_- \cap |\mathscr{C}f|$ is open-and-closed in $|\mathscr{C}f|$. By transitivity of $\mathscr{C}f^{-1}$, (3.5) implies that A_- is $\mathscr{C}f^{-1}$ +invariant and hence A_- is a preattractor for f^{-1} by Theorem 3(a)(6). $\mathscr{C}f^{-1}$ invariance follows from $\mathscr{C}f^{-1}(A_-) \supset \mathscr{C}f^{-1}(A_- \cap |\mathscr{C}f|) = A_-$. So A_- is a repellor.

Also, by Proposition 8 applied to f^{-1} a repellor is characterized by its trace on $|\mathscr{C}f| = |\mathscr{C}f^{-1}|$ and so A_- is characterized by $A_- \cap |\mathscr{C}f| = |\mathscr{C}f| - (A_+ \cap |\mathscr{C}f|)$. Uniqueness of the dual repellor follows.

Finally, $\mathscr{C}f(x) \cap A_- \neq \varnothing$ implies $x \in (\mathscr{C}f)^{-1}(y)$ for some $y \in A_-$ and so $x \in A_-$. So $x \in X - (A_+ \cup A_-)$ implies $\mathscr{C}f(x) \cap |\mathscr{C}f| \subset A_+$. Hence $\Omega\mathscr{C}f(x) = \mathscr{C}f(\mathscr{C}f(x) \cap |\mathscr{C}f|) \subset A_+$. Similarly, $\Omega\mathscr{C}f^{-1}(x) \subset A_-$. □

An inward set for f^{-1} is called an *outward set*, i.e., $\tilde{U}$ is outward if $f^{-1}(\tilde{U}) \subset \operatorname{Int}\tilde{U}$. If U is inward and G is open with $f(U) \subset G \subset U$ then $\tilde{U} = X - G$ is outward. To see this note that U is inward if and only if there exists a closed set $\tilde{U}$ with $U \cup \tilde{U} = X$ and $f(U) \cap \tilde{U} = \varnothing$. Since the latter condition is equivalent to $U \cap f^{-1}(\tilde{U}) = \varnothing$, it follows that $\tilde{U}$ is outward.

10. PROPOSITION. *Let* A_+, A_- *be an attractor-repellor pair for a closed relation* f. *There exists a* $\mathscr{C}f$ *Lyapunov function* $L\colon X \to [0, 1]$ *with* $L^{-1}(0) = A_-$, $L^{-1}(1) = A_+$, *and* $|L| = A_+ \cup A_-$.

PROOF. This result follows easily from Proposition 2.10 but we use a simpler approach which avoids the Urysohn Lemma construction.

If U is an inward neighborhood of a closed set A and L_U is *any* continuous real valued function satisfying

$$L_U(x) \begin{cases} = 1 & \text{for } x \in f(U) \cup A, \\ = 0 & \text{for } x \in X - \operatorname{Int} U, \\ \text{is in } [0, 1] & \text{for all } x, \end{cases}$$

then L_U is a $\mathscr{C}f$ Lyapunov function. Recall that U is $\mathscr{C}f$ + invariant and so $\mathscr{C}f(U) = f(U)$ (cf. (3.1)). L_U is a $\mathscr{C}f$ Lyapunov function because $y \in \mathscr{C}f(x)$ implies either $L_U(x) = 0$ or $L_U(y) = 1$ and L_U is between 0 and 1 in any case. So $L_U(y) \geq L_U(x)$.

Now let $x \in X - (A_+ \cup A_-)$. $A_+ \cup \mathscr{C}f(x)$ is a closed $\mathscr{C}f$ +invariant set disjoint from $A_- \cup x \cup \mathscr{C}f^{-1}(x)$. So by Theorem 3(c)(3) there exists an inward neighborhood U of $A_+ \cup \mathscr{C}f(x)$, disjoint from $A_- \cup x \cup \mathscr{C}f^{-1}(x)$. We can define a $\mathscr{C}f$ Lyapunov function $L_U \colon X \to [0, 1]$ with $L_U = 1$ on $A_+ \cup \mathscr{C}f(x)$ and $= 0$ on $A_- \cup x \cup \mathscr{C}f^{-1}(x)$. Similarly, we can define a $\mathscr{C}f^{-1}$ Lyapunov function $L_{\tilde{U}}$ with $L_{\tilde{U}} = 1$ on $A_- \cup \mathscr{C}f^{-1}(x)$ and $= 0$ on $A_+ \cup x \cup \mathscr{C}f(x)$. Then define

$$L_x = 1/2[L_U + (1 - L_{\tilde{U}})].$$

L_x is a $\mathscr{C}f$ Lyapunov function with $L_x = 0$ on $A_- \cup \mathscr{C}f^{-1}(x)$, $L_x = 1$ on $A_+ \cup \mathscr{C}f(x)$, and $L_x = 1/2$ on x. So x is a regular point of L_x. Thus, while A_+ and A_- are contained in $|L_x|$ by f and f^{-1} invariance respectively, x lies in the complementary open set.

By the Lindelöf property for the metric space $X - (A_+ \cup A_-)$ we can find a sequence $\{x_n\}$ in $X - (A_+ \cup A_-)$ such that $\bigcap |L_{x_n}| = A_+ \cup A_-$. Define $L(x) = \sum_n a_n L_{x_n}(x) / \sum_n a_n$ for any positive sequence $\{a_n\}$ with convergent sum. Apply Proposition 2.9(a). □

This proposition is the key step in the construction of general $\mathscr{C}f$ Lyapunov functions because, at least for points in $|\mathscr{C}f|$, the relation $\mathscr{C}f$ can be characterized by the attractor structure.

11. PROPOSITION. *Let f be a closed relation on X.*

(a) *The following are equivalent for $x, y \in X$.*

(1) *$y = x$ or $y \in \mathscr{C}f(x)$.*

(2) *For every preattractor B, $x \in B$ implies $y \in B$.*

(3) *For every inward set U, $x \in U$ implies $y \in U$.*

If in addition, $x \in |\mathscr{C}f|$, then these are equivalent to:

(4) *For every attractor A, $x \in A$ implies $y \in A$.*

(b) *If $x \in X - |\mathscr{C}f|$, then there exists an attractor-repellor pair $\{A_\pm\}$ with $x \notin A_+ \cup A_-$.*

PROOF. (a) (1) $\Rightarrow$ (2). A preattractor is $\mathscr{C}f$ +invariant.

(2) $\Rightarrow$ (3) and (2) $\Rightarrow$ (4). Inward sets and attractors are preattractors.

(3) $\Rightarrow$ (1) and, when $x \in |\mathscr{C}f|$, (4) $\Rightarrow$ (1). Assume $y \notin \mathscr{C}f(x)$ and $y \neq x$. Then $x \cup \mathscr{C}f(x)$ is a closed $\mathscr{C}f$ +invariant set disjoint from y. So by (3) of Theorem 3(c) there is an inward set U with $y \notin U$ and $x \cup \mathscr{C}f(x) \subset U$. If, in addition, $x \in |\mathscr{C}f|$ then x is in the associated attractor $\omega f[U] \subset U$ as well.

(b) If $x \notin |\mathscr{C}f|$ choose an inward set U containing $\mathscr{C}f(x)$ and disjoint from x (apply (3) of Theorem 3(c)). Let A_+ be the associated attractor $\omega f[U] \subset U$. Clearly $x \notin A_+$. But x does not lie in the dual repellor $A_- = \mathscr{C}f^{-1}(|\mathscr{C}f| - (A_+ \cap |\mathscr{C}f|))$ either. Assume it did. Then there would exist $y \in |\mathscr{C}f| - (A_+ \cap |\mathscr{C}f|)$ with $x \in \mathscr{C}f^{-1}(y)$, i.e., $y \in \mathscr{C}f(x)$. So $y \in U \cap |\mathscr{C}f|$. But $U \cap |\mathscr{C}f| = A_+ \cap |\mathscr{C}f|$. □

The restriction of $\mathscr{C}f$ to points of $|\mathscr{C}f|$ is a closed quasiorder, i.e., reflexive and transitive relation with $\mathscr{C}f \cap \mathscr{C}f^{-1}$ the associated equivalence relation. $\mathscr{C}f$ induces on the set of equivalence classes a partial order, i.e., a reflexive, transitive, and antisymmetric relation. For $x \in X$ the set $\{y : y \in \mathscr{C}f(x) \text{ and } x \in \mathscr{C}f(y)\}$ is nonempty precisely when $x \in |\mathscr{C}f|$ (recall Exercise 1.10), in which case the set is the $\mathscr{C}f \cap \mathscr{C}f^{-1}$ equivalence class of x which we will call the *basic set* containing x. Thus, for $x \in X$,

$$\mathscr{C}f(x) \cap \mathscr{C}f^{-1}(x) = \begin{cases} \varnothing, & x \notin |\mathscr{C}f|, \\ \text{the basic set containing } x, & x \in |\mathscr{C}f|. \end{cases} \tag{3.6}$$

Since $\mathscr{C}f \cap \mathscr{C}f^{-1}$ is closed the basic sets form a closed partition of $|\mathscr{C}f|$, i.e., a covering of $|\mathscr{C}f|$ by pairwise disjoint closed sets. Notice that Proposition 11 implies that two points x, y of $|\mathscr{C}f|$ lie in the same basic set if and only if they are contained in exactly the same attractors.

Now if L is any $\mathscr{C}f$ Lyapunov function then $L(x) = L(y)$ when x and y lie in the same basic set (because $y \in \mathscr{C}f(x)$ and $x \in \mathscr{C}f(y)$), i.e., L is constant on each basic set.

12. THEOREM. *Let f be a closed relation on X. There exists a $\mathscr{C}f$ Lyapunov function $L: X \to [0, 1]$ such that $|L| = |\mathscr{C}f|$ and L takes distinct basic sets to distinct values, i.e., $L(x) = L(y)$ for x, y in $|\mathscr{C}f|$ if and only if $x \in \mathscr{C}f(y)$ and $y \in \mathscr{C}f(x)$. We call such a map L a* complete $\mathscr{C}f$ Lyapunov function.

PROOF. Let $\{A_+^1, A_+^2, \dots\}$ be a finite or infinite listing of all the attractors for f. Recall that by Proposition 8 the set of attractors is countable. For each $n > 0$ let A_-^n be the dual repellor for A_+^n. Apply Proposition 10 to define a $\mathscr{C}f$ Lyapunov function $L_n: X \to [0, 1]$ with

$$|L_n| = A_+^n \cup A_-^n, \qquad L_n^{-1}(1) = A_+^n, \quad \text{and} \quad L_n^{-1}(0) = A_-^n.$$

For any x in $|\mathscr{C}f|$ the sequence of numbers $\{L_n(x)\}$ are all 0's or 1's because $|\mathscr{C}f| \subset A_+^n \cup A_-^n$ for all n. Think of the sequence as a coding which tells which attractors contain x. Proposition 11 says that x, $y \in |\mathscr{C}f|$ lie in the same basic set if and only if $L_n(x) = L_n(y)$ for all n. Now define

$$L(x) = \sum_n 2 \cdot 3^{-n} L_n(x). \tag{3.7}$$

As the series converges uniformly, L is a $\mathscr{C}f$ Lyapunov with $|L| \subset \bigcap_n \{A_+^n \cup A_-^n\}$ by Proposition 2.9(a). By Proposition 11(b) $\bigcap_n \{A_+^n \cup A_-^n\} = |\mathscr{C}f|$.

Finally, for $x \in |\mathscr{C}f|$ formula (3.7) replaces the 1's by 2's in the list $\{L_n(x)\}$ and then interprets the list as the ternary expansion of a point of the classical Cantor set. If $L(x) = L(y)$ for $x, y \in |\mathscr{C}f|$, then the ternary expansions agree. So $L_n(x) = L_n(y)$ for all n and x and y lie in the same basic set. □

13. Exercise. *For L given by the theorem prove that when $L(x) = L(y)$ for $x, y \in X$ either both $x \in \mathscr{C}f(y)$ and $y \in \mathscr{C}f(x)$ or neither. What about when $x, y \in |\mathscr{C}f|$?* □

We denote by $\mathscr{B}_f$ the space $|\mathscr{C}f|/\mathscr{C}f \cap \mathscr{C}f^{-1}$ whose elements are the basic sets and by $q_f: |\mathscr{C}f| \to \mathscr{B}_f$ the quotient map associating to each point x of $|\mathscr{C}f|$ the basic set $\mathscr{C}f(x) \cap \mathscr{C}f^{-1}(x)$ containing x. Because $\mathscr{C}f \cap \mathscr{C}f^{-1}$ is a closed equivalence relation, the quotient topology induced upon $\mathscr{B}_f$ is metrizable as well as compact.

It is easy to check directly that this space is totally disconnected, i.e., the open-and-closed sets form a base for the topology (cf. (3) of Theorem 3(c) applied to f and f^{-1}). However, this also follows from the construction of L in Theorem 12. Because any $\mathscr{C}f$ Lyapunov function is constant on basic sets the restriction to $|\mathscr{C}f|$ factors through q_f to define a continuous real valued map on $\mathscr{B}_f$. The complete $\mathscr{C}f$ Lyapunov function L of Theorem 12 distinguishes distinct points of $\mathscr{B}_f$ and so restricts to a homeomorphism of $\mathscr{B}_f$ onto a subset of the Cantor set which is thus a closed, nowhere dense subset of $[0, 1]$. Because the set of critical values is $L(|\mathscr{C}f|)$ it follows that the complementary set of regular values is open and dense.

Recall that in the proof of Proposition 10 we emphasized that we avoided the Urysohn Lemma construction. The reason that this is important is because with the alternate construction we can get smooth Lyapunov functions in the case when X is a smooth manifold. This is an application of the classical *Borel Lemma* whose proof we sketch.

14. Lemma. *Let $\{L_n\}$ be a sequence of C^∞ real-valued functions defined on a compact C^∞ manifold X. There exists a positive sequence $\{a_n\}$ such that $\sum_n a_n$ converges and $L = \sum a_n L_n$ is also C^∞.*

Proof. This is based upon the idea that we can use a C^∞ atlas on X to defined a Banach space structure on the vector space of C^r real valued functions on X, $\mathscr{C}^r(X; \mathbb{R})$, with the inclusion $\mathscr{C}^{r+1}(X, \mathbb{R}) \subset \mathscr{C}^r(X, \mathbb{R})$ of norm ≤ 1 $(r = 0, 1, 2, \ldots)$. That is, we have a C^r norm $\| \; \|_r$ defined on C^r maps so that $\|L\|_{r+1} \geq \|L\|_r$ if L is a C^{r+1} map.

Assuming these results, see, e.g. Akin (1978), let $a_n^{-1} = 2^n \|L_n\|_n$. Then for $n \geq r$

$$\|a_n L_n\|_r \leq 2^{-n}\|L_n\|_r / \|L_n\|_n \leq 2^{-n}.$$

So for each $r = 0, 1, \ldots$ the series $\sum a_n L_n$ converges uniformly in $\mathscr{C}^r(X; \mathbb{R})$ and the limit L is C^r. Since this is true for all r, L is C^∞. □

15. COROLLARY. *Let f be a closed relation on a compact C^∞ manifold X. There exists a C^∞ real valued function L such that*

(1) *L is a $\mathscr{C}f$ Lyapunov function with $|L| = |\mathscr{C}f|$;*
(2) *L distinguishes distinct basic sets;*

i.e., L is a smooth, complete $\mathscr{C}f$ Lyapunov function.

PROOF. By the use of smooth partitions of unity we can find, given disjoint closed sets B_1 and B_2 in X, a smooth function L mapping to $[0, 1]$ with $L = 0$ on B_1 and $= 1$ on B_2. This means that the functions L_U and $L_{\tilde{U}}$ in the proof of Proposition 10 can be chosen smooth. Applying the Borel Lemma we can choose the sequence $\{a_n\}$ so that the Lyapunov function L of Proposition 10 can be assumed smooth.

Now proceed to the proof of Theorem 12 and choose each L_n smooth by the above emendations to Proposition 10. By the Borel Lemma again there exists a positive sequence $\{a_n\}$ such that $\sum a_n L_n$ converges to a C^∞ function. Clearly, any other positive sequence $\{b_n\}$ with $b_n \leq a_n$ will work as well. Now choose $\{k_n\}$ an increasing sequence of integers such that $b_n = 2 \cdot 3^{-k_n} < a_n$. Replace (3.7) by

$$L(x) = \sum 2 \cdot 3^{-k_n} L_n(x). \tag{3.8}$$

Again each point x of $|\mathscr{C}f|$ is mapped to a point of the Cantor set. This time the ternary expansion of $L(x)$ will consist of zeros in all of the non k_n places. But no matter. The proof is completed as before. □

The analogue of this smoothness result for $\mathscr{C}f$ Lyapunov functions does not hold for $\mathscr{G}f$ Lyapunov functions. To see this let us return to the puzzle which was raised in Chapter 2. Let K be the Cantor set and $\tilde{f}$ be the homeomorphism of the circle associated to a positive interval example. $\mathscr{C}\tilde{f} = S \times S$ and so the entire circle consists of a single basic set and the only $\mathscr{C}\tilde{f}$ Lyapunov functions are constants. But $|\mathscr{G}\tilde{f}| = K$ (in S) and so by Corollary 2.13 there must exist a continuous $\mathscr{G}\tilde{f}$ Lyapunov function L with $|L| = K$. What is it? For the answer recall the existence of the classical Cantor function $C: I \to [0, 1]$. This is a continuous function with $C(0) = 0$ and $C(1) = 1$ but which is, paradoxically, constant on each middle third interval. C is the distribution function of a singular measure concentrated on the Cantor set. If $\sum_{n=1}^\infty a_n 3^{-n}$ ($a_n = 0, 1, 2$) is a ternary expansion for x, then define the binary expansion of $C(x)$ to be $\sum_{n=1}^N (a_n/2)2^{-n} + 2^{-N-1}$ where $a_k \neq 1$ for $k = 1, \ldots, N$ but $a_{N+1} = 1$ (if $a_n \neq 1$ for all n, i.e., x

is in the Cantor set, then the sum is infinite and the term 2^{-N-1} does not appear). A Lyapunov function for $\tilde{f}$ is given by $L(x) = x - C(x)$. L is strictly increasing on each middle third interval but $L(0) = L(1) = 0$ and so L can be defined on the quotient circle, S. This phenomenon cannot occur in a C^1 way. If L is a C^1 function on I which is nondecreasing on each middle third interval then $dL/dx \geq 0$ on all of I. So $L(0) = L(1)$ implies $dL/dx = 0$ on all of I and L is constant. From this it can be shown that the only C^1 Lyapunov functions for $\tilde{f}$ are constants and so $|L| = S$ not $|\mathscr{G}f|$.

Using this example we can also answer the following question: Suppose L is an f Lyapunov function which is constant on each basic set. Does it follow that L is a $\mathscr{C}f$ Lyapunov function? The answer is no. In the above example if we do not identify 0 with 1 then $|\mathscr{G}f| = |\mathscr{C}f| =$ Cantor set and the basic sets are the points of the Cantor set. $L(x) = x - C(x)$ is still an f Lyapunov function but is not a $\mathscr{C}f$ Lyapunov function because $1 \in \mathscr{C}f(t)$ for all t in I. On the other hand, $L_1(x) = C(x)$ is a $\mathscr{C}f$ Lyapunov function because it is nondecreasing. Notice that L_1 is constant on every f orbit, i.e, $|L_1| = I$ and so the critical values consist of the entire image of I which is $[0, 1]$. Thus, a $\mathscr{C}f$ Lyapunov function need not have a dense set of regular values.

However, we do have:

16. EXERCISE. *If L is an f Lyapunov function and the regular values of L are dense in $\mathbb{R}$, then L is a $\mathscr{C}f$ Lyapunov function.* (*Hint*: *If c is regular, then $\{L \geq c\}$ is inward. For c critical $\{L \geq c\}$ is the intersection of $\{L \geq c_n\}$ when $\{c_n\}$ increases toward c.*) *In particular, it follows that L is constant on each basic set.* □

Supplementary exercises

17. $\mathscr{G}f \cap \mathscr{G}f^{-1}$ is a closed equivalence relation on $|\mathscr{G}F|$ and so the quotient space $|\mathscr{G}f|/\mathscr{G}f \cap \mathscr{G}f^{-1}$ has the structure of a compact metric space. Since $\mathscr{G}f \subset \mathscr{C}f$ each $\mathscr{G}f \cap \mathscr{G}f^{-1}$ equivalence class lies in a unique $\mathscr{C}f \cap \mathscr{C}f^{-1}$ equivalence class and this inclusion defines a continuous map $|\mathscr{G}f|/\mathscr{G}f \cap \mathscr{G}f^{-1} \to \mathscr{B}_f = |\mathscr{C}f|/\mathscr{C}f \cap \mathscr{C}f^{-1}$. Prove that this map is onto by showing that $x \in |\mathscr{C}f|$ implies there exist $y_1, y_2 \in |\mathscr{G}f|$ with $x \in \mathscr{G}f(y_1)$, $y_2 \in \mathscr{G}f(x)$, and $y_1 \in \mathscr{C}f(y_2)$. (*Hint*: By Proposition 2.15 $\Omega\mathscr{C}f = \mathscr{C}f \circ \Omega\mathscr{G}f$. So $x \in \Omega\mathscr{C}f(x)$ implies $x \in \mathscr{C}f(\mathscr{G}f(x) \cap |\mathscr{G}f|)$.) Show that in contrast to $\mathscr{B}_f$, $|\mathscr{G}f|/\mathscr{G}f \cap \mathscr{G}f^{-1}$ need not be totally disconnected and even when it is the map to $\mathscr{B}_f$ need not be one-to-one.
18. Prove that $f: X \to X$ is nilpotent (cf. Exercise 2.16) if and only if there exists a continuous function $L: X \to \mathbb{R}$ with $L(y) > L(x)$ whenever $y \in f(x)$, i.e., a Lyapunov function L with $|L| = \varnothing$.

19. (a) Let E be a closed equivalence relation on X and let X/E denote the set of equivalence classes. This set is given the structure of a compact metric space by using the quotient map $q\colon X \to X/E$ associating to each point its equivalence class. In the notation of Exercise 1.14 show that $E = E_q (\equiv q^{-1} \circ q)$. If f is a closed relation on X we denote by f_E the induced closed relation $q_* f = q \circ f \circ q^{-1}$ on X/E. Prove that for $\mathscr{A} = \mathscr{O}, \omega, \Omega, \mathscr{R}, \mathscr{N}, \mathscr{G}, \mathscr{C}$

$$(\mathscr{A} f)_E \subset \mathscr{A}(f_E) \quad \text{and} \quad E \circ \mathscr{A} f \circ E \subset \mathscr{A}(E \circ f \circ E),$$
$$(\mathscr{A} f)_E = \mathscr{A}(f_E) \quad \text{if and only if} \quad E \circ \mathscr{A} f \circ E = \mathscr{A}(E \circ f \circ E).$$

Finally, if f is a continuous map on X show that f_E is a map on X/E if and only if $f \circ E = E \circ f$, i.e., $xE\tilde{x}$ implies $f(x)Ef(\tilde{x})$. (For all of these results see Exercise 1.20.)

(b) Let f be a closed relation on X and define $E^f = 1_X \cup (\mathscr{C} f \cap \mathscr{C} f^{-1})$. Prove that E^f is a closed equivalence relation and that the equivalence class of x is x for $x \notin |\mathscr{C} f|$ and the basic set containing x for $x \in |\mathscr{C} f|$. With $E = E^f$ show that $f \subset E \circ f \circ E \subset \mathscr{C} f$ and use this to prove $\mathscr{C}(f_E) = (\mathscr{C} f)_E$. In particular, the basic sets with respect to f_E are the points corresponding to the basic sets of f. If L is a $\mathscr{C} f$ Lyapunov function, prove that L factors through the quotient map to define a $\mathscr{C} f_E$ Lyapunov function L_E. If f is a continuous map then f_E is a continuous map. (This uses f invariance of the basic sets which will be proved in the next chapter.)

Show that the topology on X/E is generated by the $\mathscr{C} f_E$ Lyapunov functions, i.e., there is a finite or infinite sequence $\{L_n\}$ of $\mathscr{C} f$ Lyapunov functions such that $L_n(x_1) = L_n(x_2)$ for all n implies $x_1 E x_2$. Hence, $x \to \{L_n(x)\}$ embeds X/E in the product of intervals.

Prove that if A is an attractor for f_E then $q^{-1}(A)$ is an attractor for f and every attractor for f is of this form.

Thus, taking the quotient by E^f trivializes the recurrence, reducing f to a gradient-like relation f_E with all chain recurrent points fixed. The destiny of the transient points of f_E can be described by Lyapunov functions or the attractor structure. And conversely it is really only the quotient dynamic f_E that is captured by these devices. Further analysis of f requires an understanding of the behavior in a neighborhood of each basic set. This is where hyperbolicity conditions come in as we will see in Chapter 11.

20. For a closed relation $f\colon X \to X$ prove that B is the trace of an attractor, i.e., $B = A \cap |\mathscr{C} f|$ for some attractor A, if and only if B is a closed subset of $|\mathscr{C} f|$, open relative to the topology of $|\mathscr{C} f|$ and $\mathscr{C} f$ +invariant relative to $|\mathscr{C} f|$, i.e., $x \in B$ and $y \in \mathscr{C} f(x) \cap |\mathscr{C} f|$ implies $y \in B$. In that case the associated attractor A is uniquely defined by $A = \mathscr{C} f(B)$ in X.

21. Following Exercise 2.16, for a closed relation $f\colon X \to X$ define
$$\begin{aligned} X_+ &= \omega f[X] = \mathrm{Dom}(\omega(f^{-1})) = \bigcap_{n\geq 0} f^n(X) \\ &= \Omega\mathscr{C}f(X) = \mathrm{Dom}(\Omega\mathscr{C}f^{-1}), \\ X_- &= \omega(f^{-1})[X] = \mathrm{Dom}(\omega f) = \bigcap_{n\geq 0} f^{-n}(X) \\ &= \Omega\mathscr{C}f^{-1}(X) = \mathrm{Dom}(\Omega\mathscr{C}f). \end{aligned}$$
Prove that X_+ is an attractor and X_- is a repellor for f. (*Hint*: X is a preattractor for f and f^{-1}). Every f invariant subset and a fortiori every attractor for f is contained in X_+, with dual results for X_-.
$$|\Omega\mathscr{C}f| = |\mathscr{C}f| \subset X_+ \cap X_- \equiv X_0.$$
Define the restriction $f_{X_0}\colon X_0 \to X_0$ by $f_{X_0} = f \cap X_0 \times X_0$. Prove that $\mathrm{Dom}(f_{X_0}) = \mathrm{Dom}(f_{X_0}^{-1}) = X_0$, and $|\mathscr{C}f_{X_0}| = |\mathscr{C}f|$. ($\mathscr{C}(f_{X_0}) \subset \mathscr{C}f$ is clear from $f_{X_0} \subset f$. For the reverse use $|\Omega\mathscr{C}f| = |\mathscr{C}f|$, the characterizations of Lemma 2.14(3), and Exercise 2.16.)

Prove if A is an attractor for f then $A_0 = A \cap X_0$ is an attractor for f_{X_0} and conversely if A_0 is an attractor for f_{X_0} then $A_0 = A \cap X_0$ for a unique attractor A of f with $A = \mathscr{C}f(A_0)$ (use Exercise 20); $\{A_+, A_-\}$ is an attractor repellor pair for f if and only if $\{A_+ \cap X_0, A_- \cap X_0\}$ is an attractor repellor pair for f_{X_0}.

22. Use Proposition 11 to prove directly (without using a complete $\mathscr{C}f$ Lyapunov function) that the space of basic sets $|\mathscr{C}f|/\mathscr{C}f \cap \mathscr{C}f^{-1}$ is totally disconnected. In fact, if $q\colon |\mathscr{C}f| \to |\mathscr{C}f|/\mathscr{C}f \cap \mathscr{C}f^{-1}$ is the projection map then for T the trace of an attractor or a repellor there exists $\tilde{T}$ in $|\mathscr{C}f|/\mathscr{C}f \cap \mathscr{C}f^{-1}$ such that $q^{-1}(\tilde{T}) = T$ and $\tilde{T}$ is open and closed in the quotient topology. Furthermore, $\{\tilde{T}_+ \cap \tilde{T}_- : T_+, T_-$ are the traces of attractors and repellors$\}$ is a base for the topology of the quotient space.

23. Prove that each basic set of $|\mathscr{C}f|$ is a closed union of components of $|\mathscr{C}f|$. In particular, if $|\mathscr{C}f| = X$ then each component of X is contained in a single basic set. (Apply the results of 22. For the latter one can, alternatively, use $1_X \subset \mathscr{C}f$ and Exercise 1.9(b)).

24. Let $f\colon X \to X$ be a homeomorphism and let $\{A_+, A_-\}$ be an attractor-repellor pair for X. Prove that there exists $L\colon X \to [-\infty, \infty]$ a continuous extended real valued function with $L^{-1}(\pm\infty) = A_\pm$ and $L(f(x)) = L(x) + 1$ for $x \in X - (A_+ \cup A_-)$. (*Hint*: With U an inward neighborhood of A_+ disjoint from A_- first define $L_0\colon U - \mathrm{Int}\, f(U) \to [0, 1]$ so that $L^{-1}(0) = \mathrm{Bdry}\, U (\equiv U - \mathrm{Int}\, U)$ and $L^{-1}(1) = f(\mathrm{Bdry}\, U) = \mathrm{Bdry}\, f(U)$. Then for $x \in X - (A_+ \cup A_-)$, $f^N(x) \in U - f(U)$ for a unique integer N. Define $L(x) = L_0(f^N(x)) - N$.

Chapter 4. Mappings—Invariant Subsets and Transitivity Concepts

So far we have interpreted a dynamical system on a compact metric space as a closed relation on the space. In passing we pointed out various properties which hold when the closed relation is, in fact, a continuous map. We now examine these in some detail. We have already remarked on the importance of the limit relation, ωf, when f is a map. In that case, $\mathscr{O} f(x)$ consists of a single infinite chain: $f(x)$, $f^2(x), \dots$, and $\omega f(x)$ consists of the limit points of this sequence. Upon this result we will build a rich collection of invariant sets beginning with the limit sets $\omega f(x)$ themselves. In addition, as remarked after Proposition 1.11, the points of $\omega f(x)$ are all nonwandering and, a fortiori, they are chain recurrent. In the next chapter, we will compute the entire chain recurrent set by understanding the dynamical system on the closure of the set of such limit points.

1. Exercise. *With X the closed interval $[0, 2]$, sketch a picture of the closed relation:*

$$f = \{(2y(1-y), y) : 0 \leq y \leq 1\} \cup \{(x, 0) : 1/2 \leq x \leq 1\} \cup \{(x, 2) : 1 \leq x \leq 2\}.$$

Prove that when $0 < x < 1/2$, $\omega f(x) = \{0, 1\}$. Check that $\{0, 1\}$ is neither f +invariant nor f^{-1} +invariant. Prove that $\mathscr{C} f(1) = f(1) = \{0, 2\}$ and so observe that $1 \notin |\mathscr{C} f|$. Finally, prove that if $\{x_0, x_1, \dots\}$ is an infinite chain, i.e., $x_{i+1} \in f(x_i)$ for all i, with $0 < x_0 < 1/2$ then $\lim\{x_i\} = 0$. Consequently, 1 is not the limit point of any infinite chain. □

The invariance results that we need are implicit in formal relationships already derived:

2. Lemma. *Let $f : X \to X$ be a continuous map. For $F = \omega f$, $\Omega \mathscr{G} f$ or $\Omega \mathscr{C} f$ we have*

(a) $f \circ F = F \overset{*}{=} F \circ f$,
(b) $F \circ f^{-1} \subset F \subset f^{-1} \circ F \overset{*}{=} f^{-1} \circ F \circ f$,
(c) $F^{-1} \circ f^{-1} = F^{-1} \overset{*}{=} f^{-1} \circ F^{-1}$,

(d) $f \circ F^{-1} \subset F^{-1} \subset F^{-1} \circ f \overset{*}{=} f^{-1} \circ F^{-1} \circ f$,

(e) $F^{-1} \subset F^{-1} \circ \omega f \quad (F \neq \omega f)$.

With $F = \Omega f$ *the same relationships hold provided that the equalities marked by* $*$ *are replaced by inclusions* $\subset$.

PROOF. (a) For $F = \omega f$ or Ωf the results are contained in Proposition 1.12(a). For $F = \Omega\mathscr{C} f$ or $\Omega\mathscr{G} f$ they are contained in Proposition 2.4(c).

(b) follows from (a) by composing with f^{-1} on the left and right and using the inclusions (1.16) $1_X \subset f^{-1} \circ f$ and $f \circ f^{-1} \subset 1_X$ which hold when f is a map.

(c) follows from (a) and (d) follows from (b) by inverting and using $(F \circ G)^{-1} = G^{-1} \circ F^{-1}$.

(e) From (d) $F^{-1}(x) \subset F^{-1}(f^n(x))$ for all n. If F^{-1} is closed, i.e., $F = \Omega f$, $\Omega\mathscr{G} f$, or $\Omega\mathscr{C} f$, we can take the $\limsup$ and apply (e) of Exercise 1.5.

REMARK. Recall that $(\Omega f)^{-1} = \Omega(f^{-1})$, $(\Omega\mathscr{G} f)^{-1} = \Omega(\mathscr{G} f)^{-1} = \Omega\mathscr{G}(f^{-1})$, and $(\Omega\mathscr{C} f)^{-1} = \Omega(\mathscr{C} f)^{-1} = \Omega\mathscr{C}(f^{-1})$. However, $(\omega f)^{-1} \neq \omega(f^{-1})$. □

3. COROLLARY. *Let* $f \colon X \to X$ *be a continuous map. Let* A, B *be closed subsets of* X.

(a) *The sets* $\omega f[A]$, $\overline{\omega f(A)}$, $\Omega f(A)$, $\Omega\mathscr{G} f(A)$, *and* $\Omega\mathscr{C} f(A)$ *are* f *invariant.*

(b) *The sets* $\Omega\mathscr{G} f^{-1}(B)$ *and* $\Omega\mathscr{C} f^{-1}(B)$ *are* f^{-1} *invariant and* f *+invariant.*

PROOF. (a) $f(\omega f[A]) = f(\limsup\{f^n(A)\}) = \limsup\{f^{n+1}(A)\} = \omega f[A]$ by (f) of Exercise 1.5. $F(A)$ is f invariant for $F = \omega f$, Ωf, $\Omega\mathscr{G} f$, and $\Omega\mathscr{C} f$ because $f \circ F = F$ ((a) in the lemma). For a map f the closure of an f invariant set is f invariant (c.f. Exercise 3.5(c)). So $\overline{\omega f(A)}$ is f invariant.

(b) For $F = \Omega\mathscr{C} f$ or $\Omega\mathscr{G} f$, $f^{-1} \circ F^{-1} = F^{-1}$ and $f \circ F^{-1} \subset F^{-1}$ by (c) and (d) of the lemma.

REMARK. (a) Recall from (d) of Exercise 2.3 that the intersection of an f invariant set with a set both f +invariant and f^{-1} +invariant is f invariant. So we obtain f invariant sets by intersecting any set of (a) with a set from (b).

(b) Recall that for $\widetilde{F} = \mathscr{G} f$ or $\mathscr{C} f$, $|\widetilde{F}| = |\Omega\widetilde{F}|$. It follows from this (and $\widetilde{F} \circ \Omega\widetilde{F} = \Omega\widetilde{F} = \Omega\widetilde{F} \circ \widetilde{F}$) that for $A \subset |\widetilde{F}|$, $\widetilde{F}(A) = \Omega\widetilde{F}(A)$ and $\widetilde{F}^{-1}(A) = \Omega\widetilde{F}^{-1}(A)$. In particular, if $x \in |\mathscr{C} f|$ the basic set $\mathscr{C} f(x) \cap \mathscr{C} f^{-1}(x) = \Omega\mathscr{C} f(x) \cap \mathscr{C} f^{-1}(x)$ is f invariant. □

The plethora of f invariant subsets for a mapping is important because it is often useful to consider the dynamical system obtained by restriction to an invariant subset. Let us pause to consider this idea of the restricted system.

If f is a relation on X and B is any closed subset of X we define the restriction of f to B, $f_B\colon B \to B$, by

$$f_B = f \cap B \times B. \tag{4.1}$$

Notice that for $h\colon B \to X$ the inclusion map, f_B is h^*f (cf. (1.22)). f_B is always a relation on the compact metric space B, closed if f is closed, but the restriction of a mapping will be a mapping precisely when B is f +invariant because $\mathrm{Dom}(f_B) = B$ only in that case. Clearly (4.1) implies $(f^{-1})_B = (f_B)^{-1}$ and so we can drop the parentheses and write f_B^{-1}. However, for $n \neq \pm 1$ the obvious inclusion $(f_B)^n \subset (f^n)_B$ might be proper. In general, it is easy to check that (see also Exercise 1.20):

$$\begin{gathered}\mathscr{A}(f_B) \subset \mathscr{A}(f)_B \\ \text{for } \mathscr{A} = \mathscr{O},\ \omega,\ \Omega,\ \mathscr{R},\ \mathscr{N},\ \mathscr{G} \text{ and } \mathscr{C}.\end{gathered} \tag{4.2}$$

Now define a subset B of X to be f *semi-invariant* if $\mathscr{O}f(B) \cap \mathscr{O}f^{-1}(B) \subset B$, i.e., if the existence of positive integers n_1, n_2 such that $f^{n_1}(x)$ and $f^{-n_2}(x)$ intersect B implies $x \in B$. Equivalently, if an f chain begins and ends in B, then the intermediate terms lie in B. Of course, if B is either f +invariant or f^{-1} +invariant, then B is f semi-invariant. (In the literature an f semi-invariant set for a map f is sometimes called *unrevisited* because if the orbit of a point enters such a set and then leaves, it never returns, i.e., $\{n : f^n(x) \in B\}$ is an interval in the set of positive integers).

Notice that the concepts of f and f^{-1} semi-invariance coincide. Check that the intersection of a family of semi-invariant sets is semi-invariant.

The following illustrates the use of semi-invariance and its relationship with restriction.

4. EXERCISE. *Let f be a closed relation on X and B be an f semi-invariant set. Prove*:

$$\begin{aligned}(f_B)^n &= (f^n)_B, &&(n \text{ any integer}),\\ \mathscr{O}(f_B) &= \mathscr{O}(f)_B, &&\mathscr{O}(f_B^{-1}) = \mathscr{O}(f)_B^{-1}.\end{aligned}$$

If B is a closed f +invariant set, then for $x \in B$ prove that $f^n(x) = (f_B)^n(x)$ and so $\omega f(x) = \omega(f_B)(x)$. Hence,

$$\omega(f_B) = \omega(f)_B, \qquad \mathscr{R}(f_B) = \mathscr{R}(f)_B. \quad \square$$

Much more delicate and important is

5. THEOREM. *Let f be a closed relation on X and B let be a closed, $\mathscr{C}f$ semi-invariant subset ($\mathscr{C}f(B) \cap \mathscr{C}f^{-1}(B) \subset B$). Then*

$$\mathscr{C}(f_B) = (\mathscr{C}f)_B \quad \textit{and} \quad |\mathscr{C}(f_B)| = |\mathscr{C}f| \cap B.$$

PROOF. Suppose y, $z \in B$ with $z \in \mathscr{C}f(y)$. Given $\varepsilon > 0$ we will construct an ε chain for f_B beginning ε close to y and ending at z. By Proposition 1.8 this will show that $\mathscr{C}(f)_B \subset \mathscr{C}(f_B)$, the reverse inclusion to (4.2).

First, we can choose $\varepsilon_1 > 0$ such that

$$V_{\varepsilon/2} \circ f_B \circ V_{\varepsilon/2} \supset (\overline{V}_{\varepsilon_1} \circ f) \cap \overline{V}_{\varepsilon_1}(B) \times \overline{V}_{\varepsilon_1}(B) \tag{4.3}$$

because, as usual, the right side decreases to f_B as ε_1 tends to 0.

Now define $B_+ = B \cup \mathscr{C}f(B)$ and $B_- = B \cup \mathscr{C}f^{-1}(B)$. B_+ is $\mathscr{C}f$ +invariant, B_- is $\mathscr{C}f^{-1}$ +invariant, and $B_+ \cap B_- = B$ by $\mathscr{C}f$ semi-invariance. By Theorem 3.3(c) we can choose U_+ an inward neighborhood of B_+ an U_- an outward neighborhood of B_- so that

$$U_+ \cap U_- \subset V_{\varepsilon_1}(B).$$

By inwardness of U_+ (and of U_- for f^{-1}) and Corollary 1.2, we can choose $\delta > 0$ with $\delta < \varepsilon_1$ and

$$\overline{V}_\delta \circ f(U_+) \subset U_+, \qquad (\overline{V}_\delta \circ f)^{-1}(U_-) \subset U_-.$$

Now let $x_0, x_1, \dots, x_n$ be a δ chain connecting $x_0 = y$ to $x_n = z$. As $x_0 \in U_+$ and $(x_i, x_{i+1}) \in \overline{V}_\delta \circ f$ we have, by induction, that all $x_i \in U_+$. Similarly, $x_n \in U_-$ and $(x_{i+1}, x_i) \in (\overline{V}_\delta \circ f)^{-1}$ imply that all $x_i \in U_-$. Consequently, the δ chain lies entirely in $V_{\varepsilon_1}(B)$.

We can now apply (4.3) and find for each pair (x_i, x_{i+1}), $(i = 0, \dots, n-1)$ a pair (u_i, v_i) in f_B with $d(u_i, x_i)$ and $d(v_i, x_{i+1}) < \varepsilon/2$. So

$$d(v_i, u_{i+1}) \le d(v_i, x_{i+1}) + d(x_{i+1}, u_{i+1}) < \varepsilon \qquad (i = 0, \dots, n-1).$$

So $u_0, u_1, \dots, u_{n-1}, x_n$ is the required ε chain for f_B beginning ε close to $x_0 = y$ and ending at $x_n = z$.

In particular, $(x, x) \in \mathscr{C}(f_B)$ if and only if $x \in B$ and $(x, x) \in \mathscr{C}f$. □

6. COROLLARY. *Let f be a closed relation on X and B be a preattractor for f. For any subset A of B, $\mathscr{C}(f_B)(A) = \mathscr{C}f(A)$. Furthermore, A is a (pre)attractor for f_B if and only if A is a (pre)attractor for f.*

PROOF. B is $\mathscr{C}f$ +invariant and so the theorem applies. It says that for $x \in B$, $\mathscr{C}(f_B)(x) = (\mathscr{C}f(x)) \cap B$. The latter set is $\mathscr{C}f(x)$ by $\mathscr{C}f$ +invariance of B.

Consequently, A is $\mathscr{C}(f_B)$ (+)invariant if and only if it is $\mathscr{C}f$(+)invariant. Because B is a preattractor $B \cap |\mathscr{C}f|$ is open in $|\mathscr{C}f|$ and so $A \cap |\mathscr{C}(f_B)| = A \cap |\mathscr{C}f|$ is open in $B \cap |\mathscr{C}f|$ if and only if it is open in $|\mathscr{C}f|$. The equivalence follows from Theorem 3.3 (a)(6). □

7. Exercise. *Let $f: X \to X$ be a continuous map. Prove*:

(a) *If B is f semi-invariant, then $f(B)$, $f^{-1}(B)$, and* $\operatorname{Int}(B)$ *are f semi-invariant.*

(b) *If A and B are closed subsets of X, then $\Omega\mathscr{C}f(A) \cap \Omega\mathscr{C}f^{-1}(B)$ is $\mathscr{C}f$ semi-invariant and is f invariant* (*cf. Corollary* 2 *and Exercise* 2.3(d)).
□

The basic sets for a map f are not only f invariant, they also satisfy the dynamic analogue of connectedness. A *decomposition* of an f +invariant set A is a pair of disjoint f +invariant subsets, relatively closed and with union A, i.e., $\{B_1, B_2\}$ with $f(B_i) \subset B_i$, $\overline{B}_i \cap A = B_i$ $(i = 1, 2)$, and $B_1 \cap B_2 = \varnothing$, $B_1 \cup B_2 = A$. The *trivial decomposition* of A is $\{A, \varnothing\}$. A is called *indecomposable* if the only decomposition of A is trivial. Any connected f +invariant subset is indecomposable and when $f = 1_X$ a subset is indecomposable if and only if it is connected. For a homeomorphism f, a nontrivial periodic orbit is an example of a disconnected indecomposable set. The following properties are generalizations to indecomposibility of standard properties for connectedness.

8. Exercise. *Let $f: X \to X$ be a continuous map. Prove*:

(a) *If A is indecomposable, then $f(A)$ and $\overline{A}$ are indecomposable.*

(b) *If $\{A_n\}$ is a decreasing sequence of closed, indecomposable sets, then $\bigcap\{A_n\}$ is closed and indecomposable.* (*If $A = \bigcap\{A_n\}$ is decomposable, choose open sets U_1, U_2 with $U_1 \cap A$, $U_2 \cap A$ nonempty with union A and such that $(U_1 \cup f(U_1)) \cap (U_2 \cup f(U_2)) = \varnothing$. Show that if $A_n \subset U_1 \cup U_2$, then $\{U_1 \cap A_n, U_2 \cap A_n\}$ is a decomposition of A_n*).

(c) *If $\{A_\alpha\}$ is a family of indecomposable sets such that $\overline{A}_\alpha \cap A_\beta \neq \varnothing$ for any α, β, then $\bigcup\{A_\alpha\}$ is indecomposable.*

(d) *If A is f invariant and $\{B_1, B_2\}$ is a decomposition for A, then B_1 and B_2 are f invariant.* □

9. Lemma. *Let f be a relation on X and A be a subset of X.*

(a) *If B is f^{-1} +invariant, then*

$$f(A) \cap B = f(A \cap B) \cap B.$$

(b) *If B is f^{-1} +invariant and f +invariant, then*

$$f(A) \cap B = f(A \cap B).$$

Proof. (a) Clearly, $f(A \cap B) \cap B \subset f(A) \cap B$. Conversely if $y \in f(A) \cap B$ then $y \in f(x)$ for some $x \in A$. Because $y \in B$ and $x \in f^{-1}(y)$, $x \in B$ by f^{-1} +invariance. So $x \in A \cap B$ and y lies in both $f(A \cap B)$ and B.

(b) If, in addition, B is f +invariant then $f(A \cap B) \subset B$ and so $f(A \cap B) \cap B = f(A \cap B)$. □

10. Proposition. *Let $f: X \to X$ be a continuous map. Assume that A is either a connected, closed (but not necessarily f +invariant) subset or an indecomposable, closed subset of X.*

(a) *$A \cup \mathcal{O}f(A)$ and $\omega f[A]$ are indecomposable.*

(b) *$A \cup \mathcal{N}f(A)$ are $\Omega f(A)$ are indecomposable. Furthermore, if $A \subset B$ with B closed, f +invariant, and $\mathcal{N}f^{-1}$ +invariant, then $(A \cup \mathcal{N}f(A)) \cap B$ and $\Omega f(A) \cap B$ are indecomposable.*

(c) *$A \cup \mathcal{C}f(A)$ and $\Omega\mathcal{C}f(A)$ are indecomposable. Furthermore, if $A \subset B$ with B closed, f +invariant, and $\mathcal{C}f^{-1}$ +invariant, then $(A \cup \mathcal{C}f(A)) \cap B$ and $\Omega\mathcal{C}f(A) \cap B$ are indecomposable.*

PROOF. If A_1 is an f +invariant set containing A and $\{B_1, B_2\}$ is a decomposition of A_1, then $\{A \cap B_1, A \cap B_2\}$ is a disjoint pair of closed sets with union A. If A is f +invariant, then this pair is a decomposition of A, i.e., $A \cap B_i$ is f +invariant if A is $(i = 1, 2)$. So with either hypothesis the intersection pair is a trivial partition and we can choose the labels so that $A \subset B_1$.

(a) Let $\{B_1, B_2\}$ be a decomposition of $A_1 = A \cup \mathcal{O}f(A)$. By the above remarks we can assume $A \subset B_1$, and so by f +invariance of B_1, $A_1 \subset B_1$. Thus, $B_2 = \emptyset$ and the decomposition is trivial.

$$\omega f[A] = \bigcap_n \overline{\{f^n(A) \cup \mathcal{O}f(f^n(A))\}}.$$

Consequently, $\omega f[A]$ is indecomposable by parts (a) and (b) of Exercise 8.

Notice that if A is indecomposable, and hence f +invariant, then $A = A_1$ and $\omega f[A] = \bigcap_n \{f^n(A)\}$.

(b) Let $\{B_1, B_2\}$ be a decomposition of $A_1 = (A \cup \mathcal{N}f(A)) \cap B$ (which is closed as well as f +invariant). Because B_1 and B_2 are disjoint closed sets there exist open sets U_1 and U_2 with disjoint closures with $\overline{U}_i \cap A_1 = B_i$ $(i = 1, 2)$. Assuming, as before, that $A \subset B_1$ we will derive a contradiction from the assumption that B_2 is nonempty.

Assume $y \in B_2$. For some $x \in A$, $y \in \mathcal{N}f(x)$ and so there exist sequences $\{x_k\}$ in X and positive integers $\{n_k\}$ such that $\lim\{x_k\} = x$ and $\lim\{f^{n_k}(x_k)\} = y$. By chopping off an initial segment, if necessary, we can assume $x_k \in U_1$ and $f^{n_k}(x_k) \in U_2$ for all k. Now define s_k to be the largest integer $p \leq n_k$ such that $f^p(x_k) \in U_1$. Notice that $0 \leq s_k < n_k$ because $f^{n_k}(x_k) \in U_2$. By passing to a subsequence we can assume that $\{f^{s_k}(x_k)\}$ converges to z. So $z \in x \cup \mathcal{N}f(x)$. Also, $\{f^{n_k - s_k}(f^{s_k}(x_k))\}$ converges to y and so $y \in \mathcal{N}f(z)$, i.e., $z \in \mathcal{N}f^{-1}(y)$. Because B is $\mathcal{N}f^{-1}$ invariant, and $y \in A_1$, $z \in A_1$ as well. By definition of s_k, $z \in \overline{U}_1 \cap A_1 = B_1$. By f +invariance we have $f(z) \in B_1$. But $f(z) = \lim\{f^{s_k+1}(x_k)\}$ and this sequence lies in the closed set $X - U_1$ disjoint from B_1. Contradiction.

Notice that $A_1 = (A \cup \mathcal{O}f(A) \cup \Omega f(A)) \cap B$ and so by Lemma 9(b),

$$f^n(A_1) = (f^n(A) \cup \mathcal{O}f(f^n(A)) \cup \Omega f(A)) \cap B$$

because $f^n(\Omega f(A)) = \Omega f(A)$ by Corollary 3(a). So $\omega f[A_1] = \bigcap\{f^n(A_1)\} = \Omega f(A) \cap B$ is indecomposable by the previous portion of the proposition.

The first statement in (b) is the special case of the second with $B = X$.

(c) Let $\{B_1, B_2\}$ be a decomposition of $A_1 = (A \cup \mathscr{C}f(A)) \cap B$ labeled so that $A \subset B_1$. Again we contradict the assumption that $y \in B_2$.

As B_1 and B_2 are disjoint closed sets there exists a positive ε such that $d(x_1, x_2) > \varepsilon$ for all $(x_1, x_2) \in B_1 \times B_2$. For some $x \in A$, $y \in \mathscr{C}f(x)$. Because A_1 is the intersection of $\mathscr{C}f$ and $\mathscr{C}f^{-1}$ +invariant sets, A_1 is $\mathscr{C}f$ semi-invariant and so by Theorem 5 there exists an ε chain $\{x_i : i = 0, \dots, n\}$ in A_1 with $x_0 = x$ and $x_n = y$. We obtain a contradiction by proving, inductively, that $x_i \in B_1$ for all i. $x_0 = x \in B_1$ and if $x_i \in B_1$ then $f(x_i) \in B_1$ by f +invariance. Then $d(x_{i+1}, f(x_i)) \le \varepsilon$ implies $x_{i+1} \notin B_2$ by definition of ε. So $x_{i+1} \in B_1$.

Because $\mathscr{C}f = \mathscr{O}f \cap \Omega\mathscr{C}f$ we prove, just as in (b), that $\Omega\mathscr{C}f(A) \cap B = \omega f[A_1]$ and so it is indecomposable by (a).

Again letting $B = X$ yields the first statement in (c).

REMARKS. (a) In applying the relative results of (b) and (c) notice that if B_1 is any closed subset of X then $B = \Omega\mathscr{C}f^{-1}(B_1)$ is $\mathscr{C}f^{-1}$ +invariant and is f +invariant by Corollary 3(b). A fortiori B is $\mathscr{N}f^{-1}$ +invariant.

(b) In contrast to (a) notice that $\overline{\omega f(A)}$ need not be indecomposable. For example, $l_+[f] = \overline{\omega f(X)}$ is usually not indecomposable even when X is connected. □

11. COROLLARY. *Let $f: X \to X$ be a continuous map. If $x, y \in |\mathscr{C}f|$ with $y \in \mathscr{C}f(x)$, then $\mathscr{C}f(x) \cap \mathscr{C}f^{-1}(y)$ is indecomposable and f invariant. In particular, the basic set $\mathscr{C}f(x) \cap \mathscr{C}f^{-1}(x)$ is an indecomposable, f invariant set.*

PROOF. Since $x, y \in |\mathscr{C}f| = |\Omega\mathscr{C}f|$, $\mathscr{C}f(x) = \Omega\mathscr{C}f(x)$, and $\mathscr{C}f^{-1}(y) = \Omega\mathscr{C}f^{-1}(y)$. $\Omega\mathscr{C}f(x) \cap \Omega\mathscr{C}f^{-1}(y)$ is f invariant by Corollary 3 and the remark thereafter. It is indecomposable by (c) of the proposition. □

We turn now to a concept closely related to recurrence, namely transitivity. As originally conceived, a point is recurrent if it "follows" itself in time. A system is transitive if every point "follows" every other. We have seen that the various relations F extending a continuous map $f: X \to X$ lead to different notions of recurrence. We will now see that the same is true for transitivity.

A relation F on X is called *dynamically transitive* if for every $x \in X$, $F(x) = X$; that is, if $F = X \times X$. (The word "dynamically" is included to distinguish this concept from ordinary relation transitivity, $F \circ F \subset F$. Where the context makes the meaning clear I will drop the modifier.) Clearly, a map $f: X \to X$ is dynamically transitive if and only if X consists of a single fixed point. It is easy to check that $\mathscr{O}f$ is dynamically transitive if and only if X consists of a single periodic orbit. The transitivity concepts associated with $\mathscr{C}f$, $\mathscr{N}f$, and $\mathscr{R}f$ are not so trivial. Each is important in condensing, as does the attractor concept, a number of properties which are not, at first glance, equivalent.

12. THEOREM. *Let* $f: X \to X$ *be a continuous map.*

(a) f *is called* chain transitive *if it satisfies the following equivalent conditions.*

(1) *For all* $x \in X$, $\mathscr{C}f(x) = X$, *i.e.,* $\mathscr{C}f = X \times X$.
(2) *For all* $x \in X$, $\Omega\mathscr{C}f(x) = X$, *i.e.,* $\Omega\mathscr{C}f = X \times X$.
(3) *For some* $x \in X$, $\mathscr{C}f(x) = X$ *and* $\mathscr{C}f^{-1}(x) = X$.
(4) X *is the only nonempty inward set.*
(5) X *is the only nonempty attractor.*
(6) X *is indecomposable and* $|\mathscr{C}f| = X$.

In general, if B *is a closed* f *+invariant subset of* X *we call* B *a* chain-transitive subset *if the restriction* f_B *is a chain transitive map. A chain-transitive subset is automatically* f *invariant. Any basic set of* f *is a chain transitive subset. In fact,* B *is a chain transitive subset if and only if* B *is a basic set of the restriction* f_A *for some closed,* f *+invariant subset* A *of* X.

(b) f *is called* topologically transitive *if it satisfies the following equivalent conditions*:

(1) *For all* $x \in X$, $\mathscr{N}f(x) = X$, *i.e.,* $\mathscr{N}f = X \times X$.
(2) *For all* $x \in X$ $\Omega f(x) = X$, *i.e.,* $\Omega f = X \times X$.
(3) *For some* $x \in X$, $\mathscr{R}f(x) = X$.
(4) *Any nonempty, open* f *semi-invariant subset is dense in* X.
(5) *Any nonempty, open* f^{-1} *+invariant set is dense in* X.
(6) *Any closed, proper* f *+invariant subset of* X *is nowhere dense.*

If f *is topologically transitive, then the set* $\{x : \omega f(x) = X\}$ *is residual, i.e., it is the countable intersection of dense open sets.*

In general, if B *is a closed and* f *+invariant subset of* X, *then we call* B *a* topologically transitive subset *if the restriction* f_B *is a topologically transitive map.* B *is a topologically transitive subset of* X *if and only if* $B = \omega f(x)$ *for some positively recurrent point* x, *i.e., for some* $x \in |\omega f|$. *In that case,* $D_B = \{x \in B : \omega f(x) = B\}$ *is an* f *invariant subset of* X *that is dense in* B *and contained in* $|\omega f|$.

(c) f *is called* minimal *if it satisfies the following equivalent conditions.*

(1) *For all* $x \in X$, $\mathscr{O}f(x)$ *is dense in* X.
(2) *For all* $x \in X$, $\mathscr{R}f(x) = X$, *i.e.,* $\mathscr{R}f = X \times X$.
(3) *For all* $x \in X$, $\omega f(x) = X$, *i.e.,* $\omega f = X \times X$.
(4) *For some* $x \in X$, $\mathscr{R}f(x) = X$, *and* $(\mathscr{R}f)^{-1}(x) = X$.
(5) X *is the only nonempty, closed* f *+invariant set.*
(6) X *is the only nonempty, closed* f *invariant set.*

In general, if B *is a closed,* f *+invariant subset of* X, *we call* B *a* minimal subset *for* f *if the restriction* f_B *is minimal. A minimal subset* *is* f *invariant and contains no proper closed* f *invariant subsets (whence the name). Every nonempty, closed* f $+$*invariant set contains a nonempty minimal subset.*

PROOF. (a) We will prove the equivalences by showing $(2) \Rightarrow (1) \Rightarrow (3) \Rightarrow (4) \Rightarrow (5) \Rightarrow (2)$ and $(1) \Rightarrow (6) \Rightarrow (5)$.

$(2) \Rightarrow (1)$. $\Omega\mathscr{C}f \subset \mathscr{C}f \subset X \times X$.

$(1) \Rightarrow (3)$. Obvious.

$(3) \Rightarrow (4)$. If $y \in U$ with U inward then $x \in \mathscr{C}f(y)$ and so $x \in U$ by $\mathscr{C}f$ $+$invariance. Hence, $X = \mathscr{C}f(x) \subset U$ by $\mathscr{C}f$ $+$invariance again.

$(4) \Rightarrow (5)$. An attractor is the intersection of inward sets.

$(5) \Rightarrow (2)$. If $x \in X$, the set $A = \Omega\mathscr{C}f(x)$ is nonempty (it contains $\omega f(x)$) and $\mathscr{C}f$ $+$invariant $(\mathscr{C}f \circ \Omega\mathscr{C}f = \Omega\mathscr{C}f)$. By Theorem 3.3(c) A is the intersection of a family $\{A_\alpha\}$ of preattractors. Each A_α is nonempty because A is and so the associated attractor $\omega f[A_\alpha] = \bigcap_n \{f^n(A_\alpha)\}$ is nonempty. Since X is the only such attractor, $A_\alpha = X$ for all α, and their intersection $A = X$ as well.

$(1) \Rightarrow (6)$. Clearly, $X = |\mathscr{C}f|$ and consists of a single basic set which is indecomposable by Corollary 11.

$(6) \Rightarrow (5)$. Let A_+ be an attractor. We will show that $\{A_+, A_-\}$ is a decomposition of X where A_- is the complementary repellor. This will imply $A_+ = X$ or $\varnothing$ by indecomposibility. As $X = |\mathscr{C}f|$, A_- is the complement of A_+ by (3.5). Because a repellor is $\mathscr{C}f^{-1}$ invariant, $A_- = \Omega\mathscr{C}f^{-1}(A_-)$ and so it is f $+$invariant by Corollary 3(b). Thus, $\{A_+, A_-\}$ is a decomposition of X as required.

Notice that a basic set is f invariant and so $f(X) = X$ when f is chain transitive. In particular, if B is a chain transitive subset then $B = f_B(B) = f(B)$ and so B is f invariant.

If B is a basic set for f then $(\mathscr{C}f)_B = B \times B$ by definition. But a basic set is also $\mathscr{C}f$ semi-invariant and so $(\mathscr{C}f)_B = \mathscr{C}(f_B)$ by Theorem 5. Consequently f_B is chain transitive.

If A is an f $+$invariant subset and B is a basic set for the restriction f_A then $(f_A)_B = f_B$ is chain transitive and so B is a chain transitive subset of X. Notice that the application of Proposition 5 here is to f_A. B is $\mathscr{C}(f_A)$ semi-invariant but need not be $\mathscr{C}f$ semi-invariant.

Conversely, if B is a chain transitive subset it is the unique basic set of the restriction f_B.

(b) We will show

$$\begin{array}{ccccc} (3) & \Rightarrow & (4) & \Rightarrow & \\ \Downarrow & & & & (5) \iff (3) \quad \text{and} \quad (5) \Rightarrow (6). \\ (2) & \Rightarrow & (1) & \Rightarrow & \end{array}$$

$(3) \Rightarrow (2)$, (4). Notice first that $\mathscr{R}f(x) = X$ for any fixed x is equivalent to $\omega f(x) = X$. In fact, $x \in \mathscr{R}f(x)$ implies either that $x \in \omega f(x)$.

So $\mathscr{R}f(x) = \overline{\mathscr{O}f(x)} \subset \omega f(x) \subset \mathscr{R}f(x)$ or that $x \in \mathscr{O}f(x)$ in which case x is periodic and $\mathscr{O}f(x) = \mathscr{R}f(x) = \omega f(x)$. Now by Proposition 1.11(d) $\omega f \circ (\omega f)^{-1} \subset \Omega f$ i.e., $y_1, y_2 \in \omega f(x)$ implies $y_1 \in \Omega f(y_2)$. So $\omega f(x) = X$ implies $\Omega f = X \times X$. If U is open and f is semi-invariant then for some N either $f^n(x) \in U$ for all $n \geq N$ or $f^n(x) \notin U$ for all $n \geq N$. In the first case, $X = \omega f(x) \subset \overline{U}$ and U is dense while in the second case $X = \omega f(x)$ is disjoint from U and U is empty.

(4) $\Rightarrow$ (5). f^{-1} +invariance implies f semi-invariance.

(2) $\Rightarrow$ (1). $\Omega f \subset \mathscr{N} f \subset X \times X$.

(1) $\Rightarrow$ (5). Let U be open and f^{-1} +invariant. If $y \in U$ then there exists $\varepsilon > 0$ such that $V_\varepsilon(y) \subset U$. Given any $x \in X$, $y \in \mathscr{N} f(x)$. So if δ is positive with $\delta < \varepsilon$, then there exists $x_1 \in V_\delta(x)$ and n_1 such that $f^{n_1}(x_1) \in V_\delta(y) \subset U$. So $x_1 \in f^{-n_1}(U) \subset U$. Thus, $V_\delta(x) \cap U \neq \varnothing$ for every positive δ. It follows that U is dense.

(5) $\Rightarrow$ (3). We will prove the stronger result, quoted after the equivalence, that $\{x : \omega f(x) = X\}$ is the intersection of a sequence of dense open sets and so is dense by the Baire Category Theorem.

Let $\{U_n\}$ be a countable base for the topology of X. $\mathscr{O}f^{-1}(f^{-m}(U_n)) = \bigcup_{k>m} f^{-k}(U_n)$ is for each fixed m and n an f^{-1} +invariant open set and so is dense. Let D denote the intersection of this countable family. Clearly, $x \in D$ if and only if for every pair m, n of positive integers there exists $k > m$ such that $f^k(x) \in U_n$. As $\{U_n\}$ is a base for the topology, this is equivalent to $\omega f(x) = X$. Thus, $D = \{x : \omega f(x) = X\}$.

(5) $\Leftrightarrow$ (6). The complement of an f^{-1} +invariant set is f +invariant and vice-versa.

Now if $B = \omega f(x)$ for some positive recurrent point x, then $x \in B$ and so $B = \omega f(x) = \omega(f_B)(x)$. f_B is topologically transitive by (3) and B is a topologically transitive subset.

Conversely, if f_B is topologically transitive for some closed f +invariant subset B, then $B = \omega(f_B)(x) = \omega f(x)$ for some x in B and since $x \in B$ it is positive recurrent.

Because B is f invariant (e.g. it is chain transitive) $x \in B$ with $\omega f(x) = B$ implies there exists $y \in B$ with $f(y) = x$. As $\omega f \circ f = \omega f$, it follows that $\omega f(y) = \omega f(x) = \omega f(f(x))$. So y and $f(x)$ lie in D_B, which is dense in B by the Baire Category Theorem. Clearly, as above, all such points are positive recurrent.

(c) We will show

$$(1) \Leftrightarrow (2) \quad \text{and} \quad (3) \Rightarrow (2) \Rightarrow (4) \Rightarrow (5) \Rightarrow (6) \Rightarrow (3).$$

(1) $\Leftrightarrow$ (2). $\mathscr{R}f(x)$ is the closure of $\mathscr{O}f(x)$.

(3) $\Rightarrow$ (2). $\omega f \subset \mathscr{R} f \subset X \times X$.

(2) $\Rightarrow$ (4). Obvious.

$(4) \Rightarrow (5)$. If $y \in A$ then $x \in \mathscr{R}f(y)$ and $\mathscr{R}f(y) \subset A$ if A is f +invariant and closed. So $X = \mathscr{R}f(x)$ is also contained in A and $A = X$.

$(5) \Rightarrow (6)$. f invariance implies f +invariance.

$(6) \Rightarrow (3)$. For all $x \in X$, $\omega f(x)$ is a closed, nonempty f invariant subset of X.

For B an f +invariant subset of X the restriction f_B is minimal if and only if B contains no proper closed f invariant subset (cf. (6)). In particular, $f(B) = B$ and B is itself f invariant.

If A is a closed nonempty f +invariant subset of X, then the collection of closed nonempty f +invariant subsets of A is a nonempty collection (e.g. A) and is closed under monotone intersection. Hence, by Zorn's Lemma this family contains a minimal element B (with respect to the ordering by inclusion). By (5) such a set B is minimal in the above sense. □

The basic sets of f and the chain transitive subsets are closely related. In fact, we can characterize the basic sets as chain transitive subsets which are $\mathscr{C}f$ semi-invariant:

13. Corollary. *Let $f: X \to X$ be a continuous map and let B be a closed nonempty subset of X.*

(a) *If B is a chain transitive subset, then B is contained in a unique basic set of f, i.e., $B \subset \mathscr{C}f(x) \cap \mathscr{C}f^{-1}(x)$ for $x \in B$.*

(b) *The following properties are equivalent:*

(1) *B is a basic set, i.e., $B = \mathscr{C}f(x) \cap \mathscr{C}f^{-1}(x)$ for $x \in B$.*

(2) *B is a $\mathscr{C}f$ semi-invariant, chain transitive subset of $|\mathscr{C}f|$.*

(3) *B is a $\mathscr{C}f$ semi-invariant, f +invariant, indecomposable subset of $|\mathscr{C}f|$.*

Proof. (a) $B \times B = \mathscr{C}(f_B) \subset \mathscr{C}f$ and so all points of B are $\mathscr{C}f$ related to one another.

(b) $(1) \Rightarrow (3)$. A basic set $\mathscr{C}f(x) \cap \mathscr{C}f^{-1}(x)$ is clearly $\mathscr{C}f$ semi-invariant. f invariance and indecomposibility follow from Corollary 11.

$(3) \Rightarrow (2)$. By Theorem 5, $(\mathscr{C}f)_B = \mathscr{C}(f_B)$ and so $B \subset |\mathscr{C}f|$ then implies $|\mathscr{C}(f_B)| = B$. Chain transitivity follows from (a)(6) of the Theorem 12.

$(2) \Rightarrow (1)$. By (a), $B \subset \mathscr{C}f(x) \cap \mathscr{C}f^{-1}(x)$ for x in B. The reverse inclusion follows from $\mathscr{C}f$ semi-invariance. □

14. Proposition. *Let $f: X \to X$ be a continuous map. For $x \in X$, the limit set $\omega f(x)$ is a chain transitive subset of X. If, in addition, $x \in \omega f(x)$, i.e., x is positive recurrent, then $\omega f(x)$ is a topologically transitive subset of X.*

Proof. Recall that $(\omega f)^{-1} \circ \omega f \subset \Omega f$ (Proposition 1.12(d)) which says $(\Omega f)_B = B \times B$ when $B = \omega f(x)$. A fortiori, $(\mathscr{C}f)_B = B \times B$. But B is usually not $\mathscr{C}f$ semi-invariant and so we cannot immediately conclude $\mathscr{C}(f_B) = B \times B$. We have to take another look at the proof of Theorem 5.

Given y, $z \in B = \omega f(x)$ and $\varepsilon > 0$ we choose $\varepsilon_1 > 0$ satisfying (4.3) as before. Choose N so that $f^n(x) \in V_{\varepsilon_1}(B)$ for $n \geq N$. As y and z are limit points, we can choose $n_1 \geq N$ and $n > 0$ so that both $d(f^{n_1}(x), y)$ and $d(f^{n_1+n}(x), z) < \varepsilon_1$. So $x_i = f^{n_1+i}(x)$ $(i = 0, \dots, n)$ is a chain for f lying in $V_{\varepsilon_1}(B)$ and beginning and ending ε_1 close to y and z respectively. Complete the proof as in Theorem 5.

The stronger property of topological transitivity of $\omega f(x)$ when x is positive recurrent was already proved in part (b) of Theorem 12. □

From Corollary 13 it is clear that

$$(4.4)\qquad \begin{aligned} |\mathscr{C} f| &= \bigcup\{B : B \text{ is a basic set for } f\} \\ &= \bigcup\{B : B \text{ is a chain transitive subset of } X\}. \end{aligned}$$

If x is a positive recurrent point then $x \in \omega f(x)$ a topologically transitive subset. While a topologically transitive subset B may contain points which are not positive recurrent, the points of the dense subset $\{x \in B : \omega f(x) = B\}$ are positive recurrent. Consequently

$$(4.5)\qquad |\omega f| \subset \bigcup\{B : B \text{ is a topologically transitive subset}\} \subset \overline{|\omega f|}.$$

Hence, the closure of the set of positive recurrent points is the same as the closure of the union of topologically transitive subsets. We call this closure the *center* of f.

Analogously we define

$$(4.6)\qquad m[f] \equiv \bigcup\{B : B \text{ is a minimal subset for } f\},$$

and call its closure the *min-center* of f.

Every point of a minimal subset is positive recurrent and every periodic orbit is a minimal set. Thus, we have a tower of inclusions for any continuous map f:

$$(4.7)\qquad |f| \subset |\mathscr{O} f| \subset m[f] \subset |\omega f| \subset \omega f(X) \subset |\Omega f| \subset |\mathscr{G} f| \subset |\mathscr{C} f|.$$

Recall that we denote by $l_+[f]$ the closure of $\omega f(X) = \bigcup\{\omega f(x) : x \in X\}$.

15. PROPOSITION. *Let $f: X \to X$ be a continuous map and let B be a closed and $f+$ invariant subset of X.*

(a) *The set of fixed points $|f|$ is closed and f invariant. For the restriction f_B, $|f_B| = |f| \cap B$. In particular, if $|f| \subset B$, then $|f_B| = |f|$.*

(b) *The set of periodic points $|\mathscr{O} f|$ is f invariant as is its closure. For the restriction f_B, $|\mathscr{O} f_B| = |\mathscr{O} f| \cap B$. In particular, if $|\mathscr{O} f| \subset B$, then $|\mathscr{O} f_B| = |\mathscr{O} f|$.*

(c) *The set $m[f]$ is f invariant as is its closure, the min-center. For the restriction f_B, $m[f_B] = m[f] \cap B$. In particular, if $m[f] \subset B$, then $m[f_B] = m[f]$ and the min-center of f and f_B are the same.*

(d) *The set of positive recurrent points* $|\omega f|$ *is* f *invariant as is its closure, the center. For the restriction* f_B, $|\omega(f_B)| = |\omega f| \cap B$. *In particular, if* $|\omega f| \subset B$, *then* $|\omega(f_B)| = |\omega f|$ *and the centers of* f *and* f_B *agree.*

(e) *The set of positive limit points* $\omega f(X)$ *is* f *invariant as is its closure* $l_+[f]$. *If* $\omega f(X) \subset B$, *then* $l_+[f] \subset |\mathscr{C}(f_B)|$ *and each* $\omega f(x)$ *for* $x \in X$ *is contained in a unique basic set of* f_B.

(f) *The chain recurrent set* $|\mathscr{C} f|$ *is* f *invariant. If* $|\mathscr{C} f| \subset B$, *then* $|\mathscr{C}(f_B)| = |\mathscr{C} f|$ *and the basic sets agree, i.e., for each* $x \in B$, $\mathscr{C} f(x) \cap \mathscr{C} f^{-1}(x) = \mathscr{C}(f_B)(x) \cap \mathscr{C}(f_B)^{-1}(x)$.

PROOF. (a) and (b) are obvious. For (c) note that $m[f]$ is the union of f invariant sets. If A is minimal for f, then $A \cap B$ is an f +invariant subset of A and so equals A if it is nonempty, i.e., $A \subset B$ or $A \cap B = \emptyset$. When $A \subset B$, $f_A = (f_B)_A$ and so minimality for f and for f_B agree.

The f invariance of $|\omega f|$ follows from the following sharpening of (4.5) (cf. part (b) of Theorem 12).

$$|\omega f| = \bigcup \{D_B : B \text{ is a topologically transitive subset of } X\}. \tag{4.8}$$

For $x \in B$, $\omega(f_B)(x) = \omega f(x)$ and so $x \in |\omega(f_B)|$ if and only if $x \in |\omega f|$, proving (d).

(e) $\omega f(X)$ is the union of f invariant sets $\omega f(x)$ and so is f invariant (Corollary 3(a)). Alternatively apply Lemma 2(a) $(f \circ \omega f = \omega f)$. By Proposition 14 each $\omega f(x)$ is a chain transitive subset of X. So if $\omega f(x) \subset B$ it is a chain transitive subset of B and the inclusion in a unique basic set follows from Corollary 13(a). Notice that we are not assuming $x \in B$ here.

(f) $|\mathscr{C} f|$ is the union of the basic sets of f and so is f invariant. If such a basic set $\mathscr{C} f(x) \cap \mathscr{C} f^{-1}(x)$ is contained in B, then its chain transitivity implies, just as for (e), that it is contained in a basic set of B, namely, $\mathscr{C}(f_B)(x) \cap \mathscr{C}(f_B)^{-1}(x)$ since $x \in \mathscr{C} f(x) \cap \mathscr{C} f^{-1}(x)$ lies in B. The reverse inclusion follows from $\mathscr{C}(f_B) \subset \mathscr{C} f$. □

The contrast between results in (a)–(d) and those in (e) and (f) is due to the difference between properties which are intrinsic and those which are not. Let $f: X \to X$ be a continuous map, and let A be a subset of X. We call a mapping property of A *heritable* when its holding for A with respect to f implies that it holds for A with respect to the restriction f_B whenever B is a closed f invariant subset containing A. The property is called *intrinsic* if the converse is true as well. For example, + invariance itself is intrinsic because A is f +invariant if and only if it is f_B +invariant. Similarly, invariance is intrinsic. On the other hand, A f^{-1} +invariant implies that A is f_B^{-1} +invariant but the converse need not hold. Thus, f^{-1} +invariance is a heritable property which is not intrinsic. Parts (a)–(d) of the proposition say that the properties "A is the fixed point set", "A is the set of periodic points", "A is the min-center", and "A is the center" are all heritable properties. If A is itself closed and f +invariant then an additional property is intrinsic when it depends only on the restriction f_A. Thus, by definition, "A is a chain transitive subset", "A is a topologically transitive subset" and "A is minimal" are intrinsic properties.

A mapping property of a point x is called intrinsic when, analogously, it holds for x with respect to f if and only if it holds to the restriction f_B for any closed f +invariant subset B containing x, or, equivalently, when it depends only upon the restriction f_A with $A = \{x\} \cup \mathscr{R}f(x) = \{x\} \cup \mathscr{O}f(x) \cup \omega f(x)$. So "$x$ is a fixed point", "x is a periodic point" and "x is a positive recurrent point" are intrinsic properties.

16. EXERCISE. (a) *For a continuous map f prove that $|\Omega f|$ is $f+$ invariant and $|\mathscr{G}f| = |\Omega\mathscr{G}f|$ is f invariant.* (*For f +invariance use Lemma* 2(b), *$F \subset f^{-1} \circ F \circ f$. For f invariance of $|\mathscr{G}f|$ use the $\mathscr{G}f$ analogues of the basic sets.*)

(b) *Let $g(\theta)$ be a smooth, nonnegative real valued function in $[0, 2\pi]$ with $g^{-1}(0) = [0, \pi/4] \cup [3\pi/4, 5\pi/4] \cup [7\pi/4, 2\pi]$. Let f be the time-one map for the system of differential equations in the plane* (*in polar coordinates*):

$$dr/dt = r(1-r), \qquad d\theta/dt = r[(1-r)^2 + g(\theta)].$$

Prove that $B =$ the unit circle $\{r = 1\}$ is an attractor for the system ($L(r, \theta) = 1 - (1-r)^2$ *is a Lyapunov function*). *$|\mathscr{C}f| = |\mathscr{G}f| = |\Omega f| = l_+[f]$ consist of the origin and B; and B is a basic set. $|\mathscr{O}f| = |f|$ consists of the origin and $\widetilde{B} = \{(r, \theta) : r = 1$ and $g(\theta) = 0\}$. For the restriction f_B, prove $|\mathscr{G}f_B| = |\Omega(f_B)| = l_+[f_B] = |\mathscr{O}f_B| = |f_B| = \widetilde{B}$, while $|\mathscr{C}(f_B)| = B$. So the points of $B - \widetilde{B}$ are not even generalized nonwandering points for f_B.*

(c) *Prove that "A is indecomposable" and "A is f semi-invariant" are intrinsic properties while "A is $\mathscr{C}f$ semi-invariant" is heritable but not intrinsic. Prove that "A is the set of nonwandering points" and "A is the set of generalized nonwandering points" are not even heritable. Prove that "x is a nonwandering point" and "x is a generalized nonwandering point" are not heritable properties.*

(d) *Prove that "x is a chain recurrent point" is not a heritable property. On the other hand, "A is the set of chain recurrent points" is heritable* (*but not intrinsic*). *Prove that if $A \subset B$ and B is closed and $f+$ invariant, then A is a basic set for f implies A is a basic set for f_B, while the converse does not hold. Thus, "A is a basic set" is heritable but not intrinsic. Notice from Corollary* 13 *that $\mathscr{C}f$ semi-invariance is the nonintrinsic aspect of a basic set.*
□

If we let B denote the chain recurrent set $|\mathscr{C}f|$ it follows from Proposition 15(f) that $|\mathscr{C}(f_B)| = B$, i.e., all points of B are chain recurrent with respect to the restriction f_B. This is just a way of observing that "A is the chain recurrent set" is a heritable property. On the other hand, "A is the nonwandering set" is not heritable so if we let $B_1 = |\Omega f|$ then $B_2 = |\Omega(f_{B_1})|$ may be a proper subset of B_1 albeit closed and f +invariant. Inductively, define $B_{\alpha+1} = |\Omega(f_{B_\alpha})|$ and proceed by transfinite induction, if necessary, defining $B_\alpha = \bigcap_{\beta<\alpha}\{B_\beta\}$ when α is a limit ordinal. The process stabilizes (before the first uncountable ordinal) when we reach a closed f +invariant subset B with $B = |\Omega(f_B)|$. Similarly, we can begin with the limit point set $\widetilde{B}_1 = l_+[f]$ and define $\widetilde{B}_{\alpha+1} = l_+[f_{\widetilde{B}_\alpha}]$ with $\widetilde{B}_\alpha = \bigcap_{\beta<\alpha}\{\widetilde{B}_\beta\}$ (α a limit ordinal) and arrive at a set $\widetilde{B}$ closed and f+ invariant with $\widetilde{B} = l_+[f_{\widetilde{B}}]$. It is clear that in every stage of the above process the positive recurrent points remain in $\widetilde{B}_\alpha$ and $\widetilde{B}_\alpha \subset B_\alpha$. Thus, the center of f is contained in $\widetilde{B}$ and $\widetilde{B} \subset B$. In fact, the three are the same set as follows from

17. **Proposition.** *A continuous map $f: X \to X$ is called* central *if it satisfies the following equivalent conditions.*

(1) *X is the center, i.e., $|\omega f|$ is dense in X.*
(2) *$X = l_+[f]$, i.e., $\omega f(X)$ is dense in X.*
(3) *$X = |\Omega f|$.*

When f is central the set of positive recurrent points $|\omega f|$ is residual, i.e., it is the countable intersections of dense open sets.

Proof. (1) $\Rightarrow$ (2) $\Rightarrow$ (3) are obvious. We prove that (3) implies $|\omega f|$ is residual.

Let $\mathscr{U}_n$ be a finite cover of X by open sets of diameter less than $1/n$ for $n = 1, 2, \ldots$. Define

$$G_n = \bigcup\{\mathscr{O}f^{-1}(U) \cap U : U \in \mathscr{U}_n\}.$$

Clearly, G_n is open and we show that $X = |\Omega f|$ implies it is dense. For $x \in X$ and $\varepsilon > 0$ choose $U \in \mathscr{U}_n$ and a positive $\delta < \varepsilon$ such that $V_\delta(x) \subset U$. Because $x \in |\Omega f|$ there exists a point y and a positive integer N such that y and $f^N(y)$ lie in $V_\delta(x)$. So $y \in \mathscr{O}f^{-1}(V_\delta(x)) \cap V_\delta(x) \subset V_\varepsilon(x) \cap G_n$. Thus, G_n is dense. Furthermore, if $x \in |\omega f|$, i.e., $x \in \omega f(x)$ we can choose $y = x$ and so $|\omega f| \subset G_n$. We complete the proof by showing that $\bigcap_n\{G_n\} \subset |\omega f|$.

If $x \in \bigcap_n\{G_n\}$, then for any n we can choose $U \in U_n$ and a positive integer N_n such that $x \in f^{-N_n}(U) \cap U$, i.e., x, $f^{N_n}(x) \in U$ and so $d(x, f^{N_n}(x)) < 1/n$. So as $n \to \infty$, $\{f^{N_n}(x)\}$ approaches x. Either $\{N_n\}$ approaches ∞ with n and so $x \in \omega f(x)$ or $\{N_n\}$ takes on some fixed value N infinitely often so that $f^N(x) = x$ and $x \in |\mathscr{O}f| \subset |\omega f|$. □

For a mapping f the ω limit relation has proved to be quite useful. When f is a homeomorphism the corresponding relation for f^{-1} becomes useful as well. In general, define

$$\alpha f[A] = \omega(f^{-1})[A] = \limsup\{f^{-n}(A)\}, \qquad \alpha f = \omega(f^{-1}),$$

$$l_-[f] = \overline{\alpha f(X)} = \overline{\bigcup\{\alpha f(x) : x \in X\}} = l_+[f^{-1}] \tag{4.9}$$

$$l[f] = l_+[f] \cup l_-[f] = l[f^{-1}].$$

18. Exercise. *Let* $f\colon X \to X$ *be a homeomorphism. Prove:*

(a) A *is* f *invariant if and only if* A *is* f^{-1} *invariant if and only if* A *is both* f $+$*invariant and* f^{-1} $+$*invariant*.

(b) *The family of* f *invariant subsets is closed under arbitrary intersections, arbitrary unions, difference and closure.*

(c) *If* A *is* f *invariant, then* A *is indecomposable with respect to* f *if and only if it is indecomposable with respect to* f^{-1}.

(d) *For* $\mathscr{A} = \mathscr{O}, \Omega, \mathscr{G}, \mathscr{C}$, $\mathscr{A}(f^{-1}) = (\mathscr{A}f)^{-1}$ *and so* $|\mathscr{A}f^{-1}| = |\mathscr{A}f|$, *and these sets are all* f *invariant (especially including* $|\Omega f|$).

(e) f *is chain transitive if and only if* f^{-1} *is chain transitive.*

(f) f *is topologically transitive if and only if* f^{-1} *is topologically transitive. In that case the set* $\{x : \omega f(x) = X = \alpha f(x)\}$ *is residual.*

(g) f *is minimal if and only if* f^{-1} *is minimal. So for any homeomorphism* f, $m[f] = m[f^{-1}]$.

(h) f *is central if and only if* f^{-1} *is central. In that case* $|\alpha f \cap \omega f| = |\alpha f| \cap |\omega f|$ *is residual.* □

Recall that a point x is called positive recurrent if $x \in |\omega f|$, i.e., $x \in \omega f(x)$. Similarly, we call x *negative recurrent* if $x \in |\alpha f|$ and *recurrent* if it is both positive and negative recurrent. There may exist points which are positive recurrent without being negative recurrent and vice-versa. But for a homeomorphism most points recurrent either way are recurrent both ways

$$\overline{|\alpha f|} = \overline{|\alpha f| \cap |\omega f|} = \overline{|\omega f|}. \tag{4.10}$$

This equation and the stronger result that $|\alpha f| \cap |\omega f|$ is residual in $\overline{|\omega f|}$ follows from restricting f to the center $\overline{|\omega f|}$ and applying (h) of the exercise.

19. Exercise. *For a homeomorphism* f *we adjust the definition of intrinsic property slightly by assuming that the superset* B *be* f *invariant (rather than merely* f $+$*invariant). With this adjustment show that "*x *is negative recurrent" and "*x *is recurrent" are intrinsic properties. The adjustment is motivated by the observation that the restriction of a homeomorphism* f *to a subset* B *is a homeomorphism of* B *exactly when* B *is* f *invariant.* □

For a map $f\colon X \to X$ the limit set $\omega f(x)$ is entirely contained in some basic set by (e) of Proposition 15. If f is a homeomorphism the same is true for $\alpha f(x)$ as well. if B_+, B_- are the basic sets containing $\omega f(x)$ and $\alpha f(x)$, then $\{f^n(x)\}$ eventually enters any neighborhood of B_+ as $n \to \infty$

and any neighborhood of B_- as $n \to -\infty$. We can identify these basic sets by using a complete $\mathscr{C}f$ Lyapunov function (c.f. Theorem 3.12).

20. **Proposition.** (a) *Let $f\colon X \to X$ be a continuous map and $L\colon X \to \mathbb{R}$ be a Lyapunov function for f. For $x \in X$, L is nondecreasing on the positive orbit $\{x, f(x), f^2(x), \ldots\}$ and L is constant on $\omega f(x)$ with value $\lim_{n\to\infty} L(f^n(x))$. If f is a homeomorphism then L is nonincreasing on the negative orbit $\{x, f^{-1}(x), f^{-2}(x), \ldots\}$ and L is constant on $\alpha f(x)$ with value $\lim_{n\to\infty} L(f^{-n}(x))$.*

(b) *Assume f is a homeomorphism and that L is a $\mathscr{C}f$ Lyapunov function with $|L| = |\mathscr{C}f|$ and taking different values on distinct basic sets, i.e., a complete $\mathscr{C}f$ Lyapunov function.*

(1) *If $x \in |\mathscr{C}f|$, then L is constant on $\{f^n(x) : n \in Z\}$ with values that of the basic set containing x.*

(2) *If $x \notin |\mathscr{C}f|$ then L is strictly increasing on $\{f^n(x) : n \in Z\}$. $\lim_{n\to\infty} L(f^n(x))$ is the value of L on the basic set containing $\omega f(x)$. $\operatorname{Lim}_{n\to\infty} L(f^{-n}(x))$ is the value of L on the basic set containing $\alpha f(x)$.*

Proof. $\{L(f^n(x))\}$ is nondecreasing on n because L is a Lyapunov function. So the limit as $n \to \infty$ exists. Call it a. If $y \in \omega f(x)$ then there is a sequence $n_k \to \infty$ with $f^{n_k}(x) \to y$. So $a = \lim_{k\to\infty} L(f^{n_k}(x)) = L(y)$. If $x \notin |L|$, then $L(f^{-1}(x)) < L(x) < L(f(x))$. So if $|L|$ happens to be f invariant for a homeomorphism f, then for $x \notin |L|$ the entire orbit $\{f^n(x) : n \in Z\}$ consists of regular points and so $\{L(f^n(x))\}$ is strictly increasing. In particular, when $|L| = |\mathscr{C}f|$ we have (2) of (b). (1) follows because if $x \in |\mathscr{C}f|$ then the orbit of x remains in the basic set $\mathscr{C}f(x) \cap \mathscr{C}f^{-1}(x)$ by f invariance and L is constant on basic sets because it is a $\mathscr{C}f$ Lyapunov function. □

Notice that while $\alpha f(x)$ and $\omega f(x)$ are contained in basic sets, each contains minimal subsets. Hence, for every $x \in X$ there exists $y_1, y_2 \in m[f]$ with $y_1 \in \alpha f(x)$ and $y_2 \in \omega f(x)$. Clearly, $x \in |\mathscr{C}f|$ if and only if $y_1 \in \mathscr{C}f(y_2)$ in which case y_1, y_2 and x all lie in the same basic set. Also, y_1 and y_2 are recurrent points because $m[f] \subset |\omega f|$.

Supplementary exercises

21. Let $f\colon X \to X$ be a continuous map. Prove that if f is minimal, then for every $\varepsilon > 0$ there exists a positive integer n such that for all $x \in X$ the sequence $\{x, f(x), \ldots, f^n(x)\}$ is ε-dense (i.e., $y \in X$ implies $d(y, f^k(x)) \leq \varepsilon$ for some $0 \leq k \leq n$). (*Hint*: Let $\{U_i\}$ be a finite open cover of X be sets of diameter at most ε and find $n > 0$ such that $\{U_i, f^{-1}(U_i), \ldots, f^{-n}(U_i)\}$ covers X for every i.) Conversely, f is minimal if some $x \in X$ satisfies the condition that for every $\varepsilon > 0$, then there exists a positive integer n such that $\{f^i(x), \ldots, f^{i+n}(x)\}$ is

ε dense in X for every $i \geq 0$. (*Hint*: Let B be a minimal subset of $\omega f(x)$ and suppose that B is a proper subset of X. $x \in B$ easily contradicts this hypothesis. For $V_\varepsilon(x) \cap B = \emptyset$ and any n there exists U a neighborhood of B such that $U \cup \cdots \cup f^n U \subset V_\varepsilon(B)$. The orbit of x eventually enters U.)

22. On the rectangle $\widetilde{X} = [0, 1] \times [-1, 1]$ identify $(0, y)$ with $(1, -y)$ to obtain a Möbius strip X. Let g be a smooth, nonnegative function on $\widetilde{X}$ with $g(0, y) = g(1, -y)$ and $g^{-1}(0) = 0 \times [-1, 0] \cup 1 \times [0, 1]$. Let $f\colon X \to X$ be the time-one homeomorphism of the horizontal flow defined on X by

$$dx/dt = g(x, y), \qquad dy/dt = 0$$

(see Figure 4.1). Let $A_1 = 0 \times [-1, 0] = 1 \times [0, 1]$ in X and let $A_2 = [0, 1] \times 0$ in X (the central circle of the strip). Prove that $|f| = A_1$ and $|\Omega(f)| = |\mathscr{C}(f)| = A_1 \cup A_2$. Now let U be the open subset $(0, 1/2) \times [-1, 0)$ of X and let B be its complement $X - U$. Prove that B is f +invariant and for the restriction f_B, $|f_B| = A_1$ and $|\mathscr{G}(f_B)| = |\mathscr{C}(f_B)| = A_1 \cup A_2$. However, $|\Omega(f_B)| = A_1 \cup [1/2, 1] \times 0$ and so $|\Omega(f_B)|$ is not f_B invariant.

23. For $x \in |\mathscr{C} f|$ the basic set $\mathscr{C} f(x) \cap \mathscr{C} f^{-1}(x)$ is f invariant when f is a map. Prove that it is usually not $\mathscr{C} f$ +invariant. In fact, the basic set containing x is $\mathscr{C} f$ +invariant if and only if x is $\mathscr{C} f$ maximal, i.e., $y \in \mathscr{C} f(x)$ implies $x \in \mathscr{C} f(y)$ as well.

24. Let $f\colon X \to X$ be a homeomorphism. Prove that f is chain transitive if X contains only a single basic set for f. (*Hint*: The repellor complementary to the unique nonempty attractor is empty.)

25. Let $f\colon X \to X$ be a continuous map and $\{A_n\}$ a sequence of closed subsets of X. Prove that if each A_n is f +invariant (or f invariant), then $A = \limsup\{A_n\}$ is f +invariant (resp. f invariant). If, in addition, $|\mathscr{C} f_{A_n}| = A_n$ for all n, then $|\mathscr{C} f_A| = A$. (Adapt the proof of Theorem 5 to show $x \in \mathscr{C}(f_A)(x)$.) However, show that all the A_n may be chain transitive without A being indecomposable and so chain transitive.

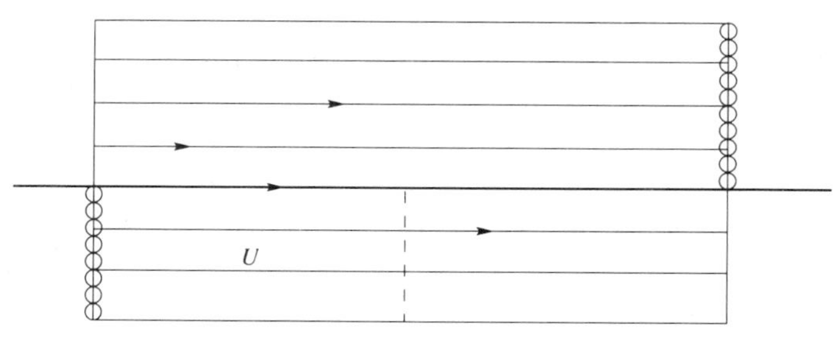

FIGURE 4.1

26. Let $f: X \to X$ be a continuous map and let B be an f +invariant, $\mathscr{C}f$ semi-invariant subset. If A is an attractor for f_B, prove there exists an attractor A_1 for f with $A_1 \cap B = A$. (*Hint*: Prove $\mathscr{C}f(A) \cap B = A$. There exists an inward set U such that $\omega f[U \cap B] = A$. Letting $A_1 = \omega f[U]$, $A_1 \cap B \subset \mathscr{C}f(U) \cap B = (\mathscr{C}f)_B(U \cap B) = \mathscr{C}(f_B)(U \cap B) = A$.)
27. Let $\{x_n\}$ be a sequence in X with limit point set A. If $\lim\{d(x_n, x_{n+1})\} = 0$ prove that A is connected. (*Hint*: if U_1 and U_2 are open sets $\varepsilon > 0$ apart, together containing the sequence, and $d(x_n, x_{n+1}) < \varepsilon$ for $n \geq N$, then $x_N \in U_1$ implies $x_n \in U_1$ for $n \geq N$). If the series $\sum_n d(x_n, x_{n+1})$ converges prove that A is a single point (the sequence is Cauchy). For $x \in X$ and $f: X \to X$ a continuous map apply these results to $\{f^n(x)\}$ to get results about $\omega f(x)$.
28. Let $X_1 = |\mathscr{C}f|$ for a continuous map f on X. Prove that $\mathscr{C}(f_{X_1}) = \bigcup\{B \times B : B$ is a basic set for $f\}$ and show by example that this is usually a proper subset of $(\mathscr{C}f)_{X_1}$.
29. (F. Takens). If $f: X \to X$ is a chain transitive map then there exists X_1 a compact metric space with $X \subset X_1$, $f_1: X_1 \to X_1$ extending f and $x \in X_1$ such that $X = \omega f_1(x)$. (*Hint*: Construct a sequence $\{x_n : n = 1, 2, \dots\}$ in X such that $d(x_{n+1}, f(x_n)) \to 0$ and $\{x_k : k > n\}$ is dense in X for all n, e.g., hook together ε chains which are ε dense. Let X_1 be the subset of $X \times [0, 1]$ consisting of $X = X \times \{0\}$ and $\{(x_n, 1/n) : n = 1, 2, \dots\}$. Define $f_1(x_n, 1/n) = (x_{n+1}, 1/n+1)$.) Compare this with Proposition 14.
30. Let $f: X \to X$ be a continuous map. When f is topologically transitive $\{x : \omega f(x) = X\}$ is residual. prove that the complement $N = \{x : \omega f(x) \neq X\}$ is either empty (in which case f is minimal) or dense. (*Hint*: $x \in N \Leftrightarrow f(x) \in N$, i.e., N is f^{-1} invariant. If $\overline{N} \neq X$, then $\omega f(x) \subset \overline{N}$ implies $\mathscr{R}f(x) \subset N$. So N is a repellor by Theorem 3.6 (b)(3) and the dual attractor is empty.)
31. Let $f: X \to X$ be a topologically transitive map. It will usually be the case that all powers f^k are topologically transitive as well, but, for example, when X is the finite set of congruence classes mod k and $s: Z_k \to Z_k$ is given by $s(i) = i+1$ then s^k is the identity. Furthermore, if $h: X \to Z_k$ is continuous and maps f to s, then f^k is not transitive. Now assume that f^k is not transitive and that $k > 0$ is the smallest such power. Fill in the details in the following steps to show that there is an f and f^{-1} invariant open subset U of X such that the restriction f_U factors through s as above.

    (a) There exists A closed and f^k +invariant with $A \neq X$ and $\operatorname{Int} A \neq \varnothing$. There exists $x_0 \in \operatorname{Int} A$ with $\omega f(x_0) = X$.

    (b) Define for $i = 0, 1, \dots, k-1$ the set A_i to be the limit point set of the sequence $\{f^{nk+i}(x_0) : n = 0, 1, \dots\}$. Observe that $A_0 = \omega(f^k)(x_0) \subset A$ and so is a proper subset of X.

(c) $A_0 \cup A_1 \cup \cdots \cup A_{k-1} = \omega f(x_0) = X$. (*Hint*: For any subsequence $\{f^{n_i}(x_0)\}$ there is a mod k congruence class containing infinitely many n_i's.)

(d) $f(A_i) = A_{i+1}$ and $f^{-1}(A_i) = A_{i-1}$ with addition and subtraction mod k. In particular, each A_i is f^k and $(f^k)^{-1}$ invariant.

(e) $f^{-1}(\operatorname{Int} A_i) \subset \operatorname{Int} A_{i-1}$ where the interior is taken with respect to X.

(f) $\operatorname{Int} A_i \neq \emptyset$ for $i = 0, \ldots, k-1$. (*Hint*: Use (c) and the Baire Category Theorem to get one A_i and (e) to get the rest.)

(g) $A_i \subset A_j$ implies $i = j$ (Otherwise A_j is f^p+ invariant for some $p < k$.)

(h) If $x \in A_i$ and $\omega f(x) = X$, then $\omega(f^k)(x) = A_i$. (For simplicity look at the $i = 0$ case: Define $\widetilde{A}_0 = \omega(f^k)(x)$ and $\widetilde{A}_i$ for $i = 1, \ldots, k-1$ as in (b) using x in place of x_0. $\widetilde{A}_i \subset A_i$ and as before $\widetilde{A}_0 \cup \cdots \cup \widetilde{A}_{k-1} = X$. So $x_0 \in \widetilde{A}_i$ for some i. By the f^k invariance of $\widetilde{A}_i$, $A_0 \subset \widetilde{A}_i \subset A_i$. So $i = 0$ by (g) and $\widetilde{A}_0 = A_0$.)

(i) f^k is topologically transitive on each A_i and $\{x \in A_i : \omega(f^k)(x) = A_i\} = A_i \cap \{x : \omega f(x) = X\}$.

(j) $A_i \cap A_j \cap \{x : \omega f(x) = X\} \neq \emptyset$ implies $i = j$.

(k) $(\operatorname{Int} A_i) \cap (\operatorname{Int} A_j) \neq \emptyset$ implies $i = j$.

(l) $\operatorname{Int} A_i$ is dense in A_i. (By (e) $\operatorname{Int} A_i$ is $(f^k)^{-1}+$ invariant.)

(m) $A_i \cap \operatorname{Int} A_j \neq \emptyset$ implies $i = j$.

(n) $\operatorname{Int} A_i = A_i - \bigcup_{j \neq i}\{A_j\}$.

(o) $f(\operatorname{Int} A_i) = \operatorname{Int} A_{i+1}$ and $f^{-1}(\operatorname{Int} A_i) = \operatorname{Int} A_{i-1}$ (*Hint*: Do the latter first.)

(p) $U = \bigcup_i\{\operatorname{Int} A_i\}$ is an f and f^{-1} invariant open subset of X and $h: U \to Z_k$ defined by $h(x) = i$ for $x \in \operatorname{Int} A_i$ maps f_U onto s.

As an application, prove that $f \times f$ is not topologically transitive on $X \times X$. (*Hint*: The invariant subset $\bigcup_i\{A_i \times A_i\}$ contains $\bigcup_i\{\operatorname{Int} A_i) \times (\operatorname{Int} A_i)\}$ and is disjoint from $(\operatorname{Int} A_i) \times (\operatorname{Int} A_j)$ for $i \neq j$.)

Similarly, show that if f is minimal and f^k is not, then h exists mapping onto s. (Use A_0 minimal for f^k. Now the A_i's themselves are pairwise disjoint.)

32. For a homeomorphism $f: X \to X$ prove that each of the following relation properties is equivalent to the condition that f is central. (a) $\mathcal{N} f$ is reflexive; (b) Ωf is reflexive; (c) $\mathcal{N} f$ is symmetric; (d) Ωf is symmetric. (*Hint*: If f is central and $y \in \mathcal{N} f(x)$, use $z \in |\omega f| \cap V_\varepsilon(x) \cap f^{-n}(V_\varepsilon(y))$ to get $x \in \Omega f(y)$. Given symmetry and $x \in X$, use $y \in \omega f(x) \Rightarrow x \in \mathcal{N} f(y) \subset \mathcal{N} f(x)$ to get $|\mathcal{N} f| = X$.)

Chapter 5. Computation of the Chain Recurrent Set

In Chapter 1 we introduced chain recurrence by looking at the errors which accumulate as you try to compute the orbit of a point. We now focus this computational view by describing how to approximate the chain recurrent set of a closed relation by using a finite cover of the space. The resulting estimates close in upon the chain recurrent set as the covers become finer. For a homeomorphism we will then describe a different procedure which extends outward from the set of limit points.

Let $\mathscr{U}$ be a finite collection of subsets whose interiors cover X. For any set U the *diameter* of U, denoted by $d(U)$, is the $\sup\{d(x, y) : x, y \in U\}$. The *mesh* of the cover $\mathscr{U}$ is defined by

$$m_1(\mathscr{U}) \equiv \max\{d(U): U \in \mathscr{U}\}. \tag{5.1}$$

Any open cover of X has a positive *Lebesgue number*, i.e., a number $\varepsilon > 0$ such that any subset A with $d(A) < \varepsilon$ is contained in some member of $\mathscr{U}$. Define the maximum Lebesgue number:

$$m_0(\mathscr{U}) \equiv \sup\{\varepsilon: d(A) < \varepsilon \text{ implies } A \subset U \text{ for some } U \in \mathscr{U}\}. \tag{5.2}$$

Define $\mathscr{T}_\varepsilon$ to be the set of open subsets of X with diameter at most ε. Then $\mathscr{U}$ refines $\mathscr{T}_\varepsilon$ when $\varepsilon > m_1(\mathscr{U})$ and $\mathscr{T}_\varepsilon$ refines $\mathscr{U}$ when $\varepsilon < m_0(\mathscr{U})$.

Regarding the set $\mathscr{U}$ as a (discrete) metric space we can think of it as a finite approximation of the space X. If $f: X \to X$ is a closed relation, then the associated relation on $\mathscr{U}$ is defined by

$$\begin{gathered} \mathscr{U}f = \{(U_1, U_2) \in \mathscr{U} \times \mathscr{U}: (U_1 \times U_2) \cap f \neq \varnothing\}, \\ \text{i.e., } U_2\, \mathscr{U}f\, U_1 \Leftrightarrow U_2 \cap f(U_1) \neq \varnothing. \end{gathered} \tag{5.3}$$

Thus, $\mathscr{U}f(U_1)$ consists of those members of $\mathscr{U}$ which intersect the image $f(U_1)$. Notice by symmetry of the condition $(U_1 \times U_2) \cap f \neq \varnothing$ we have

$$U_2 \cap f(U_1) \neq \varnothing \Leftrightarrow f^{-1}(U_2) \cap U_1 \neq \varnothing. \tag{5.4}$$

In particular, $\mathscr{U}(f^{-1}) = (\mathscr{U}f)^{-1}$ and we can omit the parentheses.

By comparing chains for $\mathscr{U}f$ in $\mathscr{U}$ with ε chains for F in X, we see how $\mathscr{U}f$ "approximates" f.

1. **Lemma.** (a) *Let* $\{U_i : i = 0, \dots, n\}$ $(n \geq 1)$ *be a* $\mathscr{U}f$ *chain in* $\mathscr{U}$ *(i.e.,* $U_{i+1}\mathscr{U}fU_i$ *for* $i = 0, \dots, n-1$*). If* $\delta_1 > m_1(\mathscr{U})$*, then there exists* $\{x_i : i = 0, \dots, n\}$ *a* δ_1 *chain for* f *in* X *with* $x_i \in U_i$ *for all* i*. Furthermore, if* $U_0 = U_n$*, then* δ_1 *chain can be chosen with* $x_0 = x_n$.

(b) *Let* $\{x_i : i = 0, \dots, n\}$ $(n \geq 1)$ *be a* δ_0 *chain for* f *in* X*. If* $\delta_0 < m_0(\mathscr{U})$*, then there exists a* $\mathscr{U}f$ *chain* $\{U_i : i = 0, \dots, n\}$ *with* $x_i \in U_i$ *for all* i*. Furthermore, if* $x_0 = x_n$*, then the* $\mathscr{U}f$ *chain can be chosen with* $U_0 = U_n$.

Proof. (a) For $i = 0, \dots, n-1$, $U_{i+1}\mathscr{U}fU_i$ implies that we can choose $x_i \in U_i$ with $f(x_i) \cap U_{i+1} \neq \varnothing$. Because $d(U_{i+1}) < \delta_1$, $U_{i+1} \subset V_{\delta_1} \circ f(x_i)$ $(i = 0, \dots, n-1)$. In particular, $x_0, \dots, x_{n-1}$ is a δ_1 chain as is the extension by choosing x_n to be an arbitrary point of U_n. If $U_n = U_0$, choose $x_n = x_0$.

(b) Because $\{x_i\}$ is a δ_0 chain there exists for $i = 0, \dots, n-1$ a point $y_{i+1} \in f(x_i)$ such that $d(x_{i+1}, y_{i+1}) \leq \delta_0$. Because $\delta_0 < m_0(\mathscr{U})$ we can choose $U_{i+1} \in \mathscr{U}$ with $x_{i+1}, y_{i+1} \in U_{i+1}$ $(i = 0, \dots, n-1)$ and so $f(x_i) \cap U_{i+1} \neq \varnothing$. In particular, $U_1, \dots, U_n$ is a $\mathscr{U}f$ chain as is the extension by choosing U_0 to be an arbitrary member of $\mathscr{U}$ containing x_0. If $x_0 = x_n$, then choose $U_0 = U_n$. □

Because $\mathscr{U}$, regarded as a space, is discrete $\mathscr{C}(\mathscr{U}f)$ is just the transitive extension $\mathscr{O}(\mathscr{U}f)$. Notice that (b) of Lemma 1 implies $\mathscr{U}(f^n) \subset (\mathscr{U}f)^n$ $(n \geq 1)$ but the inclusion is usually strict. The members of $|\mathscr{O}(\mathscr{U}f)|$ we will call the *cyclic elements* of $\mathscr{U}$. So U is cyclic when there exists a $\mathscr{U}f$ chain $\{U_i : i = 0, \dots, n\}$ $(n \geq 1)$ with $U_0 = U = U_n$. We denote by $|\mathscr{U}|_f$ the union of the cyclic elements of $\mathscr{U}$:

$$|\mathscr{U}|_f = \{x : x \in U \text{ for some } U \in |\mathscr{O}(\mathscr{U}f)|\}. \tag{5.5}$$

The set $|\mathscr{U}|_f$, which is an open (or closed) subset if $\mathscr{U}$ is an open (resp. closed) cover of X, is the desired approximation of the chain recurrent set $|\mathscr{C}f|$.

2. **Theorem.** *Let* f *be a closed relation on* X.

(a) *For every* $\varepsilon > 0$ *there exists* $\delta > 0$ *such that*

$$V_\delta(|\mathscr{O}(V_\delta \circ f)|) \subset V_\varepsilon(|\mathscr{C}f|).$$

(b) *Let* $\mathscr{U}$ *be a finite collection of subsets of* X *whose interiors cover* X*. If* $\delta_1 > m_1(\mathscr{U})$*, the mesh of* $\mathscr{U}$*, and* $\delta_0 < m_0(\mathscr{U})$*, the Lebesgue number of* $\mathscr{U}$*, then*

$$|\mathscr{C}f| \subset |\mathscr{O}(V_{\delta_0} \circ f)| \subset |\mathscr{U}|_f, \qquad |\mathscr{U}|_f \subset V_{\delta_1}(|\mathscr{O}(V_{\delta_1} \circ f)|),$$

where $|\mathscr{U}|_f$ *is the union of the* $\mathscr{U}f$ *cyclic elements as defined by* (5.5).

Proof. (a) By Proposition 1.8 the decreasing sequence of closed relations $\{\mathscr{N}(\overline{V}_\delta \circ f) : \delta > 0\}$ has intersection $\mathscr{C}f$. If $d : X \to X \times X$ is the diagonal map $d(x) = (x, x)$ then $|\mathcal{N}(\overline{V}_\delta \circ f)| = d^{-1}(\mathcal{N}(\overline{V}_\delta \circ f))$ has intersection $|\mathscr{C} f| = d^{-1}(\mathscr{C} f)$. So for δ sufficiently small $|\mathcal{N}(\overline{V}_\delta \circ f)|$ is contained in the open neighborhood $V_\varepsilon(|\mathscr{C} f|)$. By shrinking δ further still we can get $V_\delta(|\mathcal{N}(\overline{V}_\delta \circ f)|) \subset V_\varepsilon(|\mathscr{C} f|)$, which implies the inclusion of (a).

(b) If $x \in |\mathscr{C} f|$, then there exists a δ_0 chain connecting x with itself, i.e., $x \in |\mathscr{O}(V_{\delta_0} \circ f)|$. If $\{x_i : i = 0, \dots, n\}$ is any δ_0 chain with $x_0 = x_n$, then by (b) of the lemma there exists a $\mathscr{U} f$ chain $\{U_i : i = 0, \dots, n\}$ with $x_i \in U_i$ and $U_0 = U_n$. So the U_i's are cyclic elements of $\mathscr{U}$ and $x_0 = x_n \in |\mathscr{U}|_f$.

If $x \in |\mathscr{U}|_f$, then $x \in U_0$ for some $\mathscr{U} f$ chain $\{U_i : i = 0, \dots, n\}$ with $U_0 = U_n$. By part (a) of the lemma there is a δ_1 chain $\{x_i : i = 0, \dots, n\}$ with $x_i \in U_i$ and $x_0 = x_n$. So $x_o \in |\mathscr{O}(V_{\delta_1} \circ f)|$ and because x, $x_0 \in U_0$, which has diameter less than δ_1, $x \in V_{\delta_1}(|\mathscr{O}(V_{\delta_1} \circ f)|)$. □

Putting the two parts of this theorem together we see that for every $\varepsilon > 0$ there exists $\delta > 0$ such that for any finite cover with mesh less than δ, the chain recurrent set $|\mathscr{C} f|$ is contained in the computable set $|\mathscr{U}|_f$ and every point of $|\mathscr{U}|_f$ is within ε of some point in $|\mathscr{C} f|$. Notice that the flaw in this procedure is part (a) because we lack any way of estimating how small δ must be for a given ε. The only result I know which bears on this question at all is the following which allows an estimate of the intersection when a sequence decreases slowly enough. It is also helpful in generalizations of some of these results to the noncompact case.

3. EXERCISE. *Let X be a complete, but not necessarily compact, metric space. Suppose $\{A_n : n = 1, 2, \dots\}$ is a decreasing sequence of closed nonempty sets with $A_n \subset V_{\delta_n}(A_{n+1})$. If the series $\sum_{n=1}^{\infty} \delta_n$ has finite sum δ then $A = \bigcap_n \{A_n\}$ is nonempty and $A_1 \subset V_\delta(A)$. (Hint: for $x_1 \in A_1$ choose a sequence $\{x_n\}$ with $x_n \in A_n$ and $d(x_{n+1}, x_n) < \delta_n$ $(n = 1, \dots)$. $\{x_n\}$ is a Cauchy sequence with limit in A.)* □

On the cyclic set $|\mathscr{O}(\mathscr{U} f)|$ the relation $\mathscr{O}(\mathscr{U} f) \cap \mathscr{O}(\mathscr{U} f)^{-1}$ is an equivalence relation with equivalence classes the basic sets of the relation $\mathscr{U} f$ on the space $\mathscr{U}$. Just as the union $|\mathscr{U}|_f$ of $|\mathscr{O}(\mathscr{U} f)|$ approximates the chain recurrent set we would hope to approach each separate basic set via the union of such a $\mathscr{U} f$ basic set. So for $U \in \mathscr{U}$ defines

$$\widetilde{G}(U) = \bigcup \{U_1 \in \mathscr{U} : U_1 \in \mathscr{O}(\mathscr{U} f)(U) \cap \mathscr{O}(\mathscr{U} f)^{-1}(U)\}. \tag{5.6}$$

Thus, $\widetilde{G}(U)$ is empty unless U is cyclic in which case it is the union of all U_1's which are $\mathscr{O}(\mathscr{U} f) \cap \mathscr{O}(\mathscr{U} f)^{-1}$ equivalent to U. Clearly,

$$|\mathscr{U}|_f = \bigcup \{\widetilde{G}(U) : U \in \mathscr{U}\} = \bigcup \{\widetilde{G}(U) : U \in |\mathscr{O}(\mathscr{U} f)|\}. \tag{5.7}$$

If U_0 and U_1 in $|\mathscr{O}(\mathscr{U} f)|$ are not $\mathscr{O}(\mathscr{U} f) \cap \mathscr{O}(\mathscr{U} f)^{-1}$ equivalent, then of course, their basic sets are disjoint. The same need not be true of the unions $\widetilde{G}(U_0)$ and $\widetilde{G}(U_1)$.

4. Exercise. *With* $K = \{0, 1\}$ *let* $f: I \to I$ *be a homeomorphism associated with a positive interval example. For any* t *in* $(0, 1)$ *choose* a, b *with* $t < a < b < f(t)$ *and define* $U_0 = [0, b]$, $U_1 = [a, 1]$. *Check that the interiors of* $\mathscr{U} = \{U_0, U_1\}$ *cover* I *and* $\mathscr{U}f = \{(U_0, U_0), (U_0, U_1), (U_1, U_1)\}$. *Both elements of* $\mathscr{U}$ *are cyclic with* $\widetilde{G}(U_0) = U_0$ *and* $\widetilde{G}(U_1) = U_1$. *But* $U_0 \cap U_1 \neq \varnothing$. $\square$

However, we will soon find use for the following partial result along these lines.

5. Lemma. *If for* U_0, $U_1 \in \mathscr{U}$

$$\widetilde{G}(U_0) \cap \widetilde{G}(U_1) \cap f(\widetilde{G}(U_0) \cap \widetilde{G}(U_1)) \neq \varnothing,$$

then U_0 *and* U_1 *are* $\mathscr{O}(\mathscr{U}f) \cap \mathscr{O}(\mathscr{U}f)^{-1}$ *equivalent members of* $|\mathscr{O}(\mathscr{U}f)|$ *and so* $\widetilde{G}(U_0) = \widetilde{G}(U_1)$.

Proof. Because $\widetilde{G}(U_0)$ and $\widetilde{G}(U_1)$ are nonempty, U_0 and U_1 are cyclic elements of $\mathscr{U}$. Now choose points x and y with $x \in \widetilde{G}(U_0) \cap \widetilde{G}(U_1)$ and $y \in f(x) \cap \widetilde{G}(U_0) \cap \widetilde{G}(U_1)$. There exist $\widehat{U}_i$ $\mathscr{O}(\mathscr{U}f) \cap \mathscr{O}(\mathscr{U}f)^{-1}$ equivalent to U_i $(i = 0, 1)$ such that $x \in \widehat{U}_0$ and $y \in \widehat{U}_1$. Since $y \in f(\widehat{U}_0) \cap \widehat{U}_1$, we have $\widehat{U}_1 \ \mathscr{U}f \ \widehat{U}_0$. So by transitivity of $\mathscr{O}(\mathscr{U}f)$, $U_1 \ \mathscr{O}(\mathscr{U}f) \ U_0$. By a symmetric argument $U_0 \ \mathscr{O}(\mathscr{U}f) \ U_1$, and U_0 and U_1 are $\mathscr{O}(\mathscr{U}f) \cap \mathscr{O}(\mathscr{U}f)^{-1}$ equivalent. As they thus lie in the same basic set of $|\mathscr{O}(\mathscr{U}f)|$ the unions $\widetilde{G}(U_0)$ and $\widetilde{G}(U_1)$ agree. $\square$

When the closed relation f is a continuous map we can state this result somewhat differently.

6. Corollary. *Assume* $f: X \to X$ *is a continuous map. Define for each* $U \in \mathscr{U}$ *the set*

$$G(U) = \widetilde{G}(U) \cap f^{-1}(\widetilde{G}(U)). \tag{5.8}$$

If for U_0, $U_1 \in \mathscr{U}$, $G(U_0) \cap G(U_1) \neq \varnothing$, *then* U_0 *and* U_1 *are* $\mathscr{O}(\mathscr{U}f) \cap \mathscr{O}(\mathscr{U}f)^{-1}$ *equivalent members of* $|\mathscr{O}(\mathscr{U}f)|$ *and so* $G(U_0) = G(U_1)$.

Proof. For a map f, we have $f^{-1}(B_0 \cap B_1) = f^{-1}(B_0) \cap f^{-1}(B_1)$. So $G(U_0) \cap G(U_1) \neq \varnothing$ is equivalent to

$$f^{-1}(\widetilde{G}(U_0) \cap \widetilde{G}(U_1)) \cap \widetilde{G}(U_0) \cap \widetilde{G}(U_1) \neq \varnothing.$$

By (5.4) this is the same condition as the hypothesis of the lemma. $\square$

The key result that we need to sharpen Theorem 2 in the map case is

7. Lemma. *Assume* $f: X \to X$ *is a continuous map. If for* $U \in \mathscr{U}$, $x \in U \cap |\mathscr{C}f|$, *then* U *is a cyclic element of* $\mathscr{U}$ *and the basic set* $\mathscr{C}f(x) \cap \mathscr{C}f^{-1}(x)$ *is contained in* $G(U)$.

Proof. Recall that for a map f the basic sets are f invariant (e.g., Corollary 4.3 and the remarks thereafter). So if $y \in B = \mathscr{C}f(x) \cap \mathscr{C}f^{-1}(x)$, then $f(x)$, $f(y) \in B$ and there exists $z \in B$ such that $f(z) = x$. Now for any positive $\delta_0 < m_0(\mathscr{U})$ we can find δ_0 chains for f connecting $f(x)$ to y and $f(y)$ to z. Putting them together we get a δ_0 chain $\{x_i : i = 0, \dots, n+k\}$ with $x_0 = x$, $x_1 = f(x)$, $x_n = y$, $x_{n+1} = f(y)$, and $x_{n+k} = z$. By Lemma 1(b) there is a $\mathscr{U}f$ chain $\{U_1, \dots, U_{n+k}\}$ with $x_i \in U_i$ for $i = 1, \dots, n+k$. As $f(x) = x_1 \in U_1$, $U_1 \ \mathscr{U}f \ U$. As $z \in U_{n+k}$ and $f(z) = x$, $U \ \mathscr{U}f \ U_{n+k}$. Thus, $\{U, U_1, \dots, U_{n+k}, U\}$ is a $\mathscr{U}f$ chain and its elements are mutually $\mathscr{O}(\mathscr{U}f) \cap \mathscr{O}(\mathscr{U}f)^{-1}$ equivalent. So U is cyclic and $U_i \subset \tilde{G}(U)$ for all i. Because $y \in U_n$ and $f(y) \in U_{n+1}$, $y \in G(U) = \tilde{G}(U) \cap f^{-1}(\tilde{G}(U))$. □

8. THEOREM. *Let $f : X \to X$ be a continuous map and let $\mathscr{U}$ be a finite collection of subsets whose interiors cover X. With $\tilde{G}(U)$ the union of the $\mathscr{U}f$ basic set of a cyclic element U in $\mathscr{U}$, cf.* (5.6), *we shrink $\tilde{G}(U)$ by defining $G(U) = \tilde{G}(U) \cap f^{-1}(\tilde{G}(U))$. If we define*

$$\|\mathscr{U}\|_f = \bigcup\{G(U) : U \in \mathscr{U}\} = \bigcup\{G(U) : U \in |\mathscr{O}(\mathscr{U}f)|\}. \tag{5.9}$$

Then

$$|\mathscr{C}f| \subset \|\mathscr{U}\|_f \subset |\mathscr{U}|_f.$$

The set of $G(U)$'s forms a pairwise disjoint cover of $\|\mathscr{U}\|_f$, i.e., $G(U_0) \cap G(U_1) \neq \varnothing$ implies $G(U_0) = G(U_1)$. Furthermore, each f basic set in $|\mathscr{C}f|$ is entirely contained in a unique $G(U)$.

PROOF. That $\|\mathscr{U}\|_f$ is contained in $|\mathscr{U}|_f$ is obvious from (5.7). If $x \in |\mathscr{C}f|$ and $U \in \mathscr{U}$ contains x, then U is cyclic and $G(U)$ contains the basic set $\mathscr{C}f(x) \cap \mathscr{C}f^{-1}(x)$ by Lemma 7. Uniqueness follows from Corollary 6 which says that distinct $G(U)$'s are disjoint. □

9. EXERCISE. *Let $f : X \to X$ be a continuous map and $\mathscr{U}$ be a finite open cover of X. Prove that $\tilde{G}(U)$ and $G(U)$ are $\mathscr{C}f$ semi-invariant open sets for each $U \in \mathscr{U}$. (Hint: If $x_0, x_1 \in \tilde{G}(U)$ and $y\mathscr{C}fx_0$, $x_1\mathscr{C}fy$ choose U_i equivalent to U with $x_i \in U_i$ $(i = 0, 1)$ and let $\delta < m_0(\mathscr{U})$ with $V_\delta(x_1) \subset U_1$. Put together a δ chain from x_0 to y and from y to x_1 and use the proof of Lemma* 1(b) *to construct a $\mathscr{U}f$ chain from U_0 to U_1 whose union contains y.) A fortiori they are f semi-invariant.* □

For the remainder of this chapter we will assume that $f : X \to X$ is a homeomorphism. Recall from Chapter 4 that it is for mappings that the limit point relation ωf and the corresponding set of ω-limit points $l_+[f] \equiv \overline{\omega f(X)}$ become especially useful. For a homeomorphism they are supplemented by the dual relation $\alpha f = \omega(f^{-1})$ and the set of α-limit points $l_-[f] = \overline{\alpha f(X)}$. The limit point set is then defined by:

$$l[f] = l_+[f] \cup l_-[f] = \overline{\omega f(X) \cup \alpha f(X)}.$$

If $x \in X$, then each $\omega f(x)$ and $\alpha f(x)$ is a closed, f invariant, indecomposable set. In fact, by Proposition 4.14 and Exercise 4.18(e) they are chain transitive subsets. Consequently, $l_{\pm}[f]$ and $l[f]$ are closed, f invariant subsets of X.

It often happens that the limit point set is not too hard to compute. The problem is then to describe the entire chain recurrent set by using this information. Useful here is the concept of an $l[f]$ *separating set.*

If K is a closed, f invariant set then we call F a K*-separating set* or we say F *separates* K, if F, too, is a closed f invariant set and, in addition $F \cap K$ is open as well as closed in K. Since $F \cap K$ and its complement in K are f invariant, F is K-separating if and only if the pair $\{F \cap K, K-(F \cap K)\}$ forms a decomposition of K. In particular, it is clear that a closed, f invariant set F is K separating if and only if $F \cap K$ is K separating. The usefulness of the concept comes from

10. LEMMA. *Let A be the intersection of K separating sets. If B is a closed, indecomposable f invariant subset of K, then $A \cap B \neq \varnothing$ if and only if $B \subset A$.*

PROOF. If $A = \bigcap\{A_\alpha\}$ with the A_α's K separating, then $B \subset K$ implies that $\{B \cap A_\alpha, B-(B \cap A_\alpha)\}$ is a decomposition of B. By indecomposibility it is the trivial decomposition and so $B \cap A \neq \varnothing$ implies $B \subset A_\alpha$ for all α. Thus, $B \subset A$ or $B \cap A = \varnothing$.

REMARK. If B is a basic set of the restriction f_K, then B satisfies the hypotheses of Lemma 10. So A contains any basic set of f_K that it meets. □

We apply this result in the special case where $K = l[f]$ by defining, when A is the intersection of $l[f]$ separating sets, the *stable set* (or *inset*) $W^+(A)$ and the *unstable set* (or *outset*) $W^-(A)$:

$$\begin{aligned} W^+(A) &= (\omega f)^{-1}(A) = \{x\colon \omega f(x) \cap A \neq \varnothing\} = \{x\colon \omega f(x) \subset A\}, \\ W^-(A) &= (\alpha f)^{-1}(A) = \{x\colon \alpha f(x) \cap A \neq \varnothing\} = \{x\colon \alpha f(x) \subset A\}. \end{aligned} \tag{5.10}$$

The equations follow from Lemma 10 since $\omega f(x)$ and $\alpha f(x)$ are indecomposable. Because A is f invariant and closed

$$A \subset W^+(A) \cap W^-(A), \tag{5.11}$$

though the inclusion may be strict. Also, just as a closed, f invariant subset A is $l[f]$ separating if and only if $A \cap l[f]$ is, we have that

$$W^{\pm}(A) = W^{\pm}(A \cap l[f]). \tag{5.12}$$

Finally, $\omega f(f(x)) = \omega f(x)$ and $\alpha f(f(x)) = \alpha f(x)$ imply that $W^{\pm}(A)$ are f invariant subsets of X though they are usually neither open nor closed except in the case of an attractor or repellor.

11. EXERCISE. (a) *Show that if A is an attractor or repellor then A is $|\mathscr{C} f|$ separating and a fortiori $l[f]$ separating. If A is a basic set, then*

A is the intersection of $|\mathscr{C}f|$ separating sets. In each of these cases prove $W^+(A) \cap W^-(A) = A$. (For a basic set show that if $y_1 \in \omega f(x)$ and $y_2 \in \alpha f(x)$ lie in the same basic set then y_1, x, and y_2 are all $\mathscr{C}f \cap \mathscr{C}f^{-1}$ equivalent.)

(b) *For any $x \in X$ there are unique basic sets B_+ and B_- containing $\omega f(x)$ and $\alpha f(x)$ respectively, i.e., such that $x \in W^-(B_-) \cap W^+(B_+)$. If $B_+ = B_-$ show that x lies in this basic set.* □

12. **Proposition.** *Let $f\colon X \to X$ be a homeomorphism and A be a closed, f invariant subset.*

(a) *A is an attractor if and only if it satisfies the following equivalent conditions:*

(1) *A is $l[f]$ separating and $W^-(A) = A$.*
(2) *A is Ωf +invariant and $W^+(A)$ is open.*
(3) *A is Ωf +invariant and $l[f]$ separating.*

(b) *A is a repellor if and only if it satisfies the following equivalent conditions:*

(1) *A is $l[f]$ separating and $W^+(A) = A$.*
(2) *A is Ωf^{-1} +invariant and $W^-(A)$ is open.*
(3) *A is Ωf^{-1} +invariant and $l[f]$ separating.*

(c) *A pair of disjoint closed, f invariant subsets $\{A_+, A_-\}$ is an attractor-repellor pair if and only if $x \in X - (A_+ \cup A_-)$ implies $\alpha f(x) \subset A_-$ and $\omega f(x) \subset A_+$. In that case, $\{A_+ \cap l[f], A_- \cap l[f]\}$ is a decomposition of $l[f]$.*

Proof. (a) If A is an attractor, then A is $|\mathscr{C}f|$ separating, A is $\mathscr{C}f$ +invariant, and $\{x\colon \Omega\mathscr{C}f(x) \subset A\}$ is a neighborhood of A by Theorem 3.3(a). So A is $l[f]$ separating since $l[f] \subset |\mathscr{C}f|$, Ωf +invariant since $\Omega f \subset \mathscr{C}f$, and $W^+(A)$ is a neighborhood of A. Since $\omega f \subset \Omega\mathscr{C}f$. Now let U be any open subset containing A and contained in $W^+(A)$. We show that $W^+(A)$ is open by proving

$$W^+(A) = \bigcup\{f^{-n}(U)\colon n = 0, 1, 2, \ldots\}. \tag{5.13}$$

Because $W^+(A)$ is f invariant, it contains the union. If $x \in W^+(A)$, then $\omega f(x) \subset A$ implies $f^n(x) \in U$ for sufficiently large n and this if and only if A is $x \in f^{-n}(U)$.

For the converse we apply Theorem 3.6. Each of (1), (2) and (3) imply that A is an attractor by parts (b)(2) (applied to f^{-1}) (a)(3) and (a)(4), respectively. Note that A is f invariant, and Ωf +invariant iff A is $\mathscr{N}f$ +invariant.

The repellor result (b) follows from (a) applied to f^{-1}.

(c) If A_+, A_- are closed, disjoint f invariant sets and $\alpha f(x) \subset A_-$, $\omega f(x) \subset A_+$ for all $x \in X - (A_+ \cap A_-)$, then $l[f] \subset A_+ \cup A_-$ and so

$\{A_+ \cap l[f], A_- \cap l[f]\}$ forms a decomposition of $l[f]$. Thus A_+ and A_- are $l[f]$ separating. The hypothesis also implies $W^-(A_+) = A_+$ and $W^+(A_-) = A_-$. So A_+ and A_- are attractor and repellor, respectively. By Proposition 2.4(c), $\Omega\mathscr{C}f(x) = \Omega\mathscr{C}f(\omega f(x))$. For $x \in X - (A_+ \cap A_-)$ this is a subset of A_+ since $\omega f(x) \subset A_+$ and A_+ is $\mathscr{C}f$ +invariant (it is an attractor). Similarly $(\Omega\mathscr{C}f)^{-1}(x) \subset A_-$ and so $|\mathscr{C}f| = |\Omega\mathscr{C}f| \subset A_+ \cup A_-$. Thus, A_- is the repellor dual to A_+ by Proposition 3.9. The converse is clear from the same proposition because $\omega f(x) \subset \Omega\mathscr{C}f(x)$ and $\alpha f(x) \subset (\Omega\mathscr{C}f)^{-1})(x)$.

REMARK. Exercise 3.7(a) describes an example of an $l[f]$ separating subset A which is neither an attractor nor a repellor despite $W^+(A) = W^-(A) = X$. □

We now introduce the device which will allow us to construct the chain recurrent set from the limit point set. In fact, it allows us to describe the attractor-repellor structure of a homeomorphism by using the limit behavior of the points.

An *invariant decomposition* for a homeomorphism $f: X \to X$ is a finite collection of pairwise disjoint, closed, nonempty f invariant sets which covers the limit point set. Thus $\mathscr{F} = \{F_1, \dots, F_n\}$ is an invariant decomposition when $\overline{F}_i = F_i \neq \emptyset$, $f(F_i) = F_i$, $F_i \cap F_j = \emptyset$ $(i \neq j = 1, \dots, n)$, and $l[f] \subset F_1 \cup \dots \cup F_n$. Since the complement of $F_i \cap l[f]$ in $l[f]$ is the union of the remaining $F_j \cap l[f]$'s, each F_i is $l[f]$ separating. In particular, for each point $x \in X$ there are unique F_1, $F_2 \in \mathscr{F}$ such that $\alpha f(x) \subset F_1$ and $\omega f(x) \subset F_2$, or, equivalently, $x \in W^-(F_1) \cap W^+(F_2)$. Notice that by Proposition 12, an attractor-repellor pair is an example of an invariant decomposition.

Further results require the following sharpening of a lemma of Shub and Nitecki.

13. LEMMA. *Let F be an $l[f]$ separating set. Assume $x \in X$ satisfies $\omega f(x) \cap F = \emptyset$, i.e., $x \notin W^+(F)$.*

(a) *If $\Omega f(x) \cap F \neq \emptyset$, then $\Omega f(x) \cap (W^+(F) - W^-(F)) \neq \emptyset$.*

(b) *If $\Omega\mathscr{C}f(x) \cap F \neq \emptyset$, then $\Omega\mathscr{C}f(x) \cap (W^+(F) - W^-(F)) \neq \emptyset$.*

WARNING. You may want to skip the later, technical details of the proof at first reading. Also, the notation adopted for the proof is intended to illustrate its application to flows as well as to homeomorphisms.

PROOF. The pattern of the proof is the same for (a) and (b). We can find a doubly indexed sequence $\{x_{kt}: 0 \leq t \leq t_k\}$ such that $\{t_k\} \to \infty$, $\{x_{k0}\} \to x$, and $\{x_{kt_k}\} \to y \in F$. For (a) we can assume that for each fixed k $\{x_{kt}: 0 \leq t \leq t_k\}$ is a chain for f, i.e., $x_{kt} = f^t(x_{k0})$. For (b) we can assume that $\{x_{kt}: 0 \leq t \leq t_k\}$ is an $\varepsilon_k > 0$ chain for f with $\{\varepsilon_k\} \to 0$. For any sequence $\{s_k^*\}$ with $0 \leq s_k^* \leq t_k$ we can assume, by going to a subsequence, that $\{x^*_{ks_k}\}$ converges to a point y^*. If $\{s_k^*\} \to \infty$, then

$y^* \in \Omega f(x)$ in (a) and $\in \Omega\mathscr{C} f(x)$ in (b).

Let us be a bit more specific. Because $\omega f(x) \cap F = \emptyset$, F is disjoint from $\{x\} \cup \mathscr{R} f(x)$, the closure of the positive orbit of x. F is $l[f]$ separating as well and so we can find U a closed neighborhood of F such that:

$$\text{For} \quad U_0 = \bigcap_{|t|\le 1} f^t(U) \quad \text{and} \quad U_1 = \bigcup_{|t|\le 1} f^t(U),$$
$$F \subset \operatorname{Int} U_0, \qquad U \cap l[f] = F \cap l[f], \qquad U_1 \cap (\{x\} \cup \mathscr{R} f(x)) = \emptyset.$$

By throwing away initial terms if necessary we can assume, for all k: $x_{k0} \notin U_1$ and $x_{kt_k} \in \operatorname{Int} U_0$. Now define

$$\begin{aligned} s_k^1 &= \min\{s \ge 0 \colon x_{ks} \in U\}, \\ s_k^2 &= \max\{s \le t_k \colon x_{ks} \in X - \operatorname{Int} U\}. \end{aligned} \tag{5.14}$$
$$\begin{aligned} x_{ks} &\in X - U \quad \text{for } s < s_k^1, \\ x_{ks} &\in \operatorname{Int} U \quad \text{for } s > s_k^2. \end{aligned}$$

As there is thus no value of s that $s_k^2 < s < s_k^1$, we have $s_k^1 \le s_k^2$. Furthermore, $x_{ks_k^1} \in U - \operatorname{Int} U_0$ and $x_{ks_k^2} \in U_1 - \operatorname{Int} U$. So if $\{x_{ks_k^1}\} \to y^1$ and $\{x_{ks_k^2}\} \to y^2$, then $y^1 \in U - \operatorname{Int} U_0$ and $y^2 \in U_1 - \operatorname{Int} U$.

We claim that $\{s_k^1\} \to \infty$ and $\{t_k - s_k^2\} \to \infty$. Suppose s_k^1 had a finite limit point s^1. Then $y^1 = \lim x_{ks_k^1} = f^{s^1}(\lim x_{k0}) = f^{s^1}(x)$. This can't happen because $y^1 \in U$ and U is disjoint from the positive orbit of x. Similarly, if $\{t_k - s_k^2\}$ had a finite limit point $\bar{t}$, then $y^2 = \lim x_{ks_k^2} = f^{-\bar{t}}(\lim x_{kt_k}) = f^{-\bar{t}}(y)$. This can't happen since $y \in F$ and so the entire orbit of y lies in F while $y^2 \notin \operatorname{Int} U$.

Thus, y^1, $y^2 \in \Omega f(x)$ in (a) and $\in \Omega\mathscr{C} f(x)$ in (b) and each satisfies half of what we want. Let $t > 0$ be fixed. By (5.14) for k large enough $\{x_{ks_k^1 - t}\}$ lies in $X - \operatorname{Int} U$ and has limit $f^{-t}(y^1)$. So $\alpha f(y^1) \subset X - \operatorname{Int} U$, i.e., $y^1 \notin W^-(F)$. Also, $\{x_{ks_k^2 + t}\}$ lies in U and has limit $f^t(y^2)$. So $\omega f(y^2) \subset U \cap l[f] \subset F$, i.e., $y^2 \in W^+(F)$. The delicate part of the proof consists in choosing s_k^* so that the limit points of $\{x_{ks_k^*}\}$ will have both of these properties at once.

First, we restrict to a subsequence so that we can assume for all k that

$$s_k^1, \qquad t_k - s_k^2 > 2^k. \tag{5.15}$$

Define the sets In_k and Out_k by

$$t \in \mathrm{Out}_k \Leftrightarrow t \le t_k \quad \text{and} \quad x_{kt} \notin \operatorname{Int} U, \ \text{or } t \le 0,$$
$$t \in \mathrm{In}_k \Leftrightarrow t \ge 0 \quad \text{and} \quad x_{kt} \in U, \ \text{or } t \ge t_k.$$

We then use the following combinational lemma whose proof we defer until the dynamic results are completed.

CLAIM. Let $T = \mathbb{R}$ or Z. Assume In and Out are closed subsets of T with union T. Call a, $b \in T$ *k-separated* if $a+2^k < b-2^k$, $[a, a+2^k] \cap \mathrm{In} = \varnothing$ and $[b-2^k, b] \cap \mathrm{Out} = \varnothing$. Call s *k-balanced* if for $p = 1, 2, \ldots, k$ $[s + 2^{p-1} - 1, s + 2^p - 1] \cap \mathrm{In} \neq \varnothing$, $[s - 2^p + 1, s - 2^{p-1} + 1] \cap \mathrm{Out} \neq \varnothing$. If a, b are k separated, then the set $C_k(a, b)$ of k-balanced points in the interval $[a, b]$ is nonempty.

We apply the claim to Out_k and In_k with $a = 0$ and $b = t_k$ which are k-separated by (5.14) and (5.15). The claim implies that there exists $0 \leq s_k^* \leq t_k$ and u_{pk}^1, u_{pk}^2 $(p = 1, \ldots, k)$ such that

$$\begin{array}{ll} u_{pk}^1, u_{pk}^2 \in [2^{p-1} - 1, 2^p - 1], & \\ s_k^* - u_{pk}^1 \in \mathrm{Out}_k, \quad s_k^* + u_{pk}^2 \in \mathrm{In}_k & \end{array} \qquad (p = 1, \ldots, k). \tag{5.16}$$

Notice that (5.16) with $p = 1$ and (5.14) imply $s_k^* + 1 > s_k^1$ and so $s_k^* > 2^k - 1$ by (5.15).

Let y^* be the limit of a subsequence of $\{x_{ks_k^*}\}$, and for each p, let u_p^1 and u_p^2 be limit points of the subsequence (in k) $\{u_{pk}^1\}$, $\{u_{pk}^2\}$. So $2^{p-1} - 1 \leq u_p^1$, $u_p^2 \leq 2^p - 1$. By (5.16) and the definition of Out_k it follows that for all p:

$$f^{-u_p^1}(y^*) \notin \mathrm{Int}\, U \quad \text{and} \quad f^{u_p^2}(y^*) \in U.$$

So $\alpha f(y^*) \cap X - \mathrm{Int}\, U \neq \varnothing$ and $\omega f(y^*) \cap U \neq \varnothing$. But $\alpha f(y^*)$ and $\omega f(y^*)$ are indecomposable while $F \cap l[f] = \mathrm{Int}\, U \cap l[f] = U \cap l[f]$ and its complement in $l[f]$ form a decomposition for $l[f]$. So $\alpha f(y^*) \cap F = \varnothing$ and $\omega f(y^*) \subset F$ (cf. Lemma 10). Thus $y^* \in W^+(F) - W^-(F)$ while $y^* \in \Omega f(x)$ or $y^* \in \Omega \mathscr{C} f(x)$ in cases (a) and (b), respectively.

PROOF OF THE CLAIM. Clearly, empty or not $C_k(a, b)$ is closed and if a, b are k-balanced, then

$$C_p(a, b) \subset [a + 2^k - 1, b - 2^k + 1], \qquad p = 1, \ldots, k.$$

The proof is by induction on k.

$(k = 1)$. We are given $a+2 < b-2$, $[a, a+2] \cap \mathrm{In} = \varnothing$, and $[b-2, b] \cap \mathrm{Out} = \varnothing$. Define

$$Q_1 = \{t \colon a + 1 \leq t \leq b \text{ and } [t - 1, t] \subset \mathrm{Out}\}.$$

Clearly, Q_1 is closed, $[a+1, a+2] \subset Q_1$ and $Q_1 \cap [b-2, b] = \varnothing$. Let $t^* = \sup Q_1$. We have $t^* < b-2$ and $[t^* - 1, t^*] \subset \mathrm{Out}$. If $[t^*, t^* + 1] \cap \mathrm{In} = \varnothing$, then $[t^*, t^* + 1] \subset \mathrm{Out}$ and so $t^* + 1 \in Q_1$ contradicting the definition of t^*. So $[t^*, t^* + 1] \cap \mathrm{In} \neq \varnothing$ and $t^* \in C_1(a, b)$.

($k > 1$, assuming $k-1$). If α, β are $k-1$ separated, then by inductive hypothesis $C_{k-1}(\alpha, \beta)$ is a closed nonempty subset of $[\alpha - 2^{k-1} - 1, \beta - 2^{k-1} + 1]$. We show that $t_* = \inf C_{k-1}(\alpha, \beta)$ is an element of

$$Q_{k-1}(\alpha, \beta) = \{s \in C_{k-1}(\alpha, \beta) : [s - 2^k + 1, s - 2^{k-1} + 1] \cap \text{Out} \neq \varnothing\},$$

and so $Q_{k-1}(\alpha, \beta)$ is nonempty. If $t_* \notin Q_{k-1}$, then $[t_* - 2^k + 1, t_* - 2^{k-1} + 1]$ is disjoint from Out and so is contained in In.

Now $t_* - 2^{k-1} + 1 \geq \alpha$ but $[\alpha, \alpha + 2^{k-1}]$ is disjoint from In and so from $[t_* - 2^k + 1, t_* - 2^{k-1} + 1]$. So $t_* - 2^k + 1 > \alpha + 2^{k-1}$. This means that the pair α and $\tilde{\beta} = t_* - 2^{k-1} + 1$ are $k-1$ separated. So by the induction hypothesis again $C_{k-1}(\alpha, \tilde{\beta})$ is nonempty and since $\tilde{\beta} \leq \min(\beta, t_*)$ we have that $C_{k-1}(\alpha, \tilde{\beta}) \subset C_{k-1}(\alpha, \beta)$ and consists of points smaller than t_*. This contradicts the definition of t_*.

Suppose that a, b are k-separated. In particular, they are $k-1$ separated and so

$$Q_{k-1}(a, b) \neq \varnothing.$$

Let $t^* = \sup Q_{k-1}(a, b)$. We claim that $t^* \in C_k(a, b)$ and from the definition of Q_{k-1} this means precisely $[t^* + 2^{k-1}, t^* + 2^k - 1] \cap \text{In} \neq \varnothing$. We use the same trick one last time. If $[t^* + 2^{k-1}, t^* + 2^k - 1]$ were disjoint from In, then $\tilde{a} = t^* + 2^{k-1} - 1$ and b are $k-1$ separated and so $Q_{k-1}(\tilde{a}, b)$ is a nonempty subset of $Q_{k-1}(a, b)$ consisting of elements larger than t^*, contradicting the definition of t^*. □

Just as with a finite cover, we can regard an invariant decomposition $\mathscr{F} = \{F_1, \dots, F_n\}$ as a finite metric space. Corresponding to the homeomorphism f there are several possible ways of defining an associated relation $\mathscr{F}f$ on $\mathscr{F}$:

$$\begin{aligned}
&F_2 \, \mathscr{F}_1 f \, F_1 \Rightarrow W^-(F_1) \cap W^+(F_2) \neq \varnothing, \\
&F_2 \, \mathscr{F}_2 f \, F_1 \Rightarrow \overline{W^-(F_1)} \cap F_2 \neq \varnothing, \\
&F_2 \, \mathscr{F}_3 f \, F_1 \Rightarrow F_1 \cap \overline{W^+(F_2)} \neq \varnothing, \\
&F_2 \, \mathscr{F}_4 f \, F_1 \Rightarrow \Omega f(F_1) \cap F_2 \neq \varnothing, \\
&F_2 \, \mathscr{F}_5 f \, F_1 \Rightarrow \mathscr{C} f(F_1) \cap F_2 \neq \varnothing.
\end{aligned} \tag{5.17}$$

Clearly, an invariant decomposition for f is an invariant decomposition for f^{-1} and

$$\begin{aligned}
&\mathscr{F}_2(f^{-1}) = (\mathscr{F}_3 f)^{-1}, \\
&\mathscr{F}_i(f^{-1}) = (\mathscr{F}_i f)^{-1} \qquad (i = 1, 4, 5).
\end{aligned} \tag{5.18}$$

The important consequence of the Shub-Nitecki Lemma is that the transitive extensions of all five of these relations are the same.

14. COROLLARY. *The relations* $\mathscr{F}_i f$ $(i = 1, \dots, 5)$ *are all reflexive and satisfy the inclusions:*

$$\begin{array}{ccccc} & & \mathscr{F}_2 f & & \\ & \subset & & \subset & \\ \mathscr{F}_1 f & & & & \mathscr{F}_4 f \subset \mathscr{F}_5 f \subset \mathscr{O}(\mathscr{F}_1 f)\,. \\ & \subset & & \subset & \\ & & \mathscr{F}_3 f & & \end{array}$$

Consequently, the five relations have a common transitive extension which we will denote $\mathscr{O}(\mathscr{F} f)$.

PROOF. Reflexivity is obvious and $W^-(F_1) \cap W^+(F_2) \neq \varnothing$ implies $W^-(F_1) \cap W^+(F_2) \neq \varnothing$. This is equivalent to the apparently stronger condition $\overline{W^-(F_1)} \cap F_2 \neq \varnothing$ because $\overline{W^-(F_1)}$ is closed and f invariant so if it contains x, then it contains $\omega f(x)$. Thus, $\mathscr{F}_1 f \subset \mathscr{F}_2 f$.

$W^-(F_1) \subset \Omega f(F_1)$ since $\alpha f \subset \Omega f^{-1}$. So because $\Omega f(F_1)$ is closed we have

$$\overline{W^-(F_1)} \subset \Omega f(F_1) = \mathscr{N} f(F_1) \subset \mathscr{C} f(F_1) = \Omega \mathscr{C} f(F_1)\,, \tag{5.19}$$

where the equations follow from $\mathscr{N} f = \mathscr{O} f \cup \Omega f$, $\mathscr{C} f = \mathscr{O} f \cup \Omega \mathscr{C} f$, and f invariance of F_1. Thus, $\mathscr{F}_2 f \subset \mathscr{F}_4 f \subset \mathscr{F}_5 f$. For $\mathscr{F}_3 f$ apply (5.18).

Now suppose $F_2 \;\; \mathscr{F}_5 f \;\; F_1$, i.e., there exists $x \in F_1$ with $\mathscr{C} f(x) \cap F_2 \neq \varnothing$, we must find an $\mathscr{F}_1 f$ chain connecting F_1 and F_2. More generally we will show that if $x \in X$ and $\mathscr{C} f(x) \cap F_2 \neq \varnothing$ there is an $\mathscr{F}_1 f$ chain connecting F_1 with F_2 where F_1 is the member of $\mathscr{F}$ containing $\omega f(x)$. The proof is by induction on the number of elements n of the invariant decomposition $\mathscr{F}$.

If $n = 1$, i.e., $\mathscr{F}$ consists of a single closed, f invariant set containing $l[f]$, then the result is trivial.

For the inductive step, if $\omega f(x) \cap F_2 \neq \varnothing$, then $F_1 = F_2$ and by reflexivity $\{F_1, F_2\}$ is the required chain. Otherwise, $\omega f(x) \cap F_2 = \varnothing$ and $\Omega \mathscr{C} f(x) \cap F_2 \neq \varnothing$. By the Shub-Nitecki Lemma there exists $y \in \Omega \mathscr{C} f(x) \cap (W^+(F_2) - W^-(F_2))$. Let F_3 be the element of $\mathscr{F}$ containing $\alpha f(y)$. $F_3 \neq F_2$ since $y \notin W^-(F_2)$. If $F_3 = F_1$, then we are finished. Otherwise, let $\widetilde{\mathscr{F}}$ be the invariant decomposition obtained by replacing F_2 and F_3 by the union $F_2 \cup F_3$. By inductive hypothesis there exists an $\widetilde{\mathscr{F}}_1 f$ chain $\{F_{i_1}, \dots, F_{i_p}\}$ with $F_{i_1} = F_1$, $F_{i_k} \in \mathscr{F}$ $(k = 1, \dots, p-1)$ and $F_{i_p} = F_2 \cup F_3$. So $\{F_{i_1}, \dots, F_{i_{p-1}}\}$ is an $\mathscr{F}_1 f$ chain and $W^-(F_{i_{p-1}})$ intersects $W^+(F_2 \cup F_3) = W^+(F_2) \cup W^+(F_3)$. So either $F_2 \;\; \mathscr{F}_1 f \;\; F_{i_{p-1}}$ or $F_3 \;\; \mathscr{F}_1 f \;\; F_{i_{p-1}}$. In the first case the required chain is $\{F_{i_1}, \dots, F_{i_{p-1}}, F_2\}$ and in the second it is $\{F_{i_1}, \dots, F_{i_{p-1}}, F_3, F_2\}$. $\square$

Now we are ready for the main result of this section.

15. THEOREM. *Given an invariant decomposition* $\mathcal{F}$ *for a homeomorphism* $f: X \to X$, *let* $\mathcal{O}(\mathcal{F} f)$ *be the transitive extension of the reflexive relation* $\mathcal{F} f = \{(F_1, F_2) \in \mathcal{F} \times \mathcal{F}: W^-(F_1) \cap W^+(F_2) \neq \emptyset\}$.

Assume that $\mathcal{F}_+$ *is an* $\mathcal{F} f$ *invariant subset of* $\mathcal{F}$ *so that its complement* $\mathcal{F}_-$ *is* $\mathcal{F} f^{-1}$ *invariant. The sets defined by:*

$$A_+(\mathcal{F}_+) = \bigcup\{W^-(F): F \in \mathcal{F}_+\}, \qquad A_-(\mathcal{F}_-) = \bigcup\{W^+(F): F \in \mathcal{F}_-\}$$

form an attractor-repellor pair for f. *The collection of such* $A_+(\mathcal{F}_+)$*'s is called the* attractor structure *of the invariant decomposition* $\mathcal{F}$.

$\mathcal{O}(\mathcal{F} f) \cap \mathcal{O}(\mathcal{F} f)^{-1}$ *is an equivalence relation on* $\mathcal{F}$ *and for* $F \in \mathcal{F}$ *we define the* $\mathcal{F}$-basic set

$$B(F) = \bigcup\{W^-(F_1) \cap W^+(F_2): F_1, F_2 \in \mathcal{O}(\mathcal{F} f) \cap \mathcal{O}(\mathcal{F} f)^{-1}(F)\}.$$

$B(F)$ *is a closed,* f *invariant,* $\mathcal{C} f$ *semi-invariant,* $|\mathcal{C} f|$ *separating subset of* X, *containing* F. *Distinct* $\mathcal{F}$*-basic sets are disjoint, i.e.,* $B(F_1) \cap B(F_2) \neq \emptyset$ *implies* F_1 *and* F_2 *are* $\mathcal{O}(\mathcal{F} f) \cap \mathcal{O}(\mathcal{F} f)^{-1}$ *equivalent and so* $B(F_1) = B(F_2)$. *Each basic set for* f *is entirely contained in a unique* $\mathcal{F}$*-basic set. In particular,*

$$|\mathcal{C} f| \subset |\mathcal{F}| \equiv \bigcup\{B(F): F \in \mathcal{F}\}.$$

PROOF. $A_+(\mathcal{F}_+)$ and $A_-(\mathcal{F}_-)$ are f invariant sets because each $W^{\pm}(F)$ is. They are disjoint because $W^-(F_1) \cap W^+ F_2 \neq \emptyset$ for $F_1 \in \mathcal{F}_+$ implies $F_2 \in \mathcal{F}_+$ by $\mathcal{F} f$ invariance of $\mathcal{F}_+$. If $x \notin A_+ \cup A_1$ and F_1, F_2 are the members of $\mathcal{F}$ containing $\alpha f(x)$ and $\omega f(x)$ respectively then $F_2 \; \mathcal{F} f \; F_1$. This implies $F_1 \in \mathcal{F}_-$ and $F_2 \in \mathcal{F}_+$ for otherwise x would lie in A_- or A_+. Thus, $\alpha f(x) \subset F_1 \subset A_-$ and $\omega f(x) \subset F_2 \subset A_+$. Proposition 12(c) will then imply that A_+ and A_- form an attractor-repellor pair once we have proved they are closed.

If $x \in \overline{A_+}$, then $x \in \overline{W^-(F_1)}$ for some $F_1 \in \mathcal{F}_+$. Let F_2 be the member of $\mathcal{F}$ containing $\alpha f(x)$, i.e., $x \in W^-(F_2)$. Since $\overline{W^-(F_1)}$ is closed and f invariant $\alpha f(x) \subset \overline{W^-(F_1)} \cap F_2$, i.e., $F_2 \; \mathcal{F}_2 f \; F_1$. By Corollary 13 $F_2 \mathcal{O}(\mathcal{F} f) F_1$ and so by the $\mathcal{F} f$ invariance of $\mathcal{F}_+$, $F_2 \in \mathcal{F}_+$. Thus, $x \in W^-(F_2)$ implies $x \in A_+$. A_- is similarly proved closed.

Now for $F \in \mathcal{F}$, $\mathcal{O}(\mathcal{F} f)(F)$ and $\mathcal{O}(\mathcal{F} f)^{-1}(F)$ are, respectively, the smallest $\mathcal{F} f$ invariant set and the smallest $\mathcal{F} f^{-1}$ invariant set containing F. They are not, of course, disjoint. In fact, their intersection is the $\mathcal{O}(\mathcal{F} f) \cap \mathcal{O}(\mathcal{F} f)^{-1}$ equivalence class of F. So their associated attractor and repellor are not disjoint. Instead we will prove that

$$B(F) = A_+(\mathcal{O}(\mathcal{F} f)(F)) \cap A_-(\mathcal{O}(\mathcal{F} f)^{-1}(F)). \tag{5.20}$$

It is clear from the definitions that $B(F)$ is contained in $A_+ \cap A_-$. Conversely, $x \in A_+ \cap A_-$ means $x \in W^-(F_1) \cap W^+(F_2)$ with $F_1 \mathcal{O}(\mathcal{F} f) F$ and

$F\mathscr{O}(\mathscr{F}f)F_2$. But $W^-(F_1) \cap W^+(F_2) \neq \emptyset$ means $F_2 \ \mathscr{F}f \ F_1$. Thus, F_1, F_2, and F are $\mathscr{O}(\mathscr{F}f) \cap \mathscr{O}(\mathscr{F}f)^{-1}$ equivalent and so $x \in B(F)$.

As the intersection of an attractor and a repellor $B(F)$ is a closed, f invariant, $\mathscr{C}f$ semi-invariant, $|\mathscr{C}f|$ separating set. For any $x \in X$ there is a unique pair F_1, $F_2 \in \mathscr{F}$ with $x \in W^-(F_1) \cap W^+(F_2)$. $F_2 \ \mathscr{F}f \ F_1$ and either it is not true that $F_1\mathscr{O}(\mathscr{F}f)F_2$, in which case, x lies in no $\mathscr{F}$ basic set $B(F)$, or else F_1 and F_2 are $\mathscr{O}(\mathscr{F}f)$ equivalent and x lies in the $\mathscr{F}$ basic set for the equivalence class of F_1 and F_2. In particular, distinct basic sets are disjoint.

Now suppose x is a chain recurrent point and $x \in W^-(F_1) \cap W^+(F_2)$ with F_1, $F_2 \in \mathscr{F}$. Let B be the basic set, $\mathscr{C}f(x) \cap \mathscr{C}f^{-1}(x)$, containing x. Because B is closed and f invariant, $\alpha f(x) \subset B \cap F_1$ and $\omega f(x) \subset B \cap F_2$. So in addition to $F_2 \ \mathscr{F}f \ F_1$ we have $\mathscr{C}f(F_2) \cap F_1 \supset B \cap F_1 \neq \emptyset$ and so $F_1 \ \mathscr{F}_5 f \ F_2$. By Corollary 13 again, F_1 and F_2 are $\mathscr{O}(\mathscr{F}f) \cap \mathscr{O}(\mathscr{F}f)^{-1}$ equivalent and so x lies in the corresponding $\mathscr{F}$ basic set $B(F)$. Thus, $|\mathscr{C}f| \subset |\mathscr{F}|$. Also, $B \cap B(F) \neq \emptyset$ implies $B \subset B(F)$ because $B(F)$ is $|\mathscr{C}f|$ isolated and B is an indecomposable subset of $|\mathscr{C}f|$. Apply Lemma 10. □

For invariant decompositions $\mathscr{F}$ and $\widetilde{\mathscr{F}}$ we say that $\mathscr{F}$ *refines* $\widetilde{\mathscr{F}}$ if every F is $\mathscr{F}$ is contained in a (necessarily unique) $\widetilde{F}$ of $\widetilde{\mathscr{F}}$, i.e., there is a function $c: \mathscr{F} \to \widetilde{\mathscr{F}}$ defined by $F \subset c(F)$. c is called the *carrier map* of the refinement.

16. Exercise. *Let $\mathscr{F}$, $\widetilde{\mathscr{F}}$ etc. be invariant decompositions for a homeomorphism $f: X \to X$. Prove:*

(a) *Any indecomposable subset of $\bigcup\mathscr{F}$ is contained in a unique element of $\mathscr{F}$. In particular, if B is a closed, f $+$invariant subset of $\bigcup\mathscr{F}$, then each basic set of the restriction f_B is contained in an element of $\mathscr{F}$.*

(b) *For A an attractor for f, define $\mathscr{F}_+(A) = \{F \in \mathscr{F} : F \subset A)$ and $\mathscr{F}_-(A) = \{F \in \mathscr{F} : A \cap F = \emptyset\}$. $\mathscr{F}_+$ is $\mathscr{F}f$ invariant, $\mathscr{F}_-$ is $\mathscr{F}f^{-1}$ invariant and $\mathscr{F}_+ \cap \mathscr{F}_- = \emptyset$. $A_+(\mathscr{F}_+(A)) \subset A$ and $A_-(\mathscr{F}_-(A)) \cap A = \emptyset$. If $\mathscr{F}_+ \cup \mathscr{F}_- = \mathscr{F}$, then $A_+(\mathscr{F}_+(A)) = A$. Thus, A is in the attractor structure of $\mathscr{F}$ if and only if $A \cap F \neq \emptyset$ implies $F \subset A$ for $F \in \mathscr{F}$.*

(c) *The attractor structure of $\mathscr{F}$ determines the $\mathscr{F}$ basic sets.* (*Consider the collection $\{A_+ \cap A_-\}$ where A_+ lies in the attractor structure and A_- lies in the dual repellor structure. Use* (5.20) *to show that the $\mathscr{F}$ basic sets are the minimal nonempty sets in this collection.*)

(d) *Assume $\mathscr{F}$ refines $\widetilde{\mathscr{F}}$ with carrier map c. $F \cap \widetilde{F} \neq \emptyset$, $F \in \mathscr{F}$, and $\widetilde{F} \in \widetilde{\mathscr{F}}$ imply $\widetilde{F} = c(F)$. $x \in W^+(F_1) \cap W^-(F_2)$ and $x \in W^+(\widetilde{F}_1) \cap W^-(\widetilde{F}_2)$ with F_1, $F_2 \in \mathscr{F}$ and $\widetilde{F}_1$, $\widetilde{F}_2 \in \widetilde{\mathscr{F}}$ imply $\widetilde{F}_1 = c(F_1)$ and $\widetilde{F}_2 = c(F_2)$. In particular, $x \in \widetilde{F}$ and $x \in W^+(F_1) \cap W^-(F_2)$ with F_1, $F_2 \in \mathscr{F}$ and $\widetilde{F} \in \widetilde{\mathscr{F}}$ imply $c(F_1) = c(F_2) = \widetilde{F}$. Thus, the map c is onto. $F_2 \ \mathscr{F}f \ F_1$ implies $c(F_2) \ \widetilde{\mathscr{F}}f \ c(F_1)$.*

(e) *Assume* $\mathscr{F}$ *refines* $\widetilde{\mathscr{F}}$ *and* $\widetilde{\mathscr{F}}_+$ *is an* $\widetilde{\mathscr{F}}f$ *invariant subset of* $\widetilde{\mathscr{F}}$. *Then* $\mathscr{F}_+ = c^{-1}(\widetilde{\mathscr{F}}_+)$ *is an* $\mathscr{F}f$ *invariant subset of* $\mathscr{F}$ *and the corresponding attractors* $A_+(\mathscr{F}_+)$ *and* $A_+(\widetilde{\mathscr{F}}_+)$ *are equal. So the attractor structure of* $\widetilde{\mathscr{F}}$ *is a subset of the attractor structure of* $\mathscr{F}$. *If* $F \in \mathscr{F}$, *then* $B(F) \subset \widetilde{B}(c(F))$ *where* B *and* $\widetilde{B}$ *are the* $\mathscr{F}$ *and* $\widetilde{\mathscr{F}}$ *basic sets containing* F *and* $c(F)$, *respectively. In particular,* $|\mathscr{F}| \subset |\widetilde{\mathscr{F}}|$.

(f) *For the invariant decompositions* $\mathscr{F}_1$ *and* $\mathscr{F}_2$ *define* $\mathscr{F}_1 \wedge \mathscr{F}_2 = \{F_1 \cap F_2 \colon F_1 \in \mathscr{F}_1 \text{ and } F_2 \in \mathscr{F}_2\} - \{\varnothing\}$. $\mathscr{F}_1 \wedge \mathscr{F}_2$ *is an invariant decomposition refining* $\mathscr{F}_1$ *and* $\mathscr{F}_2$. $\mathscr{F}$ *refines* $\mathscr{F}_1 \wedge \mathscr{F}_2$ *if and only if it refines both* $\mathscr{F}_1$ *and* $\mathscr{F}_2$ *in which case* $c(F) = c_1(F) \cap c_2(F)$ *where* c, c_1, c_2 *are the carrier maps to* $\mathscr{F}_1 \wedge \mathscr{F}_2$, $\mathscr{F}_1$ *and* $\mathscr{F}_2$ *respectively.*

(g) *The refinement of* $\widetilde{\mathscr{F}}$ *by* $\mathscr{F}$ *is called trivial if* $c(F_2)\mathscr{O}(\widetilde{\mathscr{F}}f)c(F_1)$ *implies* $F_2\mathscr{O}(\mathscr{F}f)F_1$ (*compare* (d)). *A refinement is trivial if and only if the two attractor structures agree. For any* $\mathscr{F}$, $\mathscr{F} \wedge \{l[f]\} = \{F \cap l[f] \colon F \in \mathscr{F}\}$ *is a trivial refinement of* $\mathscr{F}$.

(h) $\mathscr{F}_b = \{B(F) \colon F \in \mathscr{F}\}$ *is an invariant decomposition for* f *and* $\mathscr{F}$ *is a trivial refinement with carrier map* $c(F) = B(F)$ *for* $F \in \mathscr{F}$.

(i) *A refinement of* $\widetilde{\mathscr{F}}$ *by* $\mathscr{F}$ *is trivial if and only if* $\mathscr{F}_b = \widetilde{\mathscr{F}}_b$, *i.e.,* $B(F) = \widetilde{B}(c(F))$ *for all* F *in* $\mathscr{F}$ (*compare* (e)). *In particular,* $(\mathscr{F}_b)_b = \mathscr{F}_b$.

(j) $\mathscr{F} = \mathscr{F}_b$ *if and only if the following equivalent conditions hold*:

(1) $B(F) = F$ *for all* F *in* $\mathscr{F}$.

(2) *Each* F *in* $\mathscr{F}$ *is* $\mathscr{C}f$ *semi-invariant.*

(3) $F_1 \in \mathscr{O}(\mathscr{F}f) \cap \mathscr{O}(\mathscr{F}f)^{-1}(F)$ *implies* $F_1 = F$ *and* $W^+(F) \cap W^-(F) = F$ *for all* F *in* $\mathscr{F}$.

(*Hint*: *Show* (1) $\Rightarrow$ (2) $\Rightarrow$ (3) $\Rightarrow$ (1). *Condition* (3) *is called the no-cycle condition for the invariant decomposition* $\mathscr{F}$.)

(k) *Let* $\mathscr{A}$ *be a finite collection of attractors of* f. *There exists an invariant decomposition whose attractor structure includes all the attractors in* $\mathscr{A}$. (*By enlarging it if necessary you can assume that* $\mathscr{A}$ *is closed under intersection and union and* $\varnothing$, $X \in \mathscr{A}$. *Let* $\widetilde{\mathscr{A}}$ *be the set of repellors dual to the attractors in* $\mathscr{A}$. *Define* $\mathscr{B} = \{A_+ \cap A_- \colon A_+ \in \mathscr{A},\ A_- \in \widetilde{\mathscr{A}}\}$, *and let* $\mathscr{F}$ *be the minimal nonempty members of* $\mathscr{B}$. *Show that* $\mathscr{F}$ *is a pairwise disjoint collection of closed,* f*-invariant,* $|\mathscr{C}f|$ *separating subsets of* X. *If* $x \in |\mathscr{C}f|$ *let* A_+, A_- *be the smallest elements of* $\mathscr{A}$ *and* $\widetilde{\mathscr{A}}$ *respectively containing* x. *Show that* $A_+ \cap A_- \in \mathscr{F}$. *Hence,* $|\mathscr{C}f| \subset \bigcup \mathscr{F}$. *Use* (b) *to show that the attractor structure of* $\mathscr{F}$ *contains* $\mathscr{A}$.) □

In some cases there exists an invariant decomposition which completely describes the recurrence behavior of f.

17. **Proposition.** *An invariant decomposition* $\mathscr{F}$ *for a homeomorphism* $f \colon X \to X$ *is called a* fine *decomposition if it satisfies the following equivalent*

conditions:

(1) *Every attractor for* f *is in the attractor structure for* $\mathscr{F}$.

(2) $x_1, x_2 \in F \cap l[f]$ *with* $F \in \mathscr{F}$ *implies* $x_2 \in \mathscr{C}f(x_1)$.

(3) *Every* $\mathscr{F}$ *basic set* $B(F)$ *is a basic set for* f.

(4) $|\mathscr{F}| = |\mathscr{C}f|$ *and each* $\mathscr{F}$ *basic set* $B(F)$ *is indecomposable.*

A homeomorphism f *admits a fine decomposition if and only if there are only finitely many basic sets for* f *or, equivalently, if and only if there are finitely many attractors for* f. *In that case, the collection of basic sets is a fine invariant decomposition for* f.

PROOF. We will first show $(1) \Rightarrow (2) \Rightarrow (3) \Rightarrow (1)$ and $(3) \Leftrightarrow (4)$.

$(1) \Rightarrow (2)$. If A is an attractor containing x_1 then $F \subset A$ because A is in the attractor structure and $x_1 \in F \cap A$. Hence, $x_2 \in A$. $x_2 \in \mathscr{C}f(x_1)$ now follows from (a) of Proposition 3.11.

$(2) \Rightarrow (3)$. By (g) and (i) of Exercise 16 we can replace $\mathscr{F}$ by the trivial refinement $\mathscr{F} \wedge \{l[f]\}$ without affecting the collection of $\mathscr{F}$ basic sets and so we can assume $F \subset l[f]$ for all F in $\mathscr{F}$. (2) then implies that an element F of $\mathscr{F}$ is contained in a, necessarily unique, basic set B. Because $\mathscr{F}f \subset \mathscr{F}_5 f$, all the members of the equivalence class $\mathscr{O}(\mathscr{F}f)(F) \cap \mathscr{O}(\mathscr{F}f)^{-1}(F)$ are contained in this same B. If for $x \in X$, $\alpha f(x)$, $\omega f(x) \subset B$, then $x \in B$ as well. So $B(F) \subset B$. The reverse inclusion is part of Theorem 15.

$(3) \Rightarrow (1)$. For any attractor A and any basic set B for f, $A \cap B \neq \emptyset$ implies $B \subset A$ by the $\mathscr{C}f$ +invariance of A. A is therefore in the attractor structure of $\mathscr{F}$ by (b) of Exercise 16.

$(3) \Leftrightarrow (4)$. $B(F)$ is always a closed, f invariant, $\mathscr{C}f$ semi-invariant subset of X. So by the characterization Corollary 4.13(b), $B(F)$ is a basic set if and only if it is an indecomposable subset of $|\mathscr{C}f|$.

In the notation of Exercise 16(h), (3) says that $\mathscr{F}$ is a fine decomposition if and only if the elements of $\mathscr{F}_b$ are the basic sets for f. This requires that there be finitely many basic sets. In that case, the family of basic sets is an invariant decomposition.

If f admits a fine decomposition, then by (1) there are only finitely many attractors. Conversely, if there are only finitely many attractors then (k) of Exercise 16 constructs a decomposition satisfying (1) and so is fine. □

18. EXERCISE. (a) *With* $X = [0, 1]$ *let* $f: X \to X$ *be the homeomorphism of a positive interval example with* K *the Cantor set. Show that if* $\mathscr{F}$ *is any invariant decomposition, then the basic sets are closed intervals of positive length and so the inclusion of* $K = |\mathscr{C}f|$ *in* $|\mathscr{F}|$ *is always proper. In particular, there is no fine decomposition.*

(b) *In the literature the term fine decomposition is sometimes used for the weaker condition that* $|\mathscr{C}f| = |\mathscr{F}|$. *Let* $X = [0, 1] \times [0, 1]$, *and let* g *be a smooth nonnegative function on* X *vanishing exactly on* $[0, 1] \times 0 \cup K \times 1/2 \cup [0, 1] \times 1$ *where* K *is the Cantor set. Sketch a phase portrait for the flow on* X *associated with the equations*

$$dy/dt = -g(x, y), \qquad dx/dt = 0.$$

Let $f: X \to X$ *be the associated time-one homeomorphism. Show that* f *has infinitely many attractors and so* f *does not admit a fine decomposition. Show that the set of basic sets is* $\{[0, 1] \times 0, [0, 1] \times 1\} \cup \{(x, 1/2): x \in K\}$. *On the other hand,* $\mathscr{F} = \{[0, 1] \times 0, K \times 1/2, [0, 1] \times 1\}$ *is an invariant decomposition with* $|\mathscr{F}| = \bigcup \mathscr{F} = |\mathscr{C} f|$. □

While a fine decomposition will not exist when f has infinitely many attractors, we can always close in upon the recurrent set of f by using a sequence of decompositions.

19. THEOREM. *A sequence* $\{\mathscr{F}_i\}$ *of invariant decompositions for a homeomorphism* $f: X \to X$ *is called a* fine sequence *if* $\mathscr{F}_{i+1}$ *is a refinement of* $\mathscr{F}_i$ *for all* i, *and the following equivalent conditions are satisfied:*

(1) *Every attractor for* f *is in the attractor structure of some* $\mathscr{F}_i$.

(2) *If* B *is a basic set for* f, *then* $B = \bigcap_i \{B(F_i)\}$, *where* $B(F_i)$ *is the* $\mathscr{F}_i$*-basic set containing* B.

In particular, for a fine sequence the chain recurrent set $|\mathscr{C} f|$ *is the intersection of the decreasing sequence of closed* f *invariant sets* $\{|\mathscr{F}_i|\}$.

Every homeomorphism f *admits a fine sequence of* f *invariant decompositions.*

PROOF. Recall that on the chain recurrent set $|\mathscr{C} f|$ we can take the quotient by the closed equivalence relation, $\mathscr{C} f \cap \mathscr{C} f^{-1}$, and so regard $\mathscr{B}_f$, the set of basic sets for f, as a compact, metrizable space $|\mathscr{C} f|/\mathscr{C} f \cap \mathscr{C} f^{-1}$. We saw as a consequence of the existence of a complete $\mathscr{C} f$ Lyapunov function, Theorem 3.12, that $\mathscr{B}_f$ is homeomorphic to a closed subset of the Cantor set and so is totally disconnected.

We replace each $\mathscr{F}_i$ by $(\mathscr{F}_i)_b$, if necessary, so as to assume that each element of $\mathscr{F}_i$ is an $\mathscr{F}_i$ basic set, i.e., $F_i = B(F_i)$ for $F_i \in \mathscr{F}_i$, and so $\bigcup \mathscr{F}_i = |\mathscr{F}_i| \supset |\mathscr{C} f|$. Since $\mathscr{F}_i$ is a trivial refinement of $(\mathscr{F}_i)_b$ this adjustment leaves the attractor structures and $\mathscr{F}$ basic sets unchanged. Also, the monotonicity of the sequence is preserved.

With this assumption there is a map $k_i: |\mathscr{C} f| \to \mathscr{F}_i$ defined by associating to each recurrent point x the element of F_i containing it. If we regard $\mathscr{F}_i$ as a finite metric space then this map is continuous because each $F_i = B(F_i)$ is $|\mathscr{C} f|$ separating and so $k_i^{-1}(F_i) = F_i \cap |\mathscr{C} f|$ is open and closed. Each basic set of f is contained in a unique $\mathscr{F}_i$ basic set and so k_i is constant on basic sets. Thus, it factors through the quotient process to define a continuous map, also denoted k_i, from $\mathscr{B}_f$ to $\mathscr{F}_i$.

From the sequence $\{\mathscr{F}_i\}$ we can construct the inverse limit $\overleftarrow{\operatorname{Lim}}_i \{\mathscr{F}_i\}$. It is the subset of the product $\prod_i \{\mathscr{F}_i\}$ consisting of those sequences $\{F_i\}$

with $F_i \in \mathscr{F}_i$ such that $c(F_{i+1}) = F_i$, where $c: \mathscr{F}_{i+1} \to \mathscr{F}_i$ is the carrier map of the refinement of $\mathscr{F}_i$ by $\mathscr{F}_{i+1}$. Thus, a sequence $\{F_i\}$ lies in $\overleftarrow{\operatorname{Lim}}_i\{\mathscr{F}_i\}$ when $F_i \in \mathscr{F}_i$ and $F_{i+1} \subset F_i$, i.e., the sequence is monotone decreasing. $\Pi_i\{\mathscr{F}_i\}$ is the product of finite, and hence compact, metric spaces and so is itself compact. $\overleftarrow{\operatorname{Lim}}_i\{\mathscr{F}_i\}$ is a closed subset and so is also compact. It is also totally disconnected. In fact, a basic open set is obtained by specifying finitely many members of the equation $\{F_i\}$ and such a set is closed as well.

If $x \in |\mathscr{C} f|$, then the sequence $\{k_i(x)\}$ is clearly monotone and so lies in the inverse limit. Thus, $k: |\mathscr{C} f| \to \overleftarrow{\operatorname{Lim}}_i\{F_i\}$ and its quotient $k: \mathscr{B}_f \to \overleftarrow{\operatorname{Lim}}_i\{\mathscr{F}_i\}$ are defined. k is the inverse limit of the maps k_i and so is continuous. It is always onto because $k^{-1}(\{F_i\}) = \bigcap_i\{F_i \cap |\mathscr{C} f|\}$ which is nonempty because each member of the decreasing sequence $\{F_i\}$ is closed and f invariant and so meets $|\mathscr{C} f|$ (e.g. $x \in F_i$ implies $\omega f(x) \subset F_i$). We will show that conditions (1) and (2) are each equivalent to

(3) $k: \mathscr{B}_f \to \overleftarrow{\operatorname{Lim}}\{\mathscr{F}_i\}$ is a homeomorphism
(when $\mathscr{F}_i = (\mathscr{F}_i)_b$ for all i).

Note that since k is an onto continuous map between compacta it is a homeomorphism if it is one-to-one.

$(1) \Rightarrow (3)$. If B_1 and B_2 are distinct points of $\mathscr{B}_f$, i.e., distinct f basic sets, then by (a) of Proposition 3.11 there exists an attractor A containing one and not the other. Say $B_1 \subset A$ and $A \cap B_2 = \varnothing$. If A is in the attractor structure of $\mathscr{F}_j$, then $k_j(B_1) \neq k_j(B_2)$ and so $k(B_1) = \{k_i(B_1)\} \neq \{k_i(B_2)\} = k(B_2)$. Thus, k is one-to-one.

$(3) \Rightarrow (1)$. An attractor A is determined by its trace on $|\mathscr{C} f|$, $A \cap |\mathscr{C} f|$, which is open and closed in $|\mathscr{C} f|$ as is its quotient in $\mathscr{B}_f$. Because k is a homeomorphism, $k(A \cap |\mathscr{C} f|)$ is open and closed in the inverse limit and so is the finite union of basic open sets each a specification of a finite number of indices. Let j be the largest index occurring in these lists. We then get that $A \cap |\mathscr{C} f|$ is the preimage, under k_j^{-1}, of a subset of $\mathscr{F}_j$. By (b) of Exercise 16, A is the attractor structure of $\mathscr{F}_j$.

$(2) \Rightarrow (3)$. If B_1 and B_2 are distinct basic sets, then $B_1 = \bigcap_i\{F_i(B_1)\}$ and $B_2 = \bigcap_i\{F_i(B_2)\}$, where $F_i(B_k)$ is the element of $\mathscr{F}_i$ (= the $\mathscr{F}_i$ basic set) containing B_k $(k = 1, 2)$. So the sequences $\{F_i(B_1)\} = k(B_1)$ and $\{F_i(B_2)\} = k(B_2)$ cannot be the same. Thus, k is one-to-one.

$(3) \Rightarrow (2)$. If $\{F_i\} \in \overleftarrow{\operatorname{Lim}}$, then because k is one-to-one the intersection $F = \bigcap_i\{F_i\}$ meets (and contains) exactly one basic set B for f. If $x \in F$, then $\alpha f(x)$, $\omega f(x) \subset F \cap l[f]$ because F is closed and f invariant. So $\alpha f(x)$, $\omega f(x) \subset B$ and so $x \in B$ as well (cf. Exercise 11).

Now if $x \in \bigcap_i\{|\mathscr{F}_i|\}$ (N.B. the latter equals $\bigcap_i(\bigcup \mathscr{F}_i)$ under our standing assumption that $(\mathscr{F}_i)_b = \mathscr{F}_i)$), then let $F_i(x)$ be the element of $\mathscr{F}_i$ containing

x. Clearly $\{F_i(x)\} \in \overleftarrow{\text{Lim}}\{\mathscr{F}_i\}$ and so $x \in \bigcap_i \{F_i(x)\}$ which is a basic set when the sequence is fine. So $x \in |\mathscr{C}f|$. The reverse inclusion always holds.

Finally, for any homeomorphism f the set of attractors is countable by Proposition 3.8. List them $\{A_1, A_2, \dots\}$ and apply (i) of Exercise 16 to construct an invariant decomposition $\widetilde{\mathscr{F}_i}$ whose attractor structure includes $\{A_1, \dots, A_i\}$. Inductively define $\mathscr{F}_1 = \widetilde{\mathscr{F}_1}$ and $\mathscr{F}_{i+1} = \widetilde{\mathscr{F}_{i+1}} \wedge \mathscr{F}_i$ to get a monotone sequence $\{\mathscr{F}_i\}$ of invariant decompositions. It is a fine sequence because (1) holds. □

20. Exercise. *Prove*:

(a) $\mathscr{F}$ *is a fine decomposition if and only if the one term sequence* $\{\mathscr{F}\}$ *is a fine sequence. A finite fine sequence terminates at a fine decomposition.*

(b) *A sequence* $\{\mathscr{F}_i\}$ *is fine if and only if for every invariant decomposition* $\widetilde{\mathscr{F}}$, $\mathscr{F}_i$ *refines* $\widetilde{\mathscr{F}}_b$ *for some* i. *In particular,* $\mathscr{F}$ *is a fine invariant decomposition if and only if, for every invariant decomposition* $\widetilde{\mathscr{F}}$, $\mathscr{F}$ *refines* $\widetilde{\mathscr{F}}_b$.
□

Supplementary exercises

21. (a) Assume f is a closed relation on X. Prove that if A is a closed $\mathscr{N}f$ semi-invariant set (i.e., $\mathscr{N}f(A) \cap \mathscr{N}f^{-1}(A) \subset A$), then every neighborhood of A contains an open f semi-invariant neighborhood of A (*Hint*: use the proof of Proposition 2.7). Let $\tilde{f}$ be the homeomorphism on the circle obtained from positive interval example with $K = \{0, 1/2, 1\}$ and let $A = 1/2$. Show that A has a base of $\tilde{f}$ semi-invariant neighborhoods but A is not $\mathscr{N}\tilde{f}$ semi-invariant.

    (b) Let $f: X \to X$ be a homeomorphism and assume A is the intersection of $l[f]$ separating sets. If A has a base of f semi-invariant neighborhoods, prove that $W^+(A) \cap W^-(A) = A$ (compare (5.11)). Prove that if A is a closed $\mathscr{C}f$ semi-invariant set, then A is the intersection of $|\mathscr{C}f|$ separating sets and $W^+(A) \cap W^-(A) = A$. (*Hint*: reduce to the cases when A is $\mathscr{C}f$ or $\mathscr{C}f^{-1}$ invariant.)
22. Let U be an inward neighborhood of an attractor A for a homeomorphism. Prove that $A = \omega f[U] = \bigcap\{f^n(U) : n = 0, 1, \dots\}$ if and only if $U \subset W^+(A)$. In that case, prove $W^+(A) = \bigcup\{f^{-n}(U) : n = 0, 1, \dots\}$. Thus $\{\dots f^{-2}(U), f^{-1}(U), U, f(U), \dots\}$ is a bi-infinite sequence of inward sets with intersection A and union $W^+(A)$. Furthermore, the interior of each term contains its successors.
23. For a homeomorphism $f: X \to X$, filtrations are often used in the literature as an alternative to invariant decomposition. We develop this alternative by a series of assertions whose proofs we leave to the reader.

    (a) Let A_+^1, A_+^2 be attractors with duals repellors A_-^1, A_-^2. If A_+^2 is a proper subset of A_+^1, then A_-^1 is a proper subset of A_-^2 and $A_+^1 \cap A_-^2$

is a nonempty, closed, f invariant, $\mathscr{C}f$ semi-invariant, $|\mathscr{C}f|$ separating subset of X (*Hint*: $x \in A^1_+ - A^2_+$ implies $\alpha f(x) \subset A^1_+ \cap A^2_-$). There exists inward sets U^1, U^2 with $\omega f[U^i] = A^i_+$ $(i = 1, 2)$ and $\operatorname{Int} U^1 \supset U^2$. Furthermore, for any $\varepsilon > 0$ we can choose the inward sets so that $U^i \subset V_\varepsilon(A^i_+)$ $(i = 1, 2)$.

(b) Let U^1, U^2 be inward sets with $\operatorname{Int} U^1 \supset U^2$. Recall that $X - \operatorname{Int} U^2$ is outward, i.e., inward for f^{-1}. Let A^i_+ be the associated attractor $\omega f[U^i] = \bigcap_{n=0}^{\infty}\{f^n(U^i)\} = \mathscr{C}f(U^i \cap |\mathscr{C}f|)$ $(i = 1, 2)$. Either $A^1_+ = A^2_+$ in which case $U^1 - \operatorname{Int} U^2$ contains no f invariant subset and $U^1 \cap |\mathscr{C}f| = (\operatorname{Int} U^2) \cap |\mathscr{C}f|$, or A^2_+ is a proper subset of A^1_+ and $A^1_+ \cap A^2_-$, nonempty, is the maximal invariant subset of $U^1 - \operatorname{Int} U^2$, i.e., $A^1_+ \cap A^2_- = \bigcap_{n=-\infty}^{+\infty} f^n(U^1 - \operatorname{Int} U^2)$. Furthermore, $A^1_+ \cap A^2_- \subset (\operatorname{Int} U^1) - U^2$.

(c) A *filtration* of length n of f is a sequence $\{U_1, \dots, U_{n+1}\}$ of inward sets with $X = U_1$, $\varnothing = U_{n+1}$ and $\operatorname{Int} U_i \supset U_{i+1}$ $(i = 1, \dots, n)$. Define $A^i_+ = \omega f[U_i]$ to be the attractor associated with U_i and $A^i_- = \alpha f[X - \operatorname{Int} U_i]$ to be the repellor dual to A^i_+. $\{X = A^1_+, A^2_+, \dots, A^{n+1}_+ = \varnothing\}$ is a nonincreasing sequence of attractors and $\{\varnothing = A^1_-, \dots, A^n_-, A^{n+1}_- = X\}$ is a nondecreasing sequence of repellors. Define $F_i = A^i_+ \cap A^{i+1}_-$ $(i = 1, \dots, n)$. F_i is the maximum invariant subset of the closed set $U_i - \operatorname{Int}(U_{i+1})$ and is contained in the open set $\operatorname{Int}(U_i) - U_{i+1}$. If $x \in |\mathscr{C}f|$ and $x \in A^i_+$ but $x \notin A^{i+1}_+$, then $x \in F_i$ and so $|\mathscr{C}f| \subset \cup\mathscr{F}$, where $\mathscr{F} = \{F_1, \dots, F_n\}$. $\mathscr{F}$ is an invariant decomposition for f satisfying the no-cycle condition of Exercise 16(j), i.e., $\mathscr{F}_b = \mathscr{F}$. Also, $F_i(\mathscr{F}f)F_j$ implies $i \geq j$. Each A^i_+ lies in the attractor structure of $\mathscr{F}$.

(d) Given an invariant decomposition $\mathscr{F}$ for f satisfying the no-cycle condition, there is an ordering of the elements $\mathscr{F} = \{F_1, \dots, F_n\}$ so that $\mathscr{F}_i(\mathscr{F}f)F_j$ implies $i \geq j$. (Use induction on n. Because $\mathscr{O}(\mathscr{F}f)$ is antisymmetric this is the finite version of the Lyapunov function Theorem 2.12.) Hence, $\mathscr{F}_i \equiv \{F_j : j \geq i\}$ $i = 1, \dots, n+1$ are $\mathscr{F}f$ + invariant subsets of $\mathscr{F}$. Let A^i_+ be the attractor $A_+(\mathscr{F}_i)$. Then $\{X = A^1_+, \dots, A^{n+1}_+ = \varnothing\}$ is a nonincreasing sequence of attractors. For any such sequence of attractors there exists a filtration $\{X = U_1, \dots, U_{n+1} = \varnothing\}$ with $\omega f[U_i] = A^i_+$. Furthermore, for any $\varepsilon > 0$ the filtration can be chosen with $U_i \subset V_\varepsilon(A^i_+)$. When the attractor sequence comes from an ordered invariant decomposition $\mathscr{F}$ as above ($\mathscr{F}_b = \mathscr{F}$ by the no-cycle condition) then $\mathscr{F}$ is the invariant decomposition obtained from the filtration via (c).

(e) Assume $L: X \to [0, 1]$ is a $\mathscr{C}f$ Lyapunov function for f. Let $a_2 < \cdots < a_n$ be a sequence of regular values for L in $(0, 1)$. Let $0 = a_1$ and define $U_i = L^{-1}([a_i, 1])$ $i = 1, \dots, n$; $U_{n+1} = \varnothing$. $\{X = U_1, \dots, U_{n+1} = \varnothing\}$ is a filtration for f. If L is a complete $\mathscr{C}f$ Lyapunov function (cf. Theorem 3.12) then the set of critical values is closed and nowhere dense in $[0, 1]$. So we can choose $\{b_2, b_3, b_4, \dots\}$ a dense sequence of regular values in $(0, 1)$. Define the nth filtration as above by rearranging $\{0, b_2, \dots, b_n\}$ in increasing order. Let $\mathscr{F}_n$ denote the associated invariant decomposition. $\{\mathscr{F}_n\}$ is a fine sequence of invariant decompositions.

24. For a homeomorphism $f: X \to X$ let $L = l[f]$. For the restriction $f_L: L \to L$ prove that every point is chain-recurrent (cf. chain transitivity of $\omega f(x)$ and $\alpha f(x)$), i.e., $|\mathscr{C}(f_L)| = L$, and so that the quotient space of $q: L \to L/\mathscr{C}(f_L) \cap \mathscr{C}(f_L)^{-1}$ is a compact, totally disconnected space. Let $\mathscr{F} = \{F_1, \dots, F_n\}$ be a family of subsets of L. Prove that $\mathscr{F}$ is an invariant decomposition for f if and only if there exists a pairwise disjoint sequence $\{L_1, \dots, L_n\}$ of closed subsets of $L/\mathscr{C}(f_L) \cap \mathscr{C}(f_L)^{-1}$ such that $F_i = q^{-1}(L_i)$ $i = 1, \dots, n$. Thus, the invariant decompositions of the form $\mathscr{F} \wedge \{l[f]\}$ are completely described by the topology of $L/\mathscr{C}(f_L) \cap \mathscr{C}(f_L)^{-1}$ (and the projection map q).

25. In applications it is sometimes important to be able to apply the results of this chapter to partially defined homeomorphisms. A closed relation $f: X \to X$ is called a *partial homeomorphism* if it satisfies the following equivalent conditions:

(1) Let $X_{-1} = \mathrm{Dom}(f) = f^{-1}(X)$ and let $X_1 = \mathrm{Dom}(f^{-1}) = f(X)$, so $f \subset X_{-1} \times X_1 \subset X \times X$. When regarded as a relation from X_{-1} to X_1, f is a homeomorphism.

(2) If $y_1 \in f(x_1)$ and $y_2 \in f(x_2)$, then $x_1 = x_2$ if and only if if $y_1 = y_2$.

(3) $f \circ f^{-1} \subset 1_X$ and $f^{-1} \circ f \subset 1_X$.

Check the equivalence of these conditions. Recall from exercises 2.16 and 3.21 the closed sets:

$$\begin{aligned} X_+ &= \omega f[X] = \mathrm{Dom}(\alpha f) = \bigcap_{n \geq 0} \{f^n(X)\} \\ &= \Omega\mathscr{C}f(X) = \mathrm{Dom}(\Omega\mathscr{C}f^{-1}), \\ X_- &= \alpha f[X] = \mathrm{Dom}(\omega f) = \bigcap_{n \geq 0} \{f^{-n}(X)\} \\ &= \Omega\mathscr{C}f^{-1}(X) = \mathrm{Dom}(\Omega\mathscr{C}f), \\ X_0 &= X_+ \cap X_-. \end{aligned}$$

Thus, for $x \in X$ either $\omega f(x) = \varnothing$, i.e., $x \notin X_-$, and for some positive integer n, $f^n(x) = \emptyset$ meaning that the sequence x, $f(x), \ldots, f^{n-1}(x)$ terminates at a point not in $X_{-1} = \mathrm{Dom}(f)$, or $x \in X_-$ meaning that the entire orbit sequence x, $f(x)$, $f^2(x), \ldots$ remains in X_{-1} and the limit point set $\omega f(x) \neq \emptyset$. In the latter case prove the generalization to this case of the homeomorphism results: $\omega f(x)$ is a closed f and f^{-1} invariant set consisting entirely of nonwandering points. Similarly, $\alpha f(x) = \emptyset$ and $x \notin X_+$ or $\alpha f(x)$ is a closed nonempty f and f^{-1} invariant set of nonwandering points.

X_+ is f invariant and f^{-1} $+$invariant, X_- is f^{-1} invariant and f $+$invariant and X_0 is both f and f^{-1} invariant. A subset A is both f and f^{-1} invariant if and only if the restriction $f_A: A \to A$ is a homeomorphism in which case $A \subset X_0$. In particular, $f_{X_0}: X_0 \to X_0$ is a homeomorphism. The assertions which follow (prove them) illustrate how the homeomorphism f_{X_0}, to which all our results apply, captures all the interesting dynamic behavior of f.

We have the inclusions (cf. Exercise 2.16)

$$l[f_{X_0}] \subset l[f] \subset |\mathscr{C}(f_{X_0})| = |\mathscr{C}f| \subset X_0.$$

The first inclusion may be proper because while $\omega f(x) \subset X_0$ for all x the points may not be $l[f_{X_0}]$ if $x \in X_- - X_0$. Of course, if $x \in X_0$, then $\omega f(x) = \omega(f_{X_0})(x)$. For a partial homeomorphism we have $\mathscr{C}(f_{X_0}) = (\mathscr{C}f)_{X_0}$, and so we can omit the parentheses (use Theorem 4.5). In particular, the basic sets of f and f_{X_0} agree. These basic sets and the limit point sets $\omega f(x)$ and $\alpha f(x)$, when nonempty, are subsets of X_0 chain transitive with respect to f_{X_0}.

If K and F are closed, f and f^{-1} invariant subsets of X (and hence of X_0), F is called K separating if $F \cap K$ is open and closed in K. When A is the intersection of $l[f]$ separating sets define:

$$\begin{aligned} W^+(A; f) &= \{x \in X : \omega f(x) \cap A \neq \emptyset\} \\ &= \{x \in X_- : \omega f(x) \subset A\}, \\ W^-(A; f) &= \{x \in X : \alpha f(x) \cap A \neq \emptyset\} \\ &= \{x \in X_+ : \alpha f(x) \subset A\}. \end{aligned}$$

Since $l[f]$ separating implies $l[f_{X_0}]$ separating, we can define $W^{\pm}(A; f_{X_0})$ by (5.10) applied to the homeomorphism f_{X_0}.

$$W^{\pm}(A; f_{X_0}) = W^{\pm}(A; f) \cap X_0.$$

In general $A_+^0 \subset X_0$ is an attractor for f_{X_0} if and only if $A_+^0 = A_+ \cap X_0$ for some attractor A_+ of f (cf. Exercise 3.21). In that case,

$$A_+ = \mathscr{C}f(A_+^0) = \Omega\mathscr{C}f(A_+^0) = W^-(A_+^0; f) \subset X_+.$$

Similarly, $A_-^0 \subset X_0$ is a repellor for f_{X_0} when $A_-^0 = A_- \cap X_0$ with A_- the repellor of f defined by

$$A_- = \mathscr{C}f^{-1}(A_-^0) = \Omega\mathscr{C}f^{-1}(A_-^0) = W^+(A_+^0\,;\,f) \subset X_-\,.$$

In particular, with $A_+^0 = A_-^0 = X_0$, $X_0 \subset X_+ = W^-(X_0\,;\,f)$ and $X_0 \subset X_- = W^+(X_0\,;\,f)$, the largest attractor and repellor respectively for f.

An invariant decomposition $\mathscr{F} = \{F_1, \ldots, F_n\}$ for f is a pairwise disjoint collection of closed f and f^{-1} invariant subsets of X with $l[f] \subset \bigcup\mathscr{F}$. $\mathscr{F}$ is then an invariant decomposition for f_{X_0} and for $F_1, F_2 \in \mathscr{F}$,

$$W^+(F_1\,;\,f) \cap W^-(F_2\,;\,f) = W^+(F_1\,;\,f_{X_0}) \cap W^-(F_2\,;\,f_{X_0})\,.$$

So in the definition of the relation $\mathscr{F}f$ we get equivalent results by using f_{X_0} or extending the definition by using f. The $\mathscr{F}$ basic sets are the same whether the f_{X_0} or f definition is used. Finally, if $\mathscr{F}_+$ is a $\mathscr{F}f$ +invariant subset of $\mathscr{F}$, then

$$A_+^0(\mathscr{F}_+) = A_+(\mathscr{F}_+) \cap X_0\,,$$

where

$$A_+^0(\mathscr{F}_+) = \bigcup\{W^-(F_i\,;\,f_{X_0})\colon f_i \in \mathscr{F}_+\}\,,$$
$$A_+(\mathscr{F}_+) = \bigcup\{W^-(F_i\,;\,f)\colon F_i \in \mathscr{F}_+)\,.$$

Chapter 6. Chain Recurrence and Lyapunov Functions for Flows

A *semiflow* on X is a continuous map $\varphi: X \times [0, \infty) \to X$ such that $f^0 = 1_X$ and $f^t \circ f^s = f^{t+s}$ for $t, s \in [0, \infty)$, where we denote by f^t the time-t map of φ given by $f^t(x) = \varphi(x, t)$. For the time-one map we write f, omitting the superscript. The semiflow extends the discrete time dynamical system $f, f^2, f^3, \ldots$ to continuous time by interpolating the functions f^t between the successive iterates. For a point x in X the φ-*orbit* of x, $\{\varphi(x, t): t \geq 0\}$, is a path connecting the points of the f-orbit of x.

We call φ a *flow* if the time-t maps are invertible. The inverse flow, φ^{-1}, is then defined by $\varphi^{-1}(x, t) = (f^t)^{-1}(x)$. As outlined in Exercise 0.1, if f is invertible, then φ is a flow and φ^{-1} is continuous. For a flow the definition of φ is extended to $X \times \mathbb{R}$ via $\varphi(x, t) = \varphi^{-1}(x, -t)$ for negative t. We write f^{-t} for $(f^t)^{-1}$ and note that the composition relation $f^t \circ f^s = f^{t+s}$ holds for all t, s in $\mathbb{R}$ when φ is a flow.

The most important examples of flows arise on smooth manifolds by integrating vectorfields. If ξ is a C^r vector field on a compact C^{r+1} manifold $(1 \leq r \leq \infty)$ then the associated flow is characterized by the differential equation

$$\frac{d\varphi(x, t)}{dt} = \xi(\varphi(x, t)), \tag{6.1}$$

i.e., with x fixed path function $\varphi(x, t)$ is the solution of $y' = \xi(y)$ with initial condition x.

In this chapter we describe the analogues for semiflows of the structures and concepts previously developed for maps. In each case we want to compare the construction for the semiflow φ with the corresponding construction for the time-one map f. It will also be convenient to use various "thickenings" of f, nonmap relations defined from φ and a subset J of $[0, \infty)$ (or of $\mathbb{R}$ in the flow case) by

$$f^J = \bigcup\{f^t : t \in J\}, \qquad \text{i.e., } f^J(x) = \{\varphi(x, t): t \in J\}. \tag{6.2}$$

The relation f^J is the image in $X \times X$ of $X \times J$ under the map $(x, t) \to (x, \varphi(x, t))$. So if J is compact f^J is a closed relation. For example, with $I = [0, 1]$, f^I is a closed relation with $f^I(x)$ the piece of the φ-orbit connecting x with $f(x)$.

In particular, the orbit relation for φ is

$$\mathcal{O}(\varphi) = f^{[0,\infty)} = \bigcup\{f^t : 0 \leq t < \infty\}, \tag{6.3}$$
$$\text{i.e., } y \in \mathcal{O}(\varphi)(x) \qquad \text{if } y = f^t(x) \text{ for some nonnegative } t.$$

$\mathcal{O}(\varphi)$ can be described using the map f and the closed relation f^I

$$\mathcal{O}(\varphi) = \mathcal{O}(f^I) = f^I \cup f^I \circ \mathcal{O}(f) = f^I \cup \mathcal{O}(f) \circ f^I. \tag{6.4}$$

These equations follow from the separation of any real into integer and fractional parts. In studying the limit relations we will need a more sophisticated version of this separation. The following lemma and its applications were shown to me by my colleague H. Onishi.

1. LEMMA. *Let $\{t_k\}$ be a real sequence tending to ∞.*

(a) *There exists $r \in I$ such that for some subsequence $\{t_{k'}\}$ and some sequence $\{n_{k'}\}$ of positive integers the difference $\{t_{k'} - (n_{k'} + r)\}$ has limit zero.*

(b) *There exist positive reals t such that for some subsequence $\{t_{k'}\}$ and some sequence $\{m_{k'}\}$ of positive integers the difference $\{t_{k'} - m_{k'}t\}$ has limit zero. Furthermore, the set of real numbers t for which this property holds is residual in $[0, \infty)$, i.e., it contains a countable intersection of dense open sets.*

PROOF. (a) Write $t_k = n_k + r_k$ with $r_k \in I$ and choose a subsequence so that $\{r_{k'}\}$ converges to r.

(b) Consider the map e_n of the unit circle S in the complex plane defined by $e_n(z) = z^n$. If U is any open subset of S then there exists a positive integer $N(U)$ such that for $n \geq N(U)$, U contains two succesive nth roots of unity and the interval between them, and so $e_n(U) = S$.

Now apply (a) to choose $r \in I$, a subsequence $\{t_{k'}\}$ and a positive integer sequence $\{n_{k'}\}$. For V open in S and n a positive integer define

$$K(V, n) = \bigcap_{k' \geq n} (e_{n_{k'}})^{-1}(S - V) = \{z : z^{n_{k'}} \notin V \text{ for any } k' \geq n\}.$$

Each $K(V, n)$ is closed. Furthermore, if U is open in S then $e_{n_{k'}}(U) \supset V$ for k' sufficiently large and so U is not a subset of $K(V, n)$. Thus, $K(V, n)$ is closed and has empty interior, i.e., it is nowhere dense.

The union of all of the $K(V, n)$'s is the same as the union with V in a countable base for the topology of S. So the complement of this union T_1 is a residual subset of S. $T_2 = \{s \in (0, \infty) : \exp(2\pi i s) \in T_1\}$ is a residual subset of $[0, \infty)$ as is $T_3 = \{t \in (0, \infty) : 1/t \in T_2\}$. We prove that the reals in T_3 satisfy the property described in (b).

Given $t \in T_3$, let $s = 1/t$ and $z = \exp(2\pi i s)$. Since $z \in T_1$, the sequence $\{z^{n_{k'}}\}$ enters every open set infinitely often. So if we let $w = \exp(2\pi i(1-rs))$ there is a subsequence $\{z^{n_{k''}}\}$ converging to w. Lifting to $[0, \infty)$ again this means that there is a sequence of positive integers $\{m_{k''}\}$ such that $\{n_{k''}s - (m_{k''} - 1)\}$ approaches $1 - rs$, i.e., $\{(n_{k''} + r)s - m_{k''}\}$ approaches zero. Recall that $\{t_{k'} - (n_{k'} + r)\}$ approaches zero and so for the subsequence $\{t_{k''}\}$

$$t_{k''} - m_{k''}t = [t_{k''} - (n_{k''} + r)] + [(n_{k''} + r)s - m_{k''}]t$$

converges to zero. □

For A a subset of X define

$$\omega\varphi[A] = \limsup\{f^t(A)\} = \bigcap_{t\geq 0}\overline{\bigcup_{s\geq t}\{f^s(A)\}} \tag{6.5}$$

and the corresponding limit relation

(6.6) $\omega\varphi(x) = \limsup\{f^t(x)\}$, i.e., $y \in \omega\varphi(x)$ if there exists a sequence $\{t_k\}$ with $\{f^{t_k}(x)\} \to y$ and $\{t_k\} \to \infty$.

The reader should check that the properties of the lim sup operator described in Exercise 1.5 carry over for the continuous index case. In particular, the closure of $\mathscr{O}(\varphi)(A)$ is $\mathscr{O}(\varphi)(A) \cup \omega\varphi[A]$ and $f^t(A) \subset U$ for t sufficiently large whenever U is a neighborhood of $\omega\varphi[A]$.

As before, we define

(6.7) $\mathscr{R}(\varphi) = \mathscr{O}(\varphi) \cup \omega\varphi$, i.e., $\mathscr{R}(\varphi)(x) = \overline{\mathscr{O}(\varphi)(x)}$ for all $x \in X$.

As usual, $\omega\varphi$ need not be a closed relation and we define

(6.8) $\Omega\varphi = \limsup\{f^t\}$, i.e., $y \in \Omega\varphi(x)$ if there exist sequences $\{t_k\}$ and $\{x_k\}$ such that $\{t_k\} \to \infty$, $\{x_k\} \to x$, and $\{f^{t_k}(x_k)\} \to y$. $\mathscr{N}\varphi = \mathscr{O}(\varphi) \cup \Omega\varphi = \overline{\mathscr{O}(\varphi)}$.

If φ is a flow, then for $\mathscr{A} = \mathscr{O}$, Ω, or $\mathscr{N}$

$$\mathscr{A}(\varphi^{-1}) = (\mathscr{A}\varphi)^{-1} \tag{6.9}$$

and because this is not true for ω, we define

$$\alpha\varphi[A] = \omega(\varphi^{-1})[A] = \limsup\{f^{-t}(A)\}, \qquad \alpha\varphi = \omega(\varphi^{-1}). \tag{6.10}$$

From the definitions it is easy to check that with $J = [1, 2]$, $\omega f^J = \omega\varphi$ and $\Omega f^J = \Omega\varphi$. Also, for any fixed $t > 0$, $\omega f^t \subset \omega\varphi$ and $\Omega f^t \subset \Omega\varphi$. These inclusions can be proper.

2. EXERCISE. (a) *On the annulus* $X = \{(r, \theta): 1 \le r \le 2\}$ *let* φ *be the flow associated with the equations (in polar coordinates):*

$$\frac{dr}{dt} = (2-r)(r-1), \qquad \frac{d\theta}{dt} = 2\pi .$$

Show that for $x = (1, 0)$, $\omega f(x) = x$, $\omega\varphi(x) = S$ *(the unit circle),* $\Omega f(x) = \{((r, 0): 1 \le r \le 2\}$ *and* $\Omega\varphi(x) = X$. *(See Exercise 2.1.) More generally, show that the inclusions* $\omega f^t(x) \subset \omega\varphi(x)$ *and* $\Omega f^t(x) \subset \Omega\varphi(x)$ *are proper for rational* t *and are equalities for irrational* t.

(b) *Let* X *be the annulus as in* (a) *and* φ *be the flow associated with*

$$\frac{dr}{dt} = 0, \qquad \frac{d\theta}{dt} = 2\pi r .$$

For $x_r = (r, 0)$ *show that* $\omega f^t(x_r) = \omega\varphi(x_r)$ *if and only if tr is irrational.*
□

The exercises suggest that for most values of t the inclusions $\omega f^t(x) \subset \omega\varphi(x)$ and $\Omega f^t \subset \Omega\varphi$ are equalities.

3. PROPOSITION. *Let* φ *be a semiflow on* X

(a) *For* A *a subset of* X *and* $I = [0, 1]$,

$$\omega\varphi[A] = f^I(\omega f[A]) = \omega f[f^I(A)],$$
$$\omega f^t[A] \subset \omega\varphi[A] \quad \text{for all } t \in (0, \infty),$$

and the set of t *such that* $\omega f^t[A] = \omega\varphi[A]$ *is residual in* $(0, \infty)$. *In particular,*

$$\omega\varphi = f^I \circ \omega f = \omega f \circ f^I , \qquad \omega f^t \subset \omega\varphi ,$$

and for every $x \in X$ *the set of* t *such that* $\omega f^t(x) = \omega\varphi(x)$ *is residual in* $(0, \infty)$.

(b) $\Omega\varphi = f^I \circ \Omega f \subset \Omega f \circ f^I$, $\Omega f^t \subset \Omega\varphi$ *for all* $t \in (0, \infty)$, *and the set of* t *such that* $\Omega f^t = \Omega\varphi$ *(i.e.,* $\Omega f^t(x) = \Omega\varphi(x)$ *for all* x *in* X*) is residual in* $(0, \infty)$.

(c) $\Omega\varphi \subset \Omega\varphi \circ \omega\varphi$, $(\omega\varphi)^{-1} \circ \omega\varphi \subset \Omega\varphi$.

(d) *For all* t,

$$f^t \circ \omega\varphi = \omega\varphi = \omega\varphi \circ f^t , \qquad f^t \circ \Omega\varphi = \Omega\varphi \subset \Omega\varphi \circ f^t .$$

PROOF. The inclusions $\omega f^t[A] \subset \omega\varphi[A]$, $\Omega f^t \subset \Omega\varphi$ $(t > 0)$, and $\omega\varphi[A] = \omega f[f^I(A)]$ are clear from the lim sup definitions. Since f^t is a continuous map we have

$$f^t(\omega\varphi[A]) = f^t \limsup\{f^s(A)\} = \limsup\{f^{t+s}(A)\} = \omega\varphi[A]$$

(cf. Exercise 1.5(f)) and $f^t \circ \Omega\varphi = \Omega\varphi$ is similar. Taking the union over $t \in I$ we get $f^I(\omega f[A]) \subset \omega\varphi[A]$ and $f^I \circ \Omega f \subset \Omega\varphi$. The remaining parts of (a) and (b) Use Lemma 1.

Assume $\{x_k\}$ converges to x, $\{t_k\}$ converges to ∞, and $\{f^{t_k}(x_k)\}$ converges to y. This says $y \in \Omega\varphi(x)$. If $\{x_k\}$ lies in A, we have $y \in \omega\varphi[A]$. In particular, if $x_k = x$ for all k $y \in \omega\varphi(x)$. Applying part (a) of the lemma we can choose a subsequence $\{t_{k'}\}$ and a positive integer sequence $\{n_{k'}\}$ such that $\delta_{k'} = t_{k'} - n_{k'}$ lies in I and $\{\delta_{k'}\}$ converges to r. We can also assume that $\{f^{n_{k'}}(x_{k'})\}$ converges to $\tilde{y}$. First, notice that

$$f^{t_{k'}}(x_{k'}) = \varphi(f^{n_{k'}}(x_{k'}), \delta_{k'})$$

and so by the continuity of φ, $y = f^r(\tilde{y})$. Thus, $y \in f^r(\Omega f(x))$, $f^r(\omega f[A])$, or $f^r(\omega f(x))$ depending upon which assumption on $\{x_k\}$ applies. Thus, $\Omega\varphi \subset f^I \circ \Omega f$, $\omega\varphi[A] \subset f^I(\omega f[A])$, and $\omega\varphi \subset f^I \circ \omega f$, reversing inclusions proved above.

Next, observe that

$$f^{t_{k'}}(x_{k'}) = \varphi(f^{n_{k'}}(f^r(x_{k'})), \delta_{k'} - r) \quad \text{if } \delta_{k'} \geq r,$$
$$\varphi(f^{t_{k'}}(x_{k'}), r - \delta_{k'}) = f^{n_{k'}}(f^r(x_{k'})) \quad \text{if } r \geq \delta_{k'}.$$

As one of these occurs infinitely often, $\{f^{n_{k'}}(f^r(x_{k'}))\}$ converges to y by continuity of φ. So y lies in $\Omega f(f^r(x))$, $\omega f[f^r(A)]$, or $\omega f(f^r(x))$ depending on the assumption for $\{x_k\}$. Thus, $\Omega\varphi \subset \Omega f \circ f^I$, $\omega\varphi[A] \subset \omega f[f^I(A)]$, and $\omega\varphi \subset \omega f \circ f^I$.

Now by part (b) of Lemma 1 there is a residual subset $T_{x,y}$ of $[0, \infty)$ such that for every $t \in T_{x,y}$ there is a subsequence $\{t_{k'}\}$ and a positive integer sequence $\{m_{k'}\}$ such that $\{\delta_{k'} = t_{k'} - m_{k'}t\}$ approaches 0. As before

$$f^{t_{k'}}(x_{k'}) = \varphi((f^t)^{m_{k'}}(x_{k'}), \delta_{k'}) \quad \text{if } \delta_{k'} \geq 0,$$
$$\varphi(f^{t_{k'}}(x_{k'}), -\delta_{k'}) = (f^t)^{m_{k'}}(x_{k'}) \quad \text{if } \delta_{k'} \leq 0,$$

and so $\{(f^t)^{m_{k'}}(x_{k'})\}$ approaches y. Thus, $y \in \Omega f^t(x)$, $y \in \omega f^t[A]$, or $y \in \omega f^t(x)$ for all t in $T_{x,y}$. Now the different cases require separate arguments.

In $\omega\varphi[A]$ choose a dense sequence $\{y_n\}$ and for each y_n choose $x_n \in \overline{A}$ the limit of the sequence $\{x_k^n\}$ in A associated with y_n. The set $\bigcap T_{x_n, y_n}$ is residual and for t in the intersection $\{y_n\} \subset \omega f^t[A]$ and so $\omega\varphi[A] = \omega f^t[A]$. In particular, with $A = x$ we get a residual set of t's for which $\omega\varphi(x) = \omega f^t(x)$.

For Ω we can choose a sequence $\{(x_n, y_n)\}$ dense in $\Omega\varphi$. For $t \in \bigcap T_{x_n, y_n}$ $\{(x_n, y_n)\} \subset \Omega f^t$ and so $\Omega\varphi = \Omega f^t$.

Finally, the proofs of (c) and (d) are easy adjustments of the proofs of the corresponding results in Proposition 1.12.

Remarks. (a) Example (b) of Exercise 2 shows that it need not be true that $\omega f^t = \omega\varphi$ for generic choice of t, i.e., the dependence upon x of the set T_x such that $\omega f^t(x) = \omega\varphi(x)$ cannot be eliminated as it was for Ω.

(b) For a flow the corresponding results are true for $\alpha\varphi$ as well (use φ^{-1}). Also, in the flow case the equality $\Omega\varphi = \Omega f \circ f^I$ follows from $\Omega\varphi = f^I \circ \Omega f$ applied to φ^{-1}. □

It clearly follows that $|\Omega\varphi|$, the set of nonwandering points for the semiflow, contains $|\Omega f^t|$ for all $t > 0$ with equality for residual t in $(0, \infty)$. On the other hand, because $\omega\varphi = \omega f \circ f^I$ the corresponding limit point sets always agree.

$$l_+[\varphi] \equiv \overline{\omega\varphi(X)} = \overline{\omega f(X)} = l_+[f]. \tag{6.11}$$

When φ is a flow we define $l_-[\varphi] = l_+[\varphi^{-1}]$ and $l[\varphi] = l_+[\varphi] \cup l_-[\varphi]$. Clearly, $l_-[\varphi] = l_-[f]$ and $l[\varphi] = l[f]$ in the flow case. A bit less obviously the concepts of recurrent point agree for φ and f in the flow case.

4. PROPOSITION. *If φ is a flow on X, then $|\omega\varphi| = |\omega f|$ and $|\alpha\varphi| = |\alpha f|$.*

PROOF. Since $\omega f \subset \omega\varphi$, $|\omega f| \subset |\omega\varphi|$. For the converse suppose $x \in \omega\varphi(x)$. Because $\omega\varphi = f^I \circ \omega f$, $f^r(x) \in \omega f(x)$ for some $r \in I$. Define $T = \{t \in \mathbb{R}: f^t(x) \in \omega f(x)\}$. T is a closed subset of $\mathbb{R}$ and f invariance of $\omega f(x)$ implies $t \pm 1 \in T$ whenever $t \in T$. Also, if t, $u \in T$, then $f^{t+u}(x) = f^u(f^t(x)) \in f^u \circ \omega f(x) = \omega f(f^u(x))$. $\omega f(f^u(x)) \subset \omega f(x)$ since $u \in T$ and $\omega f(x)$ is closed and f invariant. So $t + u \in T$. Starting with $r \in I \cap T$ we can choose sequences of whole numbers n_k, m_k such that $\{n_k r - m_k\}$ approaches 0. Because T is closed under addition, translation, and limits, $0 \in T$, i.e., $x \in \omega f(x)$. □

Now define an ε *chain* for a semiflow φ to be a sequence $\{x_0, t_0, x_1, t_1, \dots, t_{n-1} x_n\}$ with n, $t_k \geq 1$ and $d(x_{k+1}, f^{t_k}(x_k)) \leq \varepsilon$ for $k = 0, \dots, n-1$. The chain connects x_0 to x_n beginning at x_0 and ending at x_n. Its length is defined to be $\sum_k t_k$ which is at least n because for each k, $t_k \geq 1$. The latter condition is imposed to keep the jumps from occurring too frequently. Otherwise, any ε_1 chain for the identity map 1_X, with $\varepsilon_1 < \varepsilon$ would yield an ε chain for φ by finding s close to 0 so that $d(f^s(x), x) \leq \varepsilon - \varepsilon_1$ for all x and then setting all t_k's equal to s. So without some frequency restriction the concept of chain recurrence would lose all relationship with the dynamical system.

The chain relation $\mathscr{C}\varphi$ is defined by

$$y \in \mathscr{C}\varphi(x) \text{ if for every } \varepsilon > 0, \text{ there is an } \varepsilon \text{ chain connecting } x \text{ to } y. \tag{6.12}$$

In the ε chain given above we can write each $t_k = n_k + r_k$ with $r_k \in J = [1, 2]$ and n_k a nonnegative integer. If we replace the x_k, t_k, x_{k+1} portion of the chain by x_k, 1, $f(x_k)$, 1, ..., 1, $f^{n_k}(x_k)$, r_k, x_{k+1} we get an ε chain connecting x_0 to x_n of the same length though with more terms. Thus, we can always assume without loss of generality that all t_k's lie in J.

Consequently

$$\mathscr{C}\varphi = \mathscr{C}f^J \qquad (J = [1, 2]). \tag{6.13}$$

5. **Proposition.** *Let φ be a semiflow on X.*

(a) $y \in \Omega\mathscr{C}\varphi(x)$ *if and only if for every $\varepsilon > 0$ and $T > 0$ there exists an ε chain connecting x to y with length at least T.*

(b) $\Omega\mathscr{C}\varphi = \mathscr{C}\varphi \circ \Omega\varphi = \mathscr{C}\varphi \circ \omega\varphi = \mathscr{C}\Omega\varphi$.

(c) $\Omega\mathscr{C}\varphi = \Omega\mathscr{C}f^t$ *for all $t > 0$.*

(d) $\mathscr{C}\varphi = \mathscr{O}(f^J) \cup \Omega\mathscr{C}\varphi (J = [1, 2])$.

(e) *If φ is a flow, then $\mathscr{C}(\varphi^{-1}) = (\mathscr{C}\varphi)^{-1}$ and $\Omega\mathscr{C}(\varphi^{-1}) = (\Omega\mathscr{C}\varphi)^{-1}$.*

(f) $f^t \circ \Omega\mathscr{C}\varphi = \Omega\mathscr{C}\varphi = \Omega\mathscr{C}\varphi \circ f^t$ *for all $t \geq 0$, (and in the flow case for all $t \leq 0$ as well).*

(g) $f^t \circ (\Omega\mathscr{C}\varphi)^{-1} \subset (\Omega\mathscr{C}\varphi)^{-1} \subset (\Omega\mathscr{C}\varphi)^{-1} \circ f^t$ *for all $t \geq 0$ (with equalities in the flow case).*

Proof. For (a), (b), and (d) we use the equations already remarked: $\omega\varphi = \omega f^J$, $\Omega\varphi = \Omega f^J$, and $\mathscr{C}\varphi = \mathscr{C}f^J$, which implies $\Omega\mathscr{C}\varphi = \Omega\mathscr{C}f^J$ as well. For (a) apply Lemma 2.14 and notice that an f^J chain of length n yields a φ chain of length between n and $2n$. (b) follows from Proposition 2.15 applied to f^J. For (d) we have

$$\mathscr{C}\varphi = \mathscr{C}f^J = \mathscr{O}(f^J) \cup \Omega\mathscr{C}f^J = \mathscr{O}(f^J) \cup \Omega\mathscr{C}\varphi.$$

Proposition 2.15 applied to f^t and f^J implies $\Omega\mathscr{C}\varphi = \mathscr{C}\Omega\varphi$ and $\Omega\mathscr{C}f^t = \mathscr{C}\Omega f^t$. So by Proposition 3, $\Omega\mathscr{C}\varphi \supset \Omega\mathscr{C}f^t$ with equality for t in the residual subset T, where $\Omega\varphi = \Omega f^t$. For (c) we derive the reverse inclusion $\Omega\mathscr{C}f^t \supset \Omega\mathscr{C}\varphi$ for any $t > 0$.

For any $\varepsilon > 0$, $\overline{V}_\varepsilon \circ f^t \circ \overline{V}_\varepsilon$ is a neighborhood of the relation f^t and so there exists $\tilde{t} \in T$ such that $f^{\tilde{t}} \subset \overline{V}_\varepsilon \circ f^t \circ \overline{V}_\varepsilon$. Hence,

$$\Omega\mathscr{C}\varphi = \Omega\mathscr{C}f^{\tilde{t}} = \mathscr{C}\Omega f^{\tilde{t}} \subset \mathscr{C}\Omega(\overline{V}_\varepsilon \circ f^t \circ \overline{V}_\varepsilon).$$

By intersecting over all $\varepsilon > 0$, we see that Proposition 2.15 applied to f^t implies $\Omega\mathscr{C}\varphi \subset \mathscr{C}\Omega f^t = \Omega\mathscr{C}f^t$.

For (e), observe that for a flow $(f^J)^{-1} = f^{-J}$ with $-J = [-2, -1]$. So $(\mathscr{C}\varphi)^{-1} = (\mathscr{C}f^J)^{-1} = \mathscr{C}(f^{-J}) = \mathscr{C}(\varphi^{-1})$.

For (f) and (g) apply Lemma 4.2(a), (d) with f^t and $F = \Omega\mathscr{C}f^t = \Omega\mathscr{C}\varphi$ when $t > 0$. In the flow case apply the inverse f^{-t} on the left and right to get (f) for the negative time. Inverting the resulting equations yield (g) as equalities in the flow case.

Remark. $\mathscr{C}\varphi$ is by (d) the union of $\Omega\mathscr{C}\varphi$ and $\mathscr{O}(f^J) = f^I \circ \mathscr{O}(f) = \mathscr{O}(f) \circ f^I$ and $\mathscr{O}\varphi = f^I \cup \mathscr{O}(f^J)$. Thus, the initial segments of the φ orbits are usually not in $\mathscr{C}\varphi$. Also, by (c) $\Omega\mathscr{C}\varphi = \Omega\mathscr{C}f$. But clearly, $\mathscr{C}f = \mathscr{O}(f) \cup \Omega\mathscr{C}f$ is usually a proper subset of $\mathscr{C}\varphi$. For these reasons it is usually more convenient to use $\Omega\mathscr{C}\varphi$ rather than $\mathscr{C}\varphi$. □

We are now ready to compare the semiflow and map versions of the concepts of invariance, attractor, transitivity, etc. We begin with a list of elementary results whose proofs we leave to the reader.

6. Exercise. *Let φ be a semiflow on X.*

(a) *$A \subset X$ is called φ +invariant if $f^t(A) \subset A$ for all $t \geq 0$. A is* φ +invariant *if and only if it is f^I +invariant and this implies f^t +invariant for all t. If A is f + (c) invariant then $f^I(A)$ is φ +invariant.*

(b) *A is φ* invariant *if $f^t(A) = A$ for all $t \geq 0$. If A is φ +invariant then it is φ invariant if it is f^t invariant for any positive t, i.e., $f^t(A) = A$ for some $t > 0$ is sufficient (Hint: $f^t(A)$ is decreasing in t).*

(c) *If A is φ +invariant then $\omega\varphi[A] = \bigcap_{t>0}\{f^t A\} = \bigcap_n \{f^n(A)\} = \omega f[A]$ is the maximum φ invariant subset of A.*

(d) *A decomposition of a φ +invariant subset A is a pair of disjoint, relatively closed, φ +invariant subsets with union A. A is φ* indecomposable *if the trivial decomposition $\{A, \emptyset\}$ is the only one. For a φ +invariant subset the following are equivalent: (1) A is φ indecomposable, (2) A is f indecomposable, (3) A is connected (Hint: (3) $\Rightarrow$ (2) $\Rightarrow$ (1) are obvious. For (1) $\Rightarrow$ (3) note that the components of a φ +invariant set are φ +invariant.) If A is f +invariant then $f^I A$ is indecomposable—and hence connected—if and only if A is f indecomposable.*

(e) *If A is closed and φ +invariant the restriction $\varphi_A \colon A \times [0, \infty) \to A$ defines the restriction of the semiflow to A. If φ is a flow then φ_A is a flow if and only if A is φ invariant.*

(f) *For a closed, φ +invariant subset A, the following equivalent conditions define A is a* chain transitive subset (rel φ)*:*

(1) $\mathscr{C}(\varphi_A) = A \times A$.
(2) $\mathscr{C}(f_A) = A \times A$, *i.e., A is a chain transitive subset for f.*
(3) $|\mathscr{C}(\varphi_A)| = A$ *and A is connected.*

In particular, these imply that A is φ invariant. For example, each $\omega\varphi(x)$ is a chain transitive subset (rel φ). *If φ is a flow then chain transitivity* (rel φ) *and* (rel φ^{-1}) *agree and each $\alpha\varphi(x)$ is chain transitive* (rel φ).

(g) *Assume φ is a flow. For a closed, φ invariant subset A the following conditions are equivalent and define A is a* central subset (rel φ).

(1) $|\Omega(\varphi_A)| = A$.
(2) $|\Omega(f_A)| = A$.
(3) $|\omega f_A| = |\omega\varphi_A|$ *is dense in A.*

In that case $|\omega f_A| \cap |\alpha f_A| = |\omega\varphi_A| \cap |\alpha\varphi_A|$ is a residual subset of A.

(h) *A closed, φ +invariant subset A is called a* topologically transitive subset (rel φ) *if $\Omega(\varphi_A) = A \times A$. If A is a topologically transitive subset with respect to the map f^t for some $t > 0$, then A is topologically transitive* (rel φ). *Conversely, if A is topologically transitive* (rel φ), *then it is topologi-*

cally transitive with respect to f^t *for* t *in a residual subset of* $(0, \infty)$, *but not necessarily for all* t. *If* x *is positive recurrent, i.e.,* $x \in \omega\varphi(x)$, *then* $\omega\varphi(x)$ *is topologically transitive* $(\mathrm{rel}\,\varphi)$.

(i) *A closed,* φ *+invariant subset* A *is called* minimal $(\mathrm{rel}\,\varphi)$ *if it contains no proper, nonempty, closed* φ *+ invariant subset. If* B *is* f *+ invariant and minimal for* f, *then* $f^t(B)$ *is also minimal for* f *for all* $t > 0$ *and the union* $f^I(B) = \bigcup\{f^t(B) \colon t \in [0, 1]\}$ *is minimal* $(\mathrm{rel}\,\varphi)$. *Conversely, if* A *is minimal* $(\mathrm{rel}\,\varphi)$, *then* $A = f^I(B)$ *for* B *any* f*-minimal subset of* A. *In particular,* $m[\varphi] = \bigcup\{A \colon A$ *is minimal* $(\mathrm{rel}\,\varphi)\} = m[f]$. □

We can apply these results to get a simple characterization of the basic sets in the semiflow case.

7. PROPOSITION. *Let* φ *be a semiflow on* X. *The basic sets for* φ *are precisely the connected components of* $|\Omega\mathscr{C}\varphi|$. *Each basic set is a chain transitive subset* $(\mathrm{rel}\,\varphi)$.

PROOF. Because $\Omega\mathscr{C}\varphi = \Omega\mathscr{C}f$ a point x is chain recurrent for φ if and only if it is chain recurrent for f, i.e., $|\Omega\mathscr{C}\varphi| = |\Omega\mathscr{C}f|$. Furthermore, $B = \Omega\mathscr{C}\varphi(x) \cap \Omega\mathscr{C}\varphi^{-1}(x) = \Omega\mathscr{C}f(x) \cap \Omega\mathscr{C}f^{-1}(x)$ is empty unless x is chain recurrent in which case it is the basic set containing x. By Corollary 4.13 such a basic set B is chain transitive with respect to f and because $B = \Omega\mathscr{C}\varphi(x) \cap \Omega\mathscr{C}\varphi^{-1}(x)$, Proposition 5(f) and (g) imply that B is φ + invariant. Now part (f) of the exercise implies that B is chain transitive $(\mathrm{rel}\,\varphi)$ and so B is connected by φ indecomposibility. Thus, B meets a single component of $|\Omega\mathscr{C}\varphi|$.

But for any closed relation f on X the space of basic sets, i.e., the quotient space $|\mathscr{C}f|/\mathscr{C}f \cap \mathscr{C}f^{-1}$, is totally disconnected and so the image under the quotient map from $|\mathscr{C}f|$ of any connected set is a single point, i.e., each basic set is a union of components. □

For the attractor results, we will need the uniform equicontinuity of the time-t maps.

8. LEMMA. *Let* φ *be a semiflow on* X *and* J *be a compact subinterval of* $[0, \infty)$. *For every* $\varepsilon > 0$ *there exists* $\delta > 0$ *such that*

$$f^t \circ V_\delta \subset V_\varepsilon \circ f^t,$$
$$V_\delta \circ (f^t)^{-1} \subset (f^t)^{-1} \circ V_\varepsilon \quad \textit{for all } t \textit{ in } J.$$

Furthermore, if A *is a* φ *+ invariant subset of* X, *then*

$$V_\delta(A) \subset \cap\{(f^t)^{-1}(V_\varepsilon(A)) \colon t \in J\}.$$

PROOF. The existence of δ such that $f^t \circ V_\delta \subset V_\varepsilon \circ f^t$ $(t \in J)$ follows from the uniform continuity of the restriction $\varphi \colon J \times X \to X$. Compose with $(f^t)^{-1}$ on the left and right and use $1_X \subset (f^t)^{-1} \circ f^t$ and $f^t \circ (f^t)^{-1} \subset 1_X$ to get the second inclusion. If A is φ + invariant then for all $t \geq 0$,

$A \subset (f^t)^{-1}(A)$ and so $V_\delta(A) \subset V_\delta \circ (f^t)^{-1}(A) \subset (f^t)^{-1} \circ V_\varepsilon(A)$ for all t in J. □

9. PROPOSITION. *Let φ be a semiflow on X and let A be a closed subset of X.*

(a) *The following conditions are equivalent and define A is a* preattractor *for φ.*

(1) *A is φ +invariant and there exists U a closed neighborhood of A such that $\bigcap_{t\geq 0}\{f^t(U)\} \subset A$.*

(2) *A is φ +invariant and a preattractor for f.*

(3) *$A = f^I B$ for B some preattractor for f $(I = [0, 1])$.*

(4) *A is φ +invariant and $\{x: \Omega\mathscr{C}\varphi(x) \subset A\}$ is a neighborhood of A.*

(b) *A φ invariant preattractor for φ is called an* attractor *for φ. A is an attractor for φ if and only if it is an attractor for f.*

PROOF. (a) (1) $\Rightarrow$ (2). By Lemma 8 $V = \bigcap\{(f^t)^{-1}(U): t \in I\}$ is a closed neighborhood of A. It is easy to check that $\bigcap_{n\geq 0}\{f^n V\} \subset \bigcap_{t\geq 0}\{f^t U \subset A\}$. So A is a preattractor for f by Theorem 3.6(a).

(2) $\Rightarrow$ (3). Let $B = A$.

(3) $\Rightarrow$ (4). Since $\Omega\mathscr{C}\varphi = \Omega\mathscr{C}f$,

$$\{x: \Omega\mathscr{C}\varphi(x) \subset A\} \supset \{x: \Omega\mathscr{C}\varphi(x) \subset B\} = \{x: \Omega\mathscr{C}f(x) \subset B\}$$

and the latter is a neighborhood of B because B is a preattractor for f. Since $\Omega\mathscr{C}\varphi \circ f^t = \Omega\mathscr{C}\varphi$, $\{x: \Omega\mathscr{C}\varphi(x) \subset B\}$ is φ +invariant and so is its interior. Hence this interior contains $f^I(B) = A$. Notice that $f^I(B)$ is φ +invariant by Exercise 6(a).

(4) $\Rightarrow$ (1). Let U be a closed neighborhood of A contained in $\{x : \Omega\mathscr{C}\varphi(x) \subset A\}$. Let $y \in \bigcap_{t\geq 0}\{f^t U\}$. Choose a sequence $\{x_n\}$ in U with $f^n(x_n) = y$. By going to a subsequence we can assume $\{x_n\} \to x$ with x in U because U is closed. Clearly, $y \in \Omega f(x) \subset \Omega\mathscr{C}\varphi(x)$ and this is a subset of A by definition of U. So $y \in A$.

(b) An attractor for φ is clearly an attractor for f. The converse is a matter of showing that an attractor A for f is automatically φ invariant. A is f invariant and so it suffices to prove φ +invariance by Exercise 6(b). Recall equation (3.4), that an attractor is determined by its trace on the chain recurrent set

$$A = \Omega\mathscr{C}f(A \cap |\mathscr{C}f|).$$

But $\Omega\mathscr{C}f = \Omega\mathscr{C}\varphi$. Thus A is the image under $\Omega\mathscr{C}\varphi$ of its trace and so is φ +invariant.

REMARK. A repellor for f is the $(\Omega\mathscr{C}f)^{-1} = (\Omega\mathscr{C}\varphi)^{-1}$ image of its trace and so is φ invariant by Proposition 5(g). □

A is a repellor for a flow if it is an attractor for φ^{-1}. So A is a repellor for

f. $\{A_+, A_-\}$ is an attractor-repellor pair for φ if it is an attractor-repellor pair for f.

We now develop Lyapunov functions for flows.

Recall that $L: X \to \mathbb{R}$ is a Lyapunov function for a homeomorphism f on X if $L(f(x)) \geq L(x)$ for all x. x is called a critical point for L if $L(f(x)) = L(x)$ or $L(f^{-1}(x)) = L(x)$. We denote the set of critical points $|L|_f$.

If φ is a flow on X we call L a *Lyapunov function for the flow* φ if the function $L(\varphi(x, t))$ on $X \times \mathbb{R}$ is differentiable in t and the function

$$\varphi \cdot L(x) \equiv \frac{d}{ds} L(\varphi(x, s))|_{s=0} \tag{6.14}$$

is continuous and is nonnegative. Because $\varphi(x, t+s) = \varphi(\varphi(x, t), s)$ we have

$$\frac{d}{dt} L(\varphi(x, t)) = \varphi \cdot L(\varphi(x, t)), \qquad (x, t) \in X \times R. \tag{6.15}$$

Define the critical point set for L relative to φ by

$$|L|_\varphi = \{x: \varphi \cdot L(x) = 0\}. \tag{6.16}$$

Notice that (6.15) and $\varphi \cdot L \geq 0$ imply that for each x, $L(f^t(x))$ is nondecreasing in t. Hence, for $t > 0$ $L(x) = L(f^t(x))$ implies $L(x) = L(f^s(x))$ for s between 0 and t and so $\varphi \cdot L(x) = 0$. Similarly if $L(x) = L(f^{-t}(x))$ then $x \in |L|_\varphi$. This proves the first part of

10. Lemma. *Let φ be a flow on X.*

(a) *If L is a Lyapunov function for φ, then L is a Lyapunov function for f^t for all $t > 0$. Also,*

$$|L|_{f^t} \subset |L|_\varphi .$$

(b) *If L is a Lyapunov function for f^s (with $s > 0$) define*

$$\overline{L}(x) = \frac{1}{s} \int_0^s L(\varphi(x, u))\, du .$$

$\overline{L}$ is a Lyapunov function for φ with $\varphi \cdot \overline{L}(x) = \frac{1}{s}(L(f^s(x)) - L(x))$. Also, $|\overline{L}|_\varphi \subset |L|_{f^s}$.

Proof. By the change of variables $w = u + t$ we have

$$\overline{L}(\varphi(x, t)) = \frac{1}{s} \int_0^s L(\varphi(x, u+t))\, du = \frac{1}{s} \int_t^{s+t} L(\varphi(x, w))\, dw .$$

Differentiability and the formula for $\varphi \cdot \overline{L}$ follows. $\varphi \cdot \overline{L} \geq 0$ because L is a Lyapunov function for f^s. $\varphi \cdot \overline{L}(x) = 0$ if and only if $L(x) = L(f^s(x))$ which implies $x \in |L|_{f^s}$. □

Now suppose that X is a smooth manifold and φ is the flow associated with a C^1 vector field ξ on X. We call L a *Lyapunov function for the vector field* ξ if $L: X \to \mathbb{R}$ is a C^1 function and for all x

$$\text{either} \quad d_x L(\xi(x)) > 0 \quad \text{or} \quad d_x L = 0, \tag{6.17}$$

where $d_x L$ is the differential of L at x, i.e., the tangent linear map $d_x L$: $T_x X \to \mathbb{R}$.

Because $d_X L(\xi(x))$ (also denoted $\xi \cdot L(x)$) is equal to $\varphi \cdot L(x)$ for the associated flow, (6.17) implies L is a Lyapunov function for the flow. It is stronger in that it assumes that when $\varphi \cdot L(x) = 0$, i.e., $d_x L$ applied to $\xi(x)$ is 0, then the map $d_x L = 0$ when applied to all directions in $T_x X$. When L is a vector field Lyapunov function then

$$|L|_\varphi = \{x : d_x L = 0\}. \tag{6.18}$$

This is the set of critical points of the map in the usual sense.

11. Exercise. *If L is a C^∞ Lyapunov function for a vector field ξ, prove that L is a $\mathscr{C}\varphi$ Lyapunov function, i.e., $y \in \mathscr{C}\varphi(x)$ implies $L(y) \geq L(x)$. (Hint: By Sard's Theorem the set of critical values of L is nowhere dense. Apply Exercise* 3.16.) □

12. Theorem. *Let φ be a flow on ξ. There exists a Lyapunov function L for the flow such that $|L|_\varphi = |\Omega\mathscr{C}\varphi|$ and L takes distinct basic sets to distinct values. Furthermore, L is a $\mathscr{C}\varphi$ Lyapunov function. We call such an L a complete Lyapunov function for the flow.*

If φ is the flow associated with a C^r vector field ξ $(1 \leq r \leq \infty)$ on a smooth manifold X, then L can be chosen to be a C^r Lyapunov function for the vector field ξ. L is then a complete Lyapunov function for the vector field.

Proof. Apply Theorem 3.12 to get a complete $\mathscr{C}f$ Lyapunov function L_0 for f, i.e, $|L_0|_f = |\Omega\mathscr{C}f|$ and L_0 takes distinct values on distinct basic sets. Apply Lemma 10 to define the Lyapunov function $L = \overline{L}_0$ for the flow φ. $|L|_\varphi \subset |L_0|_f = |\Omega\mathscr{C}f| = |\Omega\mathscr{C}\varphi|$. For x in $|\Omega\mathscr{C}\varphi|$ the basic set $\Omega\mathscr{C}\varphi(x) \cap \Omega\mathscr{C}\varphi^{-1}(x) = \mathscr{C}f(x) \cap \mathscr{C}f^{-1}(x)$ is φ invariant. Since L_0 is constant on each basic set the averaging procedure does not change the value, i.e., $L_0 = L$ on each basic set. So $|\Omega\mathscr{C}\varphi| \subset |L|_\varphi$ and distinct basic sets take distinct values under L. If $y \in \Omega\mathscr{C}\varphi(x)$ then $\Omega\mathscr{C}\varphi = \Omega\mathscr{C}f$ and $\Omega\mathscr{C}\varphi = f^{-s} \circ \Omega\mathscr{C}\varphi \circ f^s$ imply $L_0(\varphi(y, u)) \geq L_0(\varphi(x, u))$ for $0 \leq u \leq 1$. Integrating we have $L(y) \geq L(x)$. $\mathscr{C}\varphi \subset \mathscr{O}\varphi \cup \Omega\mathscr{C}\varphi$ implies that L is a $\mathscr{C}\varphi$ Lyapunov function.

In the smooth manifold case go back to the construction for Corollary 3.15. Observe that if $\{L_k\}$ is a sequence of smooth functions and at a point x in X, $d_x L_k = 0$ for all k then $\sum a_k L_k$ convergent in C^r norm for all r (convergent C^1 is all that is needed here) implies that $d_x L = 0$ where L is the limit function. For each attractor repellor pair $\{A_+^n, A_-^n\}$ the associated Lyapunov function L_n was constructed as an infinite sum of functions constant in neighborhoods of A_+^n and A_-^n. Hence, $d_x L_n = 0$ for $x \in A_+^n \cup A_-^n$. So for L_0 defined by (3.8) $d_x L_0 = 0$ for all x in

$|\mathscr{C} f| = |\Omega\mathscr{C}\varphi|$. Furthermore, L_0 is C^∞. In averaging to define $L = \overline{L}_0$ we may lose some smoothness but L is still C^r if ξ, and hence φ is. Also, by the chain rule

$$d_x L = \int_0^1 d_{f^u(x)} L_0 \circ T_x f^u \, du .$$

If $x \in |\Omega\mathscr{C}\varphi|$, then $f^s(x) \in |\Omega\mathscr{C}\varphi|$ and so $d_{f^s(x)} L_0 = 0$. Thus, $d_x L = 0$ for $x \in |\Omega\mathscr{C}\varphi|$. If $x \notin |\Omega\mathscr{C} f|$, then $x \notin |L|_\varphi$ and so $\varphi \cdot L(x) > 0$. □

The discussion of stable and unstable sets as well as of invariant decompositions are simplified for semiflows because invariance results can be deduced from connectivity hypotheses. For example,

13. EXERCISE. *If F and K are closed subsets of X we call F K-separating if $F \cap K$ is open and closed in K. Prove that A is the intersection of K-separating sets if and only if A is closed and $A \cap K$ is a union of components of K (Hint: show that for a closed union of components $\tilde{A}$ the open-and-closed subsets containing $\tilde{A}$ form a base for the neighborhood system of $\tilde{A}$). If K is also invariant with respect to some semiflow φ on X, then A closed and $A \cap K$ a union of components of K imply $A \cap K$ is φ invariant.* □

Now assume φ is a flow on X and A is a closed subset with $A \cap l[f] = A \cap l[\varphi]$ a union of components of $l[\varphi]$. We can define the stable and unstable sets for A by

(6.19)

$$W^+(A) = \{x : \omega\varphi(x) \cap A \neq \varnothing\} = \{x : \omega f(x) \subset A\} = \{x : \omega\varphi(x) \subset A\},$$
$$W^-(A) = \{x : \alpha\varphi(x) \cap A \neq \varnothing\} = \{x : \alpha f(x) \subset A\} = \{x : \alpha\varphi(x) \subset A\}.$$

Because $\omega f(x)$ is a nonempty subset of $\omega\varphi(x)$ the conditions $\omega\varphi(x) \cap A \neq \varnothing$, $\omega f(x) \subset A$, and $\omega\varphi(x) \subset A$ are successively stronger. But if $A \cap l[\varphi]$ is a union of components then it contains any connected set $\omega\varphi(x)$ intersecting it.

An invariant decomposition $\mathscr{F}$, for the flow φ is a finite collection of pairwise disjoint, closed, φ invariant subsets which covers the limit point set $l[\varphi]$. $\mathscr{F}$ is then an invariant decomposition for the homeomorphism f and by (6.19) all of the relations and concepts like $\mathscr{F}$ attractor and $\mathscr{F}$ basic set remain unchanged when the previous definitions involving f are replaced by their analogues involving φ.

We conclude the chapter by returning to the definition of chain recurrence. For the special case where X is a manifold and φ is the solution flow of a C^1 vector field ξ on X there is a nice alternative description of the chain relations which uses the vector field directly. It requires a way of measuring the length of vectors tangent to X. So we will assume X is equipped with a Riemannian metric which provides for each $x \in X$ a norm $\| \ \|_x$ for the vectors attached to x.

A continuous map $\alpha : [0, T] \to X$ is called a *piecewise C^1 path* if for some finite sequence $0 = t_0 < t_1 < \cdots < t_{n+1} = T$ the restriction of α to each $[t_i, t_{i+1}]$ $(i = 0, \ldots, n)$ is C^1. Thus, the pieces of α fit together at the points t_i but the tangent vectors of α from the right and left at such points may disagree. For $\varepsilon \geq 0$ an ε solution for ξ is such a piecewise C^1 path which satisfies

$$\left\| \frac{d\alpha}{dt} - \xi(\alpha(t)) \right\|_{\alpha(t)} \leq \varepsilon, \qquad 0 \leq t \leq T. \tag{6.20}$$

In particular at the point t_i both tangent vectors are within ε of $\xi(\alpha(t_i))$. Clearly, when $\varepsilon = 0$ and ε solution is just a piece of the solution flow, i.e.,

$$\alpha(t) = \varphi(\alpha(0), t) = f^t(\alpha(0)) \qquad (\varepsilon = 0).$$

14. THEOREM. *Let φ be the solution flow of a C^1 vector field ξ on a compact Riemannian manifold X. $y \in \mathscr{C}\varphi(x)$ if and only if for every $\varepsilon > 0$ there exists an ε solution of ξ, $\alpha\colon [0, T] \to X$ such that $T \geq 1$, $\alpha(0) = x$, and $\alpha(T) = y$. $y \in \Omega\mathscr{C}\varphi(x)$ if and only if for every $\varepsilon > 0$ and $N > 0$ there exists an ε solution of ξ, $\alpha\colon [0, T] \to X$ such that $T \geq N$, $\alpha(0) = x$, and $\alpha(T) = y$.*

PROOF. We need some preliminary bits of folklore from differential topology. First, the smooth manifold X can be smoothly embedded in some high dimensional Euclidean space. By compactness any two metrics, d and $\tilde{d}$, on X are uniformly equivalent, i.e., for every $\varepsilon > 0$ there exists $\delta > 0$ such that $d(x_1, x_2) < \delta$ implies $\tilde{d}(x_1, x_2) < \varepsilon$ and $\tilde{d}(x_1, x_2) < \delta$ implies $d(x_1, x_2) < \varepsilon$. In particular, we can replace the original metric, d, by the metric obtained by using the Euclidean distance between the points. Also, any two Riemannian metrics, $\| \; \|$ and $| \; |$, on X are uniformly equivalent, i.e., there exists $C > 0$ such that $C^{-1}|v|_x \leq \|v\|_x \leq C|v|_x$ where v is a tangent vector based at x. So we can replace the original Riemannian metric by the ordinary Euclidean measurement of length $| \; |$. The reader should check that these replacements affect neither the original definition of the chain relation nor the proposed new one.

Now define for $\varepsilon > 0$ and $t > 0$

$$\begin{aligned} U^s_\varepsilon = \{(x, y)\colon &\text{ There exists a piecewise } C^1 \\ &\text{path } \alpha\colon [0, s] \to X \text{ with } \alpha(0) = x,\ \alpha(t) = \\ &y, \text{ and } |d\alpha/dt - \xi(\alpha(t))| < \varepsilon \text{ for } o \leq t \leq s\}. \end{aligned} \tag{6.21}$$

By using $\alpha(t) = \varphi(x, t)$ we see that

$$f^s \subset U^s_\varepsilon. \tag{6.22}$$

If α_1 and α_2 are paths on $[0, s_1]$ and $[0, s_2]$ respectively with $\alpha_1(s_1) = \alpha_2(0)$ we can define $\alpha = \alpha_2 * \alpha_1$, attaching α_2 to α_1 on $[0, s_1 + s_2]$ by

$$\alpha(s) = \begin{cases} \alpha_1(s), & 0 \leq s \leq s_1, \\ \alpha_2(s - s_1), & s_1 \leq s \leq s_1 + s_2. \end{cases}$$

Conversely, if α is a path on $[0, s_1+s_2]$ the paths α_1, α_2 such that $\alpha = \alpha_2 * \alpha_1$ are uniquely defined by $\alpha_1(s) = \alpha(s)$ and $\alpha_2(s) = \alpha(s_1+s)$.

Because these operations preserve the ε solution inequalities we obtain

$$U_\varepsilon^{s_2} \circ U_\varepsilon^{s_1} = U_\varepsilon^{s_1+s_2} \qquad (s_1, s_2, \varepsilon > 0). \tag{6.23}$$

Furthermore, it is easy to check (use local charts) that α can be varied slightly at 0 and s to obtain piecewise C^1 curves which are still ε solutions and which begin and end at arbitrary points close enough to x and y. Thus, U_ε^s is an open subset of $X \times X$.

The key step in the proof is a differential inequality argument which yields

$$\bigcap_{\varepsilon>0} \overline{U_\varepsilon^s} = f^s \tag{6.24}$$

and the strengthening, for J any compact subset of $(0, \infty)$

$$\bigcap_{\varepsilon>0} \overline{U_\varepsilon^J} = f^J, \qquad \text{where } U_\varepsilon^J = \bigcup\{U_\varepsilon^s : s \in J\}. \tag{6.25}$$

Assuming these results for the moment, we will prove that

$$\Omega\mathscr{C}\varphi = \bigcap_{\varepsilon, N>0} \left(\bigcap_{T \geq N} U_\varepsilon^T \right), \tag{6.26}$$

which is the second of our desired results.

U_ε^1 is an open neighborhood of f in $X \times X$ and so given $\varepsilon > 0$ there exists $\delta > 0$ such that $V_\delta \circ f \circ V_\delta \subset U_\varepsilon^1$. Then (6.23) and Proposition 2.15 imply (in the notation of that proposition) that

$$\Omega\mathscr{C} f \subset \mathscr{O}_n(V_\delta \circ f \circ V_\delta) \subset \bigcup_{T \geq n} U_\varepsilon^T.$$

So we have the $\Omega\mathscr{C}\varphi = \Omega\mathscr{C} f \subset \bigcap_{\varepsilon, N}$ half of (6.26).

For the other direction, given $\delta > 0$ apply (6.25) to choose $\varepsilon > 0$ so that with $J = [1, 2]$ $U_\varepsilon^J \subset V_\delta \circ f^J \circ V_\delta$. By (6.23) again we have with N any integer greater than 1,

$$\bigcup_{T \geq N} U_\varepsilon^T = \bigcup_{k \geq N-1} (U_\varepsilon^1)^k \circ U_\varepsilon^J \subset \mathscr{O}_N(V_\delta \circ f^J \circ V_\delta).$$

Applying Proposition 2.15 to f^J and intersecting on N, ε, and δ, we have $\bigcap_{\varepsilon, N} \subset \Omega\mathscr{C} f^J = \Omega\mathscr{C}\varphi$; the other half of (6.26).

The result for $\mathscr{C}\varphi$ follows from the $\Omega\mathscr{C}\varphi$ equivalence because $\mathscr{C}\varphi = f^{[1,\infty)} \cup \Omega\mathscr{C}\varphi$. If the lengths T of the paths connecting x and y remain bounded by N as ε tends to zero then with $J = [1, N]$ we have $y \in f^J(x)$ by (6.25).

We now return to (6.24) and (6.25). In essence, they are versions of the classical result that as ε tends to zero the ε solutions of differential equation approach the actual solution.

Let K and $\widetilde{K} > 0$ be Lipschitz constants for ξ on X and φ on $X \times J$, respectively. With $(x, y) \in U_\varepsilon^s$ $(s \in J)$ let $\alpha: [0, s] \to X$ be an ε solution connecting x to y, and define

$$d(t) = |\alpha(t) - \varphi(x, t)|, \qquad 0 \le t \le s.$$

The difference function $d(t)$ has left and right derivative except possibly where $d(t) = 0$. There we denote by $d'(t)$ the maximum limit point of $d(t+h)/|h|$ as h tends to 0. In any case the triangle inequality implies that

$$|d'(t)| \le \left| \frac{d\alpha}{dt} - \xi(\varphi(x, t)) \right|, \qquad 0 \le t \le s.$$

Now since

$$\frac{d\alpha}{dt} - \xi(\varphi(x, t)) = \xi(\alpha(t)) - \xi(\varphi(x, t)) + \frac{d\alpha}{dt} - \xi(\alpha(t)),$$

$$|d'(t)| \le K d(t) + \varepsilon \quad \text{or} \quad |(e^{-Kt} d(t))'| \le \varepsilon e^{-Kt}.$$

Therefore, $d(0) = 0$ implies

$$d(t) \le \frac{\varepsilon}{K}(e^{Kt} - 1), \qquad 0 \le t \le s.$$

So with $t = s$ and (x_1, y_1) within ε of (x, y)

$$\begin{aligned} |y_1 - f^s(x_1)| &\le |y_1 - y| + |\varphi(x_1, s) - \varphi(x, s)| + d(s) \\ &\le \varepsilon + \widetilde{K}\varepsilon + \varepsilon(e^{Ks} - 1)/K. \end{aligned}$$

With s fixed or varying over J this approaches 0 as ε tends to 0, proving (6.24) and (6.25). □

Supplementary exercises

15. Let φ be a flow on X and let B be a closed, f invariant subset of X. Prove that $T = \{t: f^t(B) = B\}$ is a closed additive subgroup of R containing the integers $\mathbb{Z}$ and so either $T = \mathbb{R}$ or $T = \mathbb{Z}(1/n)$ for some positive integer n. $T = \mathbb{R}$ if and only if B is φ invariant. If $T = \mathbb{Z}(1/n)$ and $I_n = [0, 1/n]$ then $A = f^{I_n}(B)$ is the smallest φ invariant subset of X containing B.

    Recall from Exercise 0.2 the suspension construction for the homeomorphism $f_B^{1/n}: B \to B$ defining $\widetilde{B}$ to be the quotient space of $B \times I$ with $(x, 1)$ identified with $(f^{1/n}(x), 0)$. When $T = \mathbb{Z}(1/n)$, show that $h: B \times I \to X$ given by $h(x, t) = \varphi(x, t/n)$ induces a continuous map $\tilde{h}$ of $\widetilde{B}$ onto A. Describe how the suspension flow on $\widetilde{B}$ is related to the restriction φ_A.

    If A is a minimal subset $(\text{rel}\,\varphi)$ and B is a subset of A with B minimal for f, prove that, with $T = \{t: f^t(B) = B\}$, either $T = \mathbb{R}$ and A is minimal for f, or $T = \mathbb{Z}(1/n)$, $A = f^{I_n}(B)$, and $\tilde{h}: \widetilde{B} \to A$ is a homeomorphism. In the latter case define a continuous map z of

A onto the unit circle S of complex numbers such that $z(\varphi(x, t)) = \exp(2\pi\, \mathrm{int}) z(x)$.

Apply the previous paragraph to f^s instead of f to prove that if A is a minimal (rel φ), then either A is minimal for all f^s with $s \neq 0$ (A = a fixed point for the flow is an example of this), or A fibers over the circle as above. In the latter case prove that A is minimal for f^s with s in a residual subset of $\mathbb{R}$.

16. Review the Mobius strip example of exercise 4.22. Let φ be the flow on $\widetilde{X}$. Now show that a positive real-valued function $\tilde{i}$ is defined on $(0, 1]$ by

$$\varphi((1/2, -y), \tilde{i}(y)) = (1/2, y),$$

i.e., $\tilde{i}(y)$ is the time required for the point $(1/2, -y)$ on $1/2\times[-1, 0)$ to flow back to $(1/2, y)$ on $1/2\times(0, 1]$. Prove that $\tilde{i}$ is continuous and as $y \to 0$, $\tilde{i}(y) \to \infty$. Hence, for some positive integer N, there exists for each $n \geq N$, a point $0 < y_n < 1$ with $\tilde{i}(y_n) = n + 1/2$, and the sequence $\{y_n\}$ approaches 0. Let $A = \bigcup\{[0, 1] \times \{y_n, -y_n\} : n \geq N\} \cup [0, 1] \times 0$. Show that in the quotient space $\widetilde{X}$, A projects to a closed φ invariant subset $\widetilde{A}$. For the restriction φ_A prove that $|\Omega(\varphi_A)|$ consists of the fixed points $\{(0, -y_n) = (1, y_n) : n \geq N\}$ and the central circle $[0, 1] \times 0$. Prove that $(1/2, 0)$ is not in $|\Omega(f_A)|$. Thus, the analogue of Proposition 4 for nonwandering points is not true.

17. Let φ be a flow on X. Assume F_1, F_2 are closed f invariant subsets of X with $F_i \cap l[f]$ a union of components in $l[f]$ $(i = 1, 2)$. Prove that $f^I(F_1) \cap f^I(F_2) \neq \emptyset$ implies $F_1 \cap F_2 \cap l[f] \neq \emptyset$. (Choose x such that $f^{r_i}(x) \in F_i$ with $0 \leq r_i \leq 1$ $(i = 1, 2)$. $\omega f(f^{r_i}(x)) \subset F_i \cap l[f]$ $(i = 1, 2)$. Let C be the component of $l[f] = l[\varphi]$ containing $\omega\varphi(x)$. Show that $C \subset F_1 \cap F_2 \cap l[f]$.) Apply this result to prove that if $\widetilde{\mathscr{F}} = \{\widetilde{F}_1, \dots, \widetilde{F}_n\}$ is an invariant decomposition for f then $\mathscr{F} = \{f^I(\widetilde{F}_1), \dots, f^I(\widetilde{F}_n)\}$ is an invariant decomposition for φ. Regarding both $\widetilde{\mathscr{F}}$ and $\mathscr{F}$ as invariant decompositions for f show that $\widetilde{\mathscr{F}}$ is a trivial refinement of $\mathscr{F}$. In fact, $\widetilde{\mathscr{F}} \wedge \{l[f]\} = \mathscr{F} \wedge \{l[f]\}$.

18. Assume φ is a flow on X. A closed subset U of X is called *inward* with respect to φ if $f^t(U) \subset \mathrm{Int}(U)$ for all $t > 0$. Prove that if L is a Lyapunov function for φ and a is a regular value of L (i.e., $a \notin L(|L|_\varphi)$) then $U = L^{-1}([a, \infty))$ is inward with respect to φ. Observe that if, in addition, φ is the solution flow of a C^r vector field ξ on a smooth manifold $(1 \leq r \leq \infty)$ of dimension n and L is a C^r Lyapunov function for the vector field then $U = L^{-1}([a, \infty))$ is a C^r submanifold of X of dimension n with boundary $L^{-1}(a)$, and the vector field ξ points inward at every point of the boundary.

If A is an attractor, prove there exists a set U inward with respect to

φ such that

$$\bigcap_{t=-\infty}^{\infty} \{f^t(U)\} = A, \qquad \bigcup_{t=-\infty}^{+\infty} \{f^t(U)\} = W^+(A).$$

Furthermore, if $\varepsilon > 0$, U can be chosen with $U \subset V_\varepsilon(A)$. (Apply Lemma 10 to the Lyapunov function of Proposition (3.10).) If $\mathscr{F} = \{F_1, \dots, F_n\}$ is an invariant decomposition for φ satisfying the no-cycle condition $\mathscr{F}_b = \mathscr{F}$ and ordered so that $F_i \ \mathscr{F}f \ F_j$ implies $i \geq j$, show that an associated filtration $\{X = U_1, \dots, U_n, U_{n+1} = \varnothing\}$ can be chosen to consist of sets inward with respect to φ (c.f. Exercise 5.23).

19. (James Murdock). Let φ be a flow on X and X_0 be a closed subset of X with $\mathscr{F} = \{e_1, \dots, e_n\}$ a family of fixed points for the flow in the interior with respect to X of X_0. Assume

(1) For every $x \in X_0$ either $\varphi(x, t) \notin X_0$ for some $t > 0$ or the limit $\lim_{t\to\infty} \varphi(x, t)$ is a point of $\mathscr{F}$, i.e., $\Omega\varphi(x)$ is one of the fixed points $e_1, \dots, e_n$, and similarly either $\varphi(x, t) \notin X_0$ for some $t < 0$ or $\lim_{t\to-\infty} \varphi(x, t)$ is a point of $\mathscr{F}$.

(2) If we define the relation $e_i > e_j$ to mean there exists $x \in X_0 - \{e_1, \dots, e_n\}$ such that $\varphi(x, t) \in X_0$ for all t and $\alpha\varphi(x) = e_j$ $\omega\varphi(x) = e_i$ then the transitive extension $\mathscr{O}(>)$ is irreflexive, i.e., there is no sequence $\{e_{i_1}, \dots, e_{i_k}\}$ with $e_{i_{j+1}} > e_{i_j}$ and with $e_{i_1} = e_{i_k}$.

Under these assumptions there exists a real-valued continuous function L on X_0 such that, for every $x \in X_0 - \{e_1, \dots, e_n\}$, $\varphi(x, t) \in X_0$ for $0 < t \leq s$ implies $L(\varphi(x, s)) > L(x)$.

We outline the proof in a sequence of steps.

Step 1. There exists X_1 with $X_0 \subset \operatorname{Int}_X X_1$ such that (1) remains true with X_0 replaced by X_1. (If $\widetilde{X}_n = \overline{V}_{1/n}(X_0)$ does not work there exists x_n with $\varphi(x_n, t) \in \widetilde{X}_n$ for all $t \geq 0$ (or $t \leq 0$) and $A_n \cap \widetilde{X}_n - X_0 \neq \varnothing$ with $A_n = \omega\varphi(x_n)$ (resp. $= \alpha\varphi(x_n)$). Let $A = \limsup\{A_n\}$. $A \subset X_0$ with $A - \operatorname{Int} X_0 \neq \varnothing$. By Exercise 4.25 the component B of A which meets $X_0 - \operatorname{Int}_X X_0$ is chain transitive with respect to φ. By (1) $\mathscr{F} \cap B$ is an invariant decomposition for B and then (2) contradicts chain transitivity of B.)

Step 2. There exists $\varepsilon > 0$ such that $f^t(X_0) \subset \operatorname{Int}_X X_1$ for all t with $|t| < \varepsilon$. Define the partial homeomorphism $g = f^\varepsilon \cap (X_1 \times X_1)$ on X_1.

Step 3. Apply exercise 5.25 to the partial homeomorphism g on X_i and show that $\mathscr{F}$ is an invariant decomposition for g satisfying the no-cycle condition. In particular, $|\mathscr{C} g| = \{e_1, \dots, e_n\}$, i.e., $\mathscr{F}$ is a fine decomposition for g.

Step 4. Let $\widetilde{L}\colon X_1 \to \mathbb{R}$ be a Lyapunov function for g with $|\widetilde{L}|_g = \{e_1, \dots, e_n\}$.

Step 5. Following Lemma 10 define $L: X_0 \to R$ by

$$L(x) = (1/\varepsilon) \int_0^\varepsilon \tilde{L}(\varphi(x, s))\, ds, \qquad x \in X_0,$$

observing that $\varphi(x, s) \in X_1$ for $0 \le s \le \varepsilon$ when $x \in X_0$. Use the proof of Lemma 10 to show that L is the required function.

In applying this kind of technique it is important to observe that we make no assumptions about returns to X_0 after an exit. Thus, by restricting to X_0 we may be simplifying the recurrence structure considerably. For example, Figure 6.1 with $\mathscr{F} = \{e\}$ satisfies the hypotheses (1) and (2).

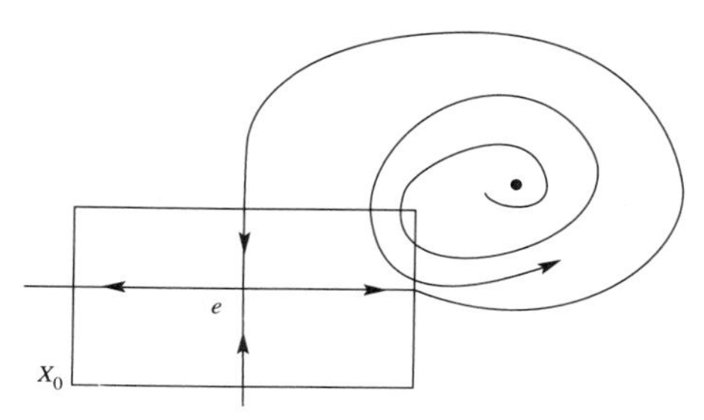

FIGURE 6.1

Chapter 7. Topologically Robust Properties of Dynamical Systems

In the preceding six chapters we developed the concepts and structures associated with a closed relation regarded as a dynamical system. We then considered the additional consequences when the relation is a map or a homeomorphism. We turn now to comparison between relations. What dynamic properties of a closed relation f on X are preserved under perturbation, i.e., the replacement of f by another relation f_1 near to f in some sense?

A homeomorphism $h: X_1 \to X_2$ is called a *conjugacy* or *topological equivalence* between relations f_1 on X_1 and f_2 on X_2 if $h \circ f_1 = f_2 \circ h$ or equivalently $h \circ f_1 \circ h^{-1} = f_2$ (cf. Definition 1.15). By Proposition 1.17 all the dynamic properties and structures we have so far considered are preserved by a conjugacy homeomorphism. A homeomorphism f on X is called *structurally stable* if every nearby homeomorphism is conjugate to f. This means that the perturbed homeomorphism is essentially the same as the original one, differing only by a change of coordinates given by the conjugacy homeomorphism, h. Of course, this concept depends upon what perturbations we allow, i.e., upon the topology used to interpret the work "nearby". For diffeomorphisms of a smooth manifold, e.g., C^r with $r \geq 2$, there is a rich class of examples structurally stable with respect to C^r perturbations. In fact, it is just such structural stability results which motivate our later concerns with hyperbolicity. However, in our current, purely topological context the concept is vacuous with essentially no examples at all.

Instead we will concern ourselves here with the weaker idea of continuity or robustness for some dynamic structure. For example, if A is a subset of X satisfying some property with respect to $f: X \to X$ we ask if for any $f_1: X \to X$, topologically close to f, there is a subset A_1 close to A with the same property. For example, suppose that U is inward for f. Then for some $\delta > 0$ we know that $\overline{V}_\delta \circ f \circ \overline{V}_\delta(U) \subset U$. Hence, if $f_1 \subset V_\delta \circ f \circ V_\delta$ U is inward for f_1. Now assume A is an attractor for f. Given $\varepsilon > 0$ there is an inward neighborhood U of A contained in $V_\varepsilon(A)$. Choose $\delta > 0$ as above, then $f_1 \subset V_\delta \circ f \circ V_\delta$ implies U is inward for f_1 and so the associated attractor for f_1, $A_1 = \omega f_1[U] = \bigcap_{n \geq 0}\{f_1^n(U)\}$ is contained in $V_\varepsilon(A)$. Thus, every point of A_1 is within ε of some point in A. However, the reverse may not be true. The attractor A map "implode" with A_1 much smaller than A. We will want to see whether such implosion possibilities are the rule or the exception.

1. EXERCISE. *Let f be a homeomorphism of* $[0, 1]$ *for a positive interval example with* $K = 0 \cup [1/2, 1]$ *so that* $A = [1/2, 1]$ *is an attractor for* f. *Show that for every* $\delta > 0$ *there exists* $f_1 \subset V_\delta \circ f \circ V_\delta$ *the homeomorphism of a positive interval example with* $K_1 = \{0, 1\}$ *and so with corresponding attractor* $A_1 = 1$. $\square$

In order to deal with these issues we must recall our initial concept of a relation f from X_1 to X_2. In Chapter 1 we rejected the view of f as a mapping from X_1 to the set of subsets of X_2 regarding f, instead, as a subset of $X_1 \times X_2$. We return to this neglected alternative, considering first the structure of the space of closed subsets of a compact metric space X, i.e.,

$$C(X) = \{A\colon A \subset X \text{ and } A \text{ is closed}\}. \tag{7.1}$$

The subset relation between elements of $C(X)$ defines

$$C = \{(A, B)\colon B \subset A\}. \tag{7.2}$$

The reversal of order in the definition is so that the set $C(A) = \{B\colon (A, B) \in C\}$ corresponds to $C(A)$ defined by (7.1) with X replaced by A. C is, of course, a partial order; a reflexive, antisymmetric, transitive relation

$$C \cap C^{-1} = 1_{C(X)} \quad \text{and} \quad C \circ C = C. \tag{7.3}$$

With respect to this partial order $C(X)$ is a complete lattice. If $\{A_\alpha\}$ is a family of elements of $C(X)$ we define

$$\bigwedge\{A_\alpha\} = \bigcap\{A_\alpha\} \quad \text{and} \quad \bigvee\{A_\alpha\} = \overline{\bigcup\{A_\alpha\}}. \tag{7.4}$$

$\bigwedge\{A_\alpha\}$ is the largest element of $C(X)$ contained in all the A_α while $\bigvee\{A_\alpha\}$ is the smallest element containing all of them. In particular, for a finite family $\bigvee$ is just the union. $\varnothing$ and X are, respectively, the smallest and largest elements of $C(X)$.

Using the metric d on X we define on $C(X)$:

$$\begin{aligned} \rho(A/B) &= \inf\{\varepsilon \geq 0\colon \overline{V}_\varepsilon(A) \supset B\}, \\ \delta(A, B) &= \max(\rho(A/B), (\rho(B/A)) \\ &= \inf\{\varepsilon \geq 0\colon \overline{V}_\varepsilon(A) \supset B \text{ and } \overline{V}_\varepsilon(B) \supset A\}. \end{aligned} \tag{7.5}$$

Clearly, ρ and δ are extended real valued functions ($\rho(A/B) = \infty$ when $A = \varnothing$ and $B \neq \varnothing$ and so $\delta(A, B) = \infty$ when exactly one of A and B is empty). Observe that for a single point x, $\rho(A/x) = d(x, A) = \inf\{d(x, y) : y \in A\}$ while $\rho(x/B) = \sup\{d(x, y) : y \in B\}$.

2. LEMMA. (a) $\rho \geq 0$ *and* $\rho(A/B) = 0$ *precisely when* $(A, B) \in C$, *i.e.,* $B \subset A$.

(b) $\rho(A/B) + \rho(B/C) \geq \rho(A/C)$.

(c) δ *is a metric on* $C(X)$ (*the Hausdorff metric*).

(d) $\rho\colon C(X) \times C(X) \to [0, \infty]$ *is monotone in each variable, decreasing in the first and increasing in the second.* ρ *is continuous when* $C(X)$ *is given the topology induced by* δ.

PROOF. (a) is obvious, while (b) follows from the triangle inequality for d. Together they easily imply (c). The monotonicity results are also clear. Finally, (b) implies $\rho(A_2/B_2) \geq \rho(A_1/B_2) - \rho(A_1/A_2)$ and $\rho(B_2/A_2) \geq \rho(A_1/A_2) - \rho(A_1/B_2)$ and so $\delta(A_2, B_2) \geq |\rho(A_1/A_2) - \rho(A_1/B_2)|$. Similarly, $\delta(A_1, B_1) \geq |\rho(A_1/B_2) - \rho(B_1/B_2)|$. Adding, we get

$$\delta(A_1, B_1) + \delta(A_2, B_2) \geq |\rho(A_1/A_2) - \rho(B_1/B_2)|, \tag{7.6}$$

which implies continuity of the function ρ. Notice that we adopt the conventions $|\infty - a| = |a - \infty| = \infty$ for finite a and $|\infty - \infty| = 0$. □

Associated with ρ and δ are the neighborhoods of the diagonal $1_{C(X)}$ in $C(X) \times C(X)$:

$$\begin{aligned} U_\varepsilon^+ &= \{(A, B)\colon \rho(A/B) < \varepsilon\}, & U_\varepsilon^- &= (U_\varepsilon^+)^{-1}, \\ \overline{U}_\varepsilon^+ &= \{(A, B)\colon \rho(A/B) \leq \varepsilon\}, & \overline{U}_\varepsilon^- &= (\overline{U}_\varepsilon^+)^{-1}, \\ U_\varepsilon &= \{(A, B)\colon \delta(A, B) < \varepsilon\} = U_\varepsilon^+ \cap U_\varepsilon^-, \\ \overline{U}_\varepsilon &= \{(A, B)\colon \delta(A, B) \leq \varepsilon\} = \overline{U}_\varepsilon^+ \cap \overline{U}_\varepsilon^-. \end{aligned} \tag{7.7}$$

We will regard $C(X)$ as topologized by the metric δ. As usual, the open balls $\{U_\varepsilon(A)\colon A \in C(X) \text{ and } \varepsilon > 0\}$ form a basis for the topology of $C(X)$. If we use (b) of the lemma in place of the triangle inequality for δ, we similarly see that $\{U_\varepsilon^+(A)\colon A \in C(X) \text{ and } \varepsilon > 0\}$ is the basis for the alternate topology on $C(X)$ and we write $C_+(X)$ for $C(X)$ equipped with this topology. Similarly, we write $C_-(X)$ for $C(X)$ equipped with the topology with basis $\{U_\varepsilon^-(A)\colon A \in C(X) \text{ and } \varepsilon > 0\}$. Notice that continuity of ρ with respect to δ (part (d)) implies that each of these topologies is coarser than the metric topology. Because $U_\varepsilon(A) = U_\varepsilon^+(A) \cap U_\varepsilon^-(A)$, the metric topology is the coarsest topology containing the other two.

3. EXERCISE. *Regarding* A *as a subset of* X *and as a point of* $C(X)$ *requires a bit of care in applying notation. Observe that* $\overline{U}_\varepsilon^+(A) = C(\overline{V}_\varepsilon(A))$ *but that* $C \circ \overline{V}_\varepsilon(A)$ *is nonsense. Also while* $\overline{U}_\varepsilon^- = (\overline{U}_\varepsilon^+)^{-1}$, *it is not true that* $\overline{U}_\varepsilon^-(A) = C^{-1}(\overline{V}_\varepsilon(A))$. □

In considering convergence for these topologies recall that for a sequence $\{A_n\}$ in $C(X)$ we have defined (Exercise 1.5)

$$\limsup\{A_n\} = \bigcap_n \overline{\bigcup_{k \geq n} \{A_k\}} = \bigwedge_n \bigvee_{k \geq n} \{A_k\}. \tag{7.8}$$

Now we define

$$\text{(7.9)} \qquad \liminf\{A_n\} = \bigcap_{\varepsilon>0} \bigcup_n \bigcap_{k\geq n} \{V_\varepsilon(A_k)\} = \bigwedge_{\varepsilon>0} \bigvee_n \bigwedge_{k\geq n} \{\overline{V}_\varepsilon(A_k)\},$$

where the equality between the two definitions follows from the easy inclusion, when $0 < \varepsilon_1 < \varepsilon$

$$\overline{\bigcup_n \bigcap_{k\geq n} \{\overline{V}_{\varepsilon_1}(A_k)\}} \subset \bigcup_n \bigcap_{k\geq n} \{V_\varepsilon(A_k)\}.$$

4. Exercise. *Recall that* $x \in \limsup\{A_n\}$ *if and only if there are sequences* $\{n_i\}$, $\{x_i\}$ *with* $\{n_i\} \to \infty$, $x_i \in A_{n_i}$ *and* $\{x_i\} \to x$. *Prove that* $x \in \liminf\{A_n\}$ *if and only if there exists* N *and a sequence* $\{x_n : n \geq N\}$ *with* $x_n \in A_n$ *and* $\{x_n\} \to x$. (*Hint: For* $x \in \liminf$ *define* N_i *to be the smallest integer such that* $x \in \bigcap_{k\geq N_i}\{V_{1/i}(A_k)\}$ *and choose* $x_n \in A_n$ *with* $d(x, x_n) < 1/i$ *for* $N_i \leq n < N_{i+1}$.) *In particular,* $\liminf\{A_n\} \subset \limsup\{A_n\}$. □

5. Lemma. *Let* $\{A_n\}$ *be a sequence in* $C(X)$.

(a) $\{A_n\}$ *converges to* A *in* $C_+(X)$ *if and only if* $\lim\{\rho(A/A_n)\} = 0$ *and if and only if* $\limsup\{A_n\} \subset A$.

(b) $\{A_n\}$ *converges to* A *in* $C_-(X)$ *if and only if* $\lim\{\rho(A_n/A)\} = 0$ *and if and only if* $\liminf\{A_n\} \supset A$.

(c) $\{A_n\}$ *is a Cauchy sequence with respect to* δ *if and only if* $\limsup\{A_n\} = \liminf\{A_n\}$ *in which case this common value is the limit of* $\{A_n\}$ *in* $C(X)$. *In particular,* $\{A_n\}$ *converges to* A *in* $C(X)$ *if and only if* $\{A_n\}$ *converges to* A *in both* $C_+(X)$ *and* $C_-(X)$.

Proof. (a) $\{U_\varepsilon^+(A) : \varepsilon > 0\}$ is a base for the neighborhood system of A in $C_+(X)$. Hence, $\{A_n\}$ converges to A in $C_+(X)$ if and only if $\{\rho(A/A_n)\} \to 0$. This, in turn, is equivalent to saying that the sets A_n eventually enter every neighborhood of A. $\limsup\{A_n\}$ is the smallest set with this property and so $\{\rho(A/A_n)\} \to 0$ if and only if $A \supset \limsup\{A_n\}$.

(b) Because $\{U_\varepsilon^-(A) : A > 0\}$ is a neighborhood base for A in $C_-(X)$, $\{A_n\}$ converges to A in $C_-(X)$ if and only if $\{\rho(A_n/A)\} \to 0$. The definition of the latter statement is

$$A \subset \bigcap_{\varepsilon>0} \bigcup_n \bigcap_{k\geq n} \{V_\varepsilon(A_k)\} = \liminf\{A_n\}.$$

(c) $\{A_n\}$ converges to A in $C(X)$ if and only if $\{\delta(A, A_n) = \max(\rho(A/A_n), \rho(A_n/A))\} \to 0$ and so if and only if $\{A_n\}$ converges to A in both $C_+(X)$ and $C_-(X)$. By (a) and (b) this is equivalent to $A \subset \liminf$ and $A \supset \limsup$. But since $\liminf \subset \limsup$ (use Exercise 4 or $\rho(\limsup / \liminf) \leq \rho(\limsup / A_n) + \rho(A_n / \liminf)$) it follows that $\liminf = \limsup = A$. Conversely, if $\{A_n\}$ is Cauchy with respect to δ and $\varepsilon > 0$ then there exists $N(\varepsilon)$ such that $m, n \geq N(\varepsilon)$ implies $\delta(A_n, A_m) < \varepsilon$. Thus, $A_m \subset \overline{V}_\varepsilon(A_n)$ and so $\rho(A_n/\limsup) \le \varepsilon$ for $n \ge N(\varepsilon)$. It follows that $\{\rho(A_n/\limsup)\} \to 0$ and so by (b), $\limsup \subset \liminf$ when the sequence is Cauchy. □

Notice that while the $\limsup$ is a pure lattice concept, the $\liminf$ defined by (7.9) is usually a larger set than the lattice version of $\liminf$, $\bigvee_n\{\bigwedge_{k\ge n}\{A_n\}\}$. Of course, when $\bigvee_n \bigwedge_{k\ge n}\{A_k\} = \bigwedge_n \bigvee_{k\ge n}\{A_k\}$ then this common value is the $C(X)$ limit of $\{A_n\}$. In particular, we have

6. COROLLARY. *If $\{A_n\}$ is an increasing (or decreasing) sequence in $C(X)$ then $\{A_n\}$ converges to $\bigvee\{A_n\}$ (respectively, to $\bigwedge\{A_n\}$) in $C(X)$.*

This lack of duality is also reflected in the special role of $\varnothing$ which is always an isolated point of $C(X)$, while X is usually not isolated. In particular, $\limsup\{A_n\} = \varnothing$ only when $A_n = \varnothing$ for n sufficiently large.

We now prove that $C(X)$ is compact. Observe that part (c) of Lemma 5 says that $C(X)$ is complete. A metric space is compact when it is complete and, in addition, it is totally bounded, that is, for every $\varepsilon > 0$ there is an ε *net*, a finite subset which is ε-dense, i.e., whose ε neighborhood is the entire space. In particular, for X there exists a finite set E such that for all $x \in X$ $d(x, y) < \varepsilon$ for some y in E. Now define the functions

$$R_\varepsilon^+, R_\varepsilon^- : C(X) \to C(E) \subset C(X) \tag{7.10}$$
$$R_\varepsilon^-(A) = V_\varepsilon(A) \cap E, \qquad R_2^+(A) = \overline{V}_\varepsilon(A) \cap E.$$

By definition, $\varepsilon \ge \rho(A/R_\varepsilon^+(A))$ and $\varepsilon > \rho(A/R_\varepsilon^-(A))$. On the other hand, because E is an ε-net for X, every point of A is within ε of some point of E and the latter lies in $R_\varepsilon^-(A)$ and a fortiori in $R_\varepsilon^+(A)$, i.e., $\varepsilon > \rho(R_\varepsilon^-(A)/A) \ge \rho(R_\varepsilon^+(A)/A)$. Thus,

$$\varepsilon \ge \delta(A, R_\varepsilon^+(A)) \quad \text{and} \quad \varepsilon > \delta(A, R_\varepsilon^-(A)). \tag{7.11}$$

This implies that the finite subset $C(E)$ is an ε-net for $C(X)$. Thus, $C(X)$ is totally bounded, proving

7. COROLLARY. *$C(X)$ is a compact metric space.*

8. EXERCISE. *Prove that $C_+(X)$ and $C_-(X)$ are compact spaces (the "identities" $C(X) \to C_\pm(X)$ are continuous). Prove that each of the topologies of $C_+(X)$ and $C_-(X)$ as well as $C(X)$ admit a countable base. (Let D be a countable dense subset of X and consider $\{U_\varepsilon^\pm(A)\}$ for ε varying over the positive rationals and A over the finite subsets of D.) This means that each of these topologies, while not Hausdorff, can be characterized by its convergent sequences.* □

If L is a subset of $C(X)$, i.e., L is a family of closed subsets of X, we denote the closure of L in $C(X)$, the closure with respect to δ, by $\overline{L}$. We denote by $Cl_+(L)$ and $Cl_-(L)$ the closure of L in $C_+(X)$ and $C_-(X)$, respectively. The latter two are related to $\overline{L}$ in a simple way.

9. **Lemma.** *For* $L \subset C(X)$,

$$Cl_+(L) = \{B : B \supset A \textit{ for some } A \in \overline{L}\} = C^{-1}(\overline{L}).$$
$$Cl_-(L) = \{B : B \subset A \textit{ for some } A \in \overline{L}\} = C(\overline{L}).$$

In particular, L *is* $C_+(X)$ *closed if and only if it is* $C(X)$ *closed and* C^{-1} *invariant, while* L *is* $C_-(X)$ *closed if and only if it is* $C(X)$ *closed and* C *invariant.* L *is* $C_+(X)$ *open if and only if it is* $C(X)$ *open and* C *invariant, while* L *is* $C_-(X)$ *open if and only if it is* $C(X)$ *open and* C^{-1} *invariant.*

Proof. $B \in Cl_+(L)$ if and only if there is some sequence $\{A_n\}$ in L such that $\{\rho(B/A_n)\} \to 0$. By compactness of $C(X)$ we can assume that $\{A_n\}$ converges to A in $C(X)$. Thus, $A \in \overline{L}$. $A = \limsup\{A_n\}$ by Lemma 5(c). By Lemma 5(a), $B \supset \limsup\{A_n\}$ and so $B \in C^{-1}(\overline{L})$. Reversing the argument we see that any B in $C^{-1}(\overline{L})$ is the $C_+(X)$ limit of some sequence in L. In particular, L is $C_+(X)$ closed precisely when $L = C^{-1}(\overline{L})$ which means $L = \overline{L}$ and $L = C^{-1}(L)$. The open set result follows because L is C invariant if and only if its complement is C^{-1} invariant.

The proof for $C_-(X)$ is similar. □

10. **Exercise.** (a) *For* $A \in C(X)$, $\{B : B \cap A = \emptyset\}$ *is* $C_+(X)$ *open. Its complement* $\{B : B \cap A \neq \emptyset\}$ *is* $C_+(X)$ *closed.*

(b) *For* $A \in C(X)$, $\{B : B \subset A\} = C(A)$ *is* $C_-(X)$ *closed and* $\{B : A \subset B\}$ *is* $C_+(X)$ *closed. For* U *open in* X, $\{B : B \subset U\}$ *is* $C_+(X)$ *open.*

(c) *For every* $\varepsilon \geq 0$, U_ε^+ *is open in the product* $C_-(X) \times C_+(X)$ *and* $\overline{U}_\varepsilon^+$ *is closed in the product* $C_+(X) \times C_-(X)$. *In particular,* C *is a closed subset of* $C_+(X) \times C_-(X)$. □

Now for the promised reinterpretation of the relation concept. It will be concept. It will be convenient to consider relations f from D to X where X is, as usual, a compact metric space but where D, still a metric space, need not be compact. We call a relation f from D to X *pointwise closed* if $f(x)$ is a closed subset X for every x in D. If f is a closed relation, i.e., a closed subset of $D \times X$, then f is pointwise closed but the converse need not hold. Recall, for example, that if f is a closed relation on X then the associated relations ωf and $\mathscr{R} f = \mathscr{O} f \cup \omega f$ are pointwise closed but usually not closed.

A pointwise closed relation $f : D \to X$ can be regarded as a function from D to $C(X)$ which we also denote by f. Conversely, every function $f : D \to C(X)$ corresponds to the pointwise closed relation $f = \{(x, y) : x \in D$ and $y \in f(x)\}$.

11. **Proposition.** *Let* f *be a pointwise closed relation from* D *to* X *with* D *metric and* X *compact metric.*

(a) *The following conditions are equivalent and define* f *is* upper semicontinuous at x (*written* f *is* usc *at* x).

(1) *Regarded as a function* $f: D \to C_+(X)$ f *is continuous at* x.
(2) *If* $\{x_n\}$ *is a sequence in* D *converging to* x, $\limsup\{f(x_n)\} \subset f(x)$.
(3) *For every* $\varepsilon > 0$ *there exists* $\delta > 0$ *such that* $d(x, x_1) < \delta$ *implies* $f(x_1) \subset V_\varepsilon(f(x))$.
(4) *If* O *is an open subset of* X *and* $f(x) \subset O$ *then* $\{x_1: f(x_1) \subset O\}$ *is a neighborhood of* x.

f *is called upper semicontinuous* (usc) *if it is upper semicontinuous at every point* $x \in D$. *So* f *is* usc *when the following equivalent conditions hold*:

(5) $f: D \to C_+(X)$ *is a continuous map.*
(6) O *an open subset of* X *implies* $\{x: f(x) \subset O\}$ *is open in* D.
(7) A *a closed subset of* X *implies* $f^{-1}(A) = \{x: f(x) \cap A \neq \varnothing\}$ *is closed in* D.
(8) *As a subset of* $D \times X$, f *is closed.*

(b) *The following conditions are equivalent and define* f *is lower semicontinuous at* x (*written* f *is* lsc *at* x).

(1) *Regarded as a function* $f: D \to C_-(X)$ f *is continuous at* x.
(2) *If* $\{x_n\}$ *is a sequence in* D *converging to* x, $f(x) \subset \liminf\{f(x_n)\}$.
(3) *For every* $\varepsilon > 0$ *there exists* $\delta > 0$ *such that* $d(x, x_1) < \delta$ *implies* $f(x) \subset V_\varepsilon(f(x_1))$.
(4) *If* O *is an open subset of* X *and* $f(x) \cap O \neq \varnothing$, *then* $\{x_1: f(x_1) \cap O \neq \varnothing\}$ *is a neighborhood of* x.

f *is called lower semicontinuous* (lsc) *if it is lower semicontinuous at every point* $x \in D$. *So* f *is* lsc *when the following equivalent conditions hold*:

(5) $f: D \to C_-(X)$ *is a continuous map.*
(6) O *an open subset of* X *implies* $f^{-1}(O) = \{x: f(x) \cap O \neq \varnothing\}$ *is open in* D.
(7) A *a closed subset of* X *implies* $\{x: f(x) \subset A\}$ *is closed in* D.

PROOF. Using Lemma 5 it is easy to check that (1) is equivalent to each of (2), (3), and (4) in parts (a) and (b). Since $\{f(x_n)\}$ converges to $f(x)$ in $C(X)$ if and only if it converges to $f(x)$ in both $C_+(X)$ and $C_-(X)$, continuity at x is equivalent to the combination of usc at x together with lsc at x. The equivalence of (5), (6), and (7) follow easily (compare (e) and (f) of Proposition 1.1).

If $\{x_n\} \to x$, then by Exercise 4, $y \in \limsup\{f(x_n)\}$ implies (x, y) is a limit point of a sequence in $f \subset D \times X$ and so (x, y) lies in f, i.e., $y \in f(x)$, when f is closed. So a closed relation is usc by (2). Conversely, suppose (x, y) is a limit of $\{(x_n, y_n)\}$ in f. Then $\{x_n\} \to x$ and $y \in \limsup\{f(x_n)\}$. If f is usc, then $y \in f(x)$, i.e., $(x, y) \in f$, and so f is closed.

REMARK. A closed relation $f: X_1 \to X_2$ is pointwise closed and usc. We call it a *continuous relation* if it is also lsc. We know from Proposition 1.1 that for closed relations $f: X_1 \to X_2$ and $g: X_2 \to X_3$ the composition $g \circ f: X_1 \to X_3$ is closed. From (6) of (b) it follows that $g \circ f$ is continuous if both f and g are, as well. □

Thus, f usc at x means f does not "explode" near x, i.e., when x_1 is near x, $f(x_1)$ is nearly a subset of $f(x)$. It may, however, "implode" near x. For x_1 arbitrarily close to x, $f(x_1)$ may be much smaller than $f(x)$. For lower semicontinuity the reverse is true: f lsc at x means f does not "implode" near x. While continuity in x means as usual that for x_1 near x, $f(x_1)$ provides a good approximation for $f(x)$ at least in the crude sense of the metric δ.

The equivalence between the concepts of usc map and closed relation provides an easy way to detect uppersemicontinuity. In particular, for X_1 compact, $f: X_1 \to C(X_2)$ is usc if and only if as a subset of $X_1 \times X_2$, $f \in C(X_1 \times X_2)$. For lower semicontinuity one is forced to rely upon the definition.

12. LEMMA. *For $\{f_\alpha\}$ a family of pointwise closed relations from D to X we define the pointwise closed relations*:

$$\bigwedge_\alpha \{f_\alpha\}(x) = \bigwedge_\alpha \{f_\alpha(x)\} = \bigcap_\alpha \{f_\alpha(x)\},$$

$$\mathrm{v}_\alpha \{f_\alpha\}(x) = \bigvee_\alpha \{f_\alpha(x)\} = \overline{\bigcup_\alpha \{f_\alpha(x)\}}.$$

If for some point x of D all the f_α's are usc *at x, then $\bigwedge\{f_\alpha\}$ is* usc *at x. If all the f_α's are* lsc *at x then $\mathrm{v}\{f_\alpha\}$ is* lsc *at x.*

PROOF. For any fixed f_β and any sequence $\{x_n\} \to x$ in D we have for the usc case

$$\limsup \left\{\bigwedge\{f_\alpha\}(x_n)\right\} \subset \limsup f_\beta(x_n) \subset f_\beta(x).$$

Then intersect over all β. For the lsc case

$$f_\beta(x) \subset \liminf f_\beta(x_n) \subset \liminf\{\mathrm{v}\{f_\alpha\}(x_n)\}.$$

Take the closure of the union over all β.

REMARK. Regarding each f_α as a subset of $D \times X$, we have $(\bigcap\{f_\alpha\})(x) = \bigcap\{f_\alpha(x)\}$. So in the usc case, i.e., each f_α is closed in $D \times X$, we obtain the closed relation $\bigwedge\{f_\alpha\}$ by intersecting in $D \times X$. Notice that when D is compact as well, this is just the $\wedge$ operation in $C(D \times X)$.

Even if D is compact and each f_α is closed, $\mathrm{v}\{f_\alpha\}$ is usually smaller than $\bigvee\{f_\alpha\}$ in $C(D \times X)$, which is why we use the symbol v instead of $\bigvee$. To see the difference, compare, when f is a closed relation on X, $\mathscr{R}f = \mathrm{v}\{f^n : n \geq 1\}$ with $\mathscr{N}f = \bigvee\{f^n : n \geq 1\}$. □

13. COROLLARY. *Let f be a pointwise closed relation from D to X. As a subset of $D \times X$ we can take the closure of f to get $\overline{f}$ a closed relation from D to X. $\overline{f}: D \to C(X)$ is the smallest* usc *function containing f. f is* usc *at x if and only if $f(x) = \overline{f}(x)$. In particular, if f is* lsc, *then f is continuous at x if and only if $f(x) = \overline{f}(x)$.*

PROOF. Consider the family of usc functions $\{f_\alpha\}$ such that $f(x) \subset f_\alpha(x)$ for all x. Then $f(x) \subset \bigwedge\{f_\alpha\}(x)$ and since $\bigwedge\{f_\alpha\}$ is usc it is the smallest member of the family. As subsets of $X_1 \times X_2$ we are just intersecting all closed sets containing f and so $\bigwedge\{f_\alpha\}$ is just the closure of $\overline{f}$. If $f(x) = \overline{f}(x)$ and $\{x_n\} \to x$, then

$$\limsup\{f(x_n)\} \subset \limsup\{\overline{f}(x_n)\} \subset \overline{f}(x) = f(x)$$

and so f is usc at x.

Conversely, suppose $y \in \overline{f}(x)$. Because $\overline{f}$ is the closure of f there is a sequence (x_n, y_n) in f converging to (x, y). By Exercise 4, $y \in \limsup\{f(x_n)\}$. If f is usc at x, then $\limsup\{f(x_n)\} \subset f(x)$ and so $y \in f(x)$, i.e., $\overline{f}(x) \subset f(x)$. As $f \subset \overline{f}$ we have $\overline{f}(x) = f(x)$ when f is usc at x.

REMARK. One can similarly consider the family $\{f_\alpha\}$ of all lsc functions with $f_\alpha(x) \subset f(x)$ for all x, and so get $\vee\{f_\alpha\} \equiv f^0$ the largest lsc function contained in x. It is easy to check as above that $f^0(x) = f(x)$ implies f is lsc at x. The converse is not true in general (see the supplementary exercises). □

There is a useful special case where infinite union preserves uppersemi-continuity as well.

14. LEMMA. *Assume X_1 and X_2 are compact and that $f: D \times X_1 \to C(X_2)$ is* usc. *Then $\vee_{X_1} f: D \to C(X_2)$ defined by $\vee_{X_1} f(d) = \bigcup\{f(d, x): x \in X_1\}$ is* usc *as well. Furthermore, $\vee_{X_1} f$ is continuous if f is continuous.*

PROOF. Recall that if A is a closed subset of $D \times X$, with X compact, then the projection of A to D is closed. For if (d_n, x_n) is a sequence in A with $\{d_n\} \to d$ we can assume by going to a subsequence that $\{x_n\} \to x$ and so $(d, x) \in A$.

As a relation, $f \subset D \times X_1 \times X_2$ is a closed subset and $\vee_{X_1} f \subset D \times X_2$ is just the projection of f into $D \times X_2$. It is therefore closed and so $\vee_{X_1} f$ is usc.

When f is continuous then $\vee_{X_1} f$ is also lsc by Lemma 12. □

When the domain as well as the range of a map is a space of subsets we can consider whether the order structure is preserved. A map $f: C(X_1) \to C(X_2)$ is called *monotone* if $A_1 \subset B_1$ in $C(X_1)$ implies $f(A_1) \subset f(B_1)$ in $C(X_2)$. For monotone maps the following result is useful for constructing semicontinuous maps by composition.

15. EXERCISE. *Let* $f\colon C(X_1) \to C(X_2)$ *be a monotone map. If* $f\colon C(X_1) \to C_+(X_2)$ *is continuous, then* $f\colon C_+(X_1) \to C_+(X_2)$ *is continuous. If* $f\colon C(X_1) \to C_-(X_2)$ *is continuous, then* $f\colon C_-(X_1) \to C_-(X_2)$ *is continuous. (Use the characterizations of Lemma* 9.) *In particular, if* $f\colon C(X_1) \to C(X_2)$ *is continuous, then* $f\colon C_+(X_1) \to C_+(X_2)$ *and* $f\colon C_-(X_1) \to C_-(X_2)$ *are both continuous. More generally, if* $f\colon C(X_1) \times C(X_2) \to C(X_3)$ *is monotone in each variable separately, then* $f\colon C(X_1)\times C(X_2) \to C_+(X_3)$ *continuous implies* $f\colon C_+(X_1) \times C_+(X_2) \to C_+(X_3)$ *is continuous and similarly for* C_-.
□

16. PROPOSITION. *Assume that* X, X_1, *etc. are compact metric spaces.*

(a) $i\colon X \to C(X)$ *defined by* $i(x) = \{x\}$ *is a metric isometry and so is continuous.*

(b) *If* $f\colon X_1 \to X_2$ *is a continuous function, then* $f_*\colon C(X_1) \to C(X_2)$ *defined by* $f_*(A) = f(A)$ *(the image of* A *under* f*) is a continuous monotone map. In particular, if* A *is a closed subset of* X *and* $i\colon A \to X$ *is the inclusion map, then* $i_*\colon C(A) \to C(X)$ *is an isometry identifying* $C(A)$ *with the corresponding subset of* $C(X)$.

(c) $\times\colon C(X_1) \times C(X_2) \to C(X_1 \times X_2)$ *defined by* $\times(A, B) = A \times B$ *is a continuous map monotone in each variable.*

(d) $\vee\colon C(X) \times C(X) \to C(X)$ *defined by* $\vee(A, B) = A \cup B$ *is a continuous map monotone in each variable.*

(e) $\wedge\colon C(X) \times C(X) \to C(X)$ *defined by* $\wedge(A, B) = A \cap B$ *is monotone in each variable and is* usc, *i.e.,* $\wedge\colon C(X) \times C(X) \to C_+(X)$ *is continuous.*

(f) $\mathrm{Im}\colon C(X_1) \times C(X_1 \times X_2) \to C(X_2)$ *defined by* $\mathrm{Im}(A, f) = f(A)$ *is monotone in each variable and is* usc.

(g) $\mathrm{In}\colon C(X_1 \times X_2) \to C(X_2 \times X_1)$ *defined by* $\mathrm{In}(f) = f^{-1}$ *is a monotone isometry.*

(h) $\mathrm{Comp}\colon C(X_1\times X_2)\times C(X_2\times X_3) \to C(X_1\times X_3)$ *defined by* $\mathrm{Comp}(f, g) = g \circ f$ *is monotone in each variable and is* usc.

(i) $|\ |\colon C(X \times X) \to C(X)$ *defined by* $|\ |(f) = |f| = \mathrm{Dom}(f \cap 1_X)$ *is monotone and* usc.

PROOF. Monotonicity throughout is obvious as are the isometry results of (a), (b), and (g), e.g., $d(x_1, x_2) = \delta(\{x_1\}\{x_2\})$ for x_1, x_2 in X. If $f\colon X_1 \to X_2$ is a continuous map, then for $\varepsilon > 0$ there exists $\delta_1 > 0$ such that $d_1(x, y) < \delta_1$ implies $d_2(f(x), f(y)) < \varepsilon$. So $\rho(A/B) < \delta_1$ implies $\rho(f(A)/f(B)) < \varepsilon$, proving (b).

For (c) use $d((x_1, x_2), (y_1, y_2)) = \max(d_1(x_1, y_1), d_2(x_2, y_2))$ on the product $X_1 \times X_2$. Then

$$\rho(A_1 \times A_2/B_1 \times B_2) \le \max(\rho_1(A_1/B_1), \rho_2(A_2/B_2)). \tag{7.12}$$

Similarly, for (d)

$$\rho(A_1 \vee A_2/B_1 \vee B_2) \le \max(\rho(A_1/B_1), \rho(A_2/B_2)). \tag{7.13}$$

In each of these cases the uniformity of the estimates allows us to get δ bounds from ρ bounds. For the remaining cases this is not true and so we only get usc results.

For (e) the family of closed sets $\{\overline{V}_\delta(A_1) \wedge \overline{V}_\delta(A_2)\}$ decreases with δ and has intersection $A_1 \wedge A_2$. So given $\varepsilon > 0$, there exists $\delta > 0$, depending upon A_1 and A_2 as well as ε, such that $\overline{V}_\delta(A_1) \wedge \overline{V}_\delta(A_2) \subset V_\varepsilon(A_1 \wedge A_2)$. Hence $\rho(A_1/B_1) < \delta$ and $\rho(A_2/B_2) < \delta$ implies $\rho(A_1 \wedge A_2/B_1 \wedge B_2) < \varepsilon$.

The uppersemiconintuity of composition, (h), is a restatement of Exercise 1.5(e). The image result then follows by the Chapter 1 trick of replacing a set by a relation, i.e., for p a single point space $C(p \times X_i)$ is isometric with $C(X_i)$ $(i = 1, 2)$ and Im is then the same as

$$C(p \times X_1) \times C(X_1 \times X_2) \xrightarrow{\text{Comp}} C(p \times X_2).$$

Finally, for (i), observe that $|f| = (f \wedge 1_X)(X)$, i.e., $|f| = \text{Im}(X, \wedge(1_X, f))$. By the previous results and Exercise 15 the composite function $\text{Im}(\ ,\wedge(\ ,\ ))$: $C_+(X) \times C_+(X \times X) \times C_+(X \times X) \to C_+(X)$ is continuous and so fixing the first and second variables we get a continuous function of the third. □

As we see, most of the natural semicontinuous functions are usc. We provide one useful lsc example.

17. LEMMA. *Let $\varepsilon > 0$ be fixed. Then the function ${}_\varepsilon\cap\colon C(X) \times C(X) \to C(X)$ defined by ${}_\varepsilon\cap(A, B) = \overline{V_\varepsilon(A) \cap B}$ is monotone in each variable and is* lsc.

PROOF. Again monotonicity is clear. Now suppose $\{A_n\} \to A$ and $\{B_n\} \to B$ in the metric topology. We must prove ${}_\varepsilon\cap(A, B) \subset \liminf\{{}_\varepsilon\cap(A_n, B_n)\}$ because the lim inf is closed it suffices to show that $x \in V_\varepsilon(A) \cap B$ is in the lim inf. $x \in B$ and there exists $y \in A$ with $d(x, y) < \varepsilon$. By Exercise 4, there exists N and sequences $\{x_n : n \geq N\}$ and $\{y_n : n \geq N\}$ with $x_n \in A_n$, $y_n \in B_n$, $\{x_n\} \to x$, and $\{y_n\} \to y$. There exists $N_1 \geq N$ so that $n \geq N_1$ implies $d(x_n, y_n) < \varepsilon$ because V_ε is open in $X \times X$. Thus, for $n \geq N_1$, $x_n \in {}_\varepsilon\cap(A_n, B_n)$ and so $x \in \liminf$ by Exercise 4 again. □

18. COROLLARY. *Let E be a finite subset of X and $\varepsilon \geq 0$. The mappings R_ε^+ and R_ε^- from $C(X)$ to $C(E)$ defined by* (7.10) *are monotone. R_ε^+ is* usc *and R_ε^- is* lsc.

PROOF. Since the inclusion of $C(E)$ into $C(X)$ preserves ρ and δ the results follow by noting that $R_\varepsilon^\pm\colon C(X) \to C(X)$ are the composites

$$R_\varepsilon^+(A) = \wedge(E, \text{Im}(A, \overline{V}_\varepsilon)), \qquad R_\varepsilon^-(A) = {}_\varepsilon\cap(A, E).$$

Notice that since E is finite, $V_\varepsilon(A) \cap E$ is closed.

Thus, R_ε^+ is obtained by fixing the first and third variables of a continuous map of $C_+(X) \times C_+(X) \times C_+(X)$ to $C_+(X)$ and R_ε^- is obtained by fixing the second variable of a continuous map of $C_{-}(X) \times C_{-}(X)$ to $C_{-}(X)$ (cf. Exercise 15). □

Notice that for a finite metric space E, $U_\varepsilon^+ = C$ for $\varepsilon > 0$ small enough. $C(E)$ is also a finite metric space and so every subset is open and closed. Lemma 9 then implies that $L \subset C(E)$ is open in $C_+(E)$ if and only if it is C invariant if and only if it is closed in $C_-(E)$. Dually, L is closed in $C_+(E)$ if and only if it is C^{-1} invariant if and only if it is open in $C_-(E)$.

Our applications of semicontinuity are all based on the following result of Takens which says that for a semicontinuous map the points of continuity are generic.

19. THEOREM. *Let $f: D \to C(X)$ be either* usc *or* lsc. *The set of continuity points $\{x \in D: f$ is continuous at $x\}$ is residual, i.e., it is the intersection of a countable family of dense open sets.*

PROOF. Consider $O_\varepsilon = \{x \in D: \delta(f(x_1), f(x_2)) < \varepsilon$ for all x_1, x_2 is some neighborhood of $x\}$. It is clear that O_ε is open in D and $\bigcap\{O_\varepsilon : \varepsilon > 0\}$ is the set of continuity points for f. We complete the proof by showing that O_ε is dense for each $\varepsilon > 0$. To be precise we fix $\varepsilon_1 > 0$ and U open in D and show $U \cap O_{\varepsilon_1} \neq \varnothing$.

Let ε be positive with $\varepsilon < \varepsilon_1/2$. Choose E an ε net for X (see the discussion leading up to Corollary 7). Assume f is usc, i.e., $f: D \to C_+(X)$ is continuous. By Corollary 18 and Exercise 15, $R_\varepsilon^+ : C_+(X) \to C_+(E)$ is continuous. Observe that if $x \in U$ such that $R_\varepsilon^+ \circ f$ is constant on a neighborhood U_0 of x in U then $x \in U \cap O_{\varepsilon_1}$. This is because x_1, $x_2 \in U_0$ implies $R_\varepsilon^+(f(x_1)) = R_\varepsilon^+(f(x_2))$, and so by (7.1)

$$\begin{aligned}\delta(f(x_1), f(x_2)) &\leq \delta(f(x_1), R_\varepsilon^+(f(x_1))) + \delta(f(x_2), R_\varepsilon^+(f(x_2))) \\ &\leq 2\varepsilon < \varepsilon_1 .\end{aligned}$$

The image of $R_\varepsilon^+ \circ f$ on U is a subset of $C(E)$. Choose a minimal element of this subset. That is, choose $x \in U$ so that with $E_1 = R_\varepsilon^+(f(x))$, $R_\varepsilon^+(f(x_1)) \subset E_1$ for $x_1 \in U$ implies $R_\varepsilon^+(f(x_1)) = E_1$. Now $C(E_1)$ is a C invariant subset of $C(E)$ and so is open in $C_+(E)$. Because the restriction $(R_\varepsilon^+ \circ f)_U : U \to C_+(E)$ is continuous, $U_0 = \{x_1 \in U: R_\varepsilon^+(f(x_1)) \subset E_1\}$ is open and contains x. Because E_1 is minimal $R_\varepsilon^+(f(x_1)) = E_1$ for all x_1 in U_0. Thus, $R_\varepsilon^+ \circ f$ is constant on the neighborhood U_0 of x and so $x \in U \cap O_{\varepsilon_1}$ as above.

When f is lsc the proof is similar. Use the lsc map R_ε^- and choose E_1 a maximal element of the image of $R_\varepsilon^- \circ f$ on U.

REMARK. D is called a *Baire Space* if any residual subset is dense in D. The Baire Category Theorem says that D is Baire if it is either a locally compact or complete metric space. □

If $f: D \to C(X)$ is usc (or lsc) and $h: D_1 \to D$ is continuous, then the composition $f \circ h: D_1 \to C(X)$ is usc (resp. lsc) and so the set of continuity points for $f \circ h$ is residual in D_1. But be careful! $f \circ h$ continuous at $x \in D_1$ does not imply f is continuous at $h(x)$. Nor is the converse true. In particular, $\{x \in D_1 : f \text{ is continuous at } h(x)\}$ can be empty. For example, let h be the inclusion of D_1 a closed nowhere dense set in the complement of the continuity set for f. Where D_1 is a subset of D (and h is the inclusion) observe the difference between the two ideas by comparing the phrase "f is continuous on D_1 (i.e., the restriction f_{D_1} is continuous)." with the much stronger property "f is continuous at every point of D_1."

In applying these results to the dynamics of a closed relation f, on X we are especially interested in the case when f is a continuous map or a homeomorphism.

20. PROPOSITION. *Assume X_1, X_2, etc. are compact metric spaces.*

(a) *Denote by $C(X_1; X_2)$ the set of continuous functions from X_1 to X_2 equipped with the sup metric*

$$d(f, g) = \sup\{d_2(f(x), g(x)) : x \in X_1\}. \tag{7.14}$$

With respect to d $C(X_1; X_2)$ is a complete metric space. Regarding $C(X_1; X_2)$ as a subset of $C(X_1 \times X_2)$ the topologies induced upon $C(X_1; X_2)$ by the metric topology on $C(X_1 \times X_2)$ and by the topology of $C_+(X_1 \times X_2)$ both agree with the topology on $C(X_1; X_2)$ given by d.

The restriction Comp: $C(X_1; X_2) \times C(X_2; X_3) \to C(X_1; X_3)$ *and* Im: $C(X_1) \times C(X_1; X_2) \to C(X_2)$ *are continuous.*

(b) *Denote by* $\mathrm{Cis}(X_1; X_2)$ *the set of homeomorphisms of X_1 onto X_2 equipped (if nonempty) with the metric*

$$\bar{d}(f, g) = \max(d(f, g), d(f^{-1}, g^{-1})). \tag{7.15}$$

With respect to $\bar{d}$ $\mathrm{Cis}(X_1; X_2)$ *is a complete metric space whose topology agrees with those induced from $(C(X_1; X_2), d)$, $(C(X_1 \times X_2), \delta)$, and $C_+(X_1 \times X_2)$.*

The inversion map $\mathrm{Cis}(X_1; X_2) \to \mathrm{Cis}(X_2; X_1)$ *is an isometry with respect to the $\bar{d}$ metric.*

PROOF. Observe that

$$\begin{aligned} d(f, g) &= \inf\{\varepsilon : g \subset \overline{V}_\varepsilon \circ f\} \\ &= \inf\{\varepsilon : g \subset \overline{V}_\varepsilon \circ f \text{ and } f \subset \overline{V}_\varepsilon \circ g\}, \end{aligned} \tag{7.16}$$

because $g \subset \overline{V}_\varepsilon \circ f$ if and only if $f \subset \overline{V}_\varepsilon \circ g$ for mappings f and g since each says $d_2(f(x), g(x)) \le \varepsilon$ for all x in X_1.

If we use the metric

$$d((x_1, x_2), (y_1, y_2)) = \max(d_1(x_1, y_1), d_2(x_2, y_2))$$

on $X_1 \times X_2$ then the ε neighborhood of f is just $\overline{V}_\varepsilon \circ f \circ \overline{V}_\varepsilon$. So (7.5) says that

$$\rho(f/g) = \inf\{\varepsilon : g \subset \overline{V}_\varepsilon \circ f \circ \overline{V}_\varepsilon\}$$

and

$$\delta(f, g) = \inf\{\varepsilon : g \subset \overline{V}_\varepsilon \circ f \circ \overline{V}_\varepsilon \text{ and } f \subset \overline{V}_\varepsilon \circ g \circ \overline{V}_\varepsilon\}.$$

So we clearly have $\rho \leq \delta \leq d$ on $C(X_1; X_2)$.

Now fix $f \in C(X_1; X_2)$ and $\varepsilon > 0$. Because f is uniformly continuous we can choose a positive $\varepsilon_1 < \varepsilon/2$ a modulus of uniform continuity for f with respect to $\varepsilon/2$, i.e., $f \circ V_{\varepsilon_1} \subset V_{\varepsilon/2} \circ f$. So if $\rho(f/g) < \varepsilon_1$, then

$$g \subset V_{\varepsilon_1} \circ f \circ V_{\varepsilon_1} \subset V_\varepsilon \circ f$$

and $d(f, g) < \varepsilon$. It follows that the three topologies agree when restricted to $C(X_1; X_2)$. Because of the dependence of ε_1 on f as well as ε the proof does not show that $C_-(X_1 \times X_2)$ restricts to this common topology as well. In fact, it is usually strictly coarser (use Exercise 21 below).

The completeness of $C(X_1; X_2)$ with respect to d is the standard result that uniform convergence preserves continuity.

Continuity of the composition follows because Comp: $C_+(X_1 \times X_2) \times C_+(X_2 \times X_3) \to C_+(X_1 \times X_3)$ is continuous but it is instructive to prove it directly

$$\begin{aligned} d(f \circ g, f_1 \circ g_1) &\leq d(f \circ g, f \circ g_1) + d(f \circ g_1, f_1 \circ g_1) \\ &\leq d(f \circ g, f \circ g_1) + d(f \circ f_1). \end{aligned}$$

As f_1 approaches f, $d(f, f_1) \to 0$. As g_1 approaches g $d(f \circ g, f \circ g_1) \to 0$ by uniform continuity of f.

For Im we similarly have

$$\begin{aligned} \delta(f(A), f_1(A_1)) &\leq \delta(f(A), f(A_1)) + \delta(f(A_1), f_1(A_1)) \\ &\leq \delta(f(A), f(A_1)) + d(f, f_1), \end{aligned}$$

and if $f \circ V_{\varepsilon_1} \subset V_\varepsilon \circ f$, then $\delta(A, A_1) < \varepsilon_1$ implies $\delta(f(A), f(A_1)) < \varepsilon$.

On $\mathrm{Cis}(X_1; X_2)$ we have $\rho \leq \delta \leq d \leq \overline{d}$ and given $f \in \mathrm{Cis}(X_1; X_2)$ and $\varepsilon > 0$ we can choose $\varepsilon_1 > 0$ as above so that $\rho(f/g) < \varepsilon_1$ and $\rho(f^{-1}/g^{-1}) < \varepsilon_1$ imply $d(f, g) < \varepsilon$ and $d(f^{-1}, g^{-1}) < \varepsilon$. But $\rho(f/g) = \rho(f^{-1}/g^{-1})$ and so $\rho(f/g) < \varepsilon_1$ implies $\overline{d}(f, g) < \varepsilon$. So all four topologies agree on $\mathrm{Cis}(X_1; X_2)$.

If $\{f_n\}$ is Cauchy for $\overline{d}$, then $\{f_n\}$ and $\{f_n^{-1}\}$ are Cauchy for d. So $\{f_n\} \to f$, $\{f_n^{-1}\} \to g$ in $C(X_1; X_2)$ and $C(X_2; X_1)$, respectively. By continuity of composition $1_{X_1} = \lim\{f_n^{-1} \circ f_n\} = g \circ f$ and similarly, $1_{X_2} = f \circ g$. So $f \in \mathrm{Cis}(X_1; X_2)$ and $\{f_n\} \to f$ with respect to $\overline{d}$.

That the map $f \to f^{-1}$ is an isometry is obvious.

REMARK. $C(X_1; X_2)$ is usually not a closed subset of $C(X_1 \times X_2)$ and so is not compact. For this reason the metrices d and δ, while inducing the same topology on $C(X_1; X_2)$, are not uniformly equivalent. In particular, on $C(X_1; X_2)$ δ is usually not complete. Similarly, $\mathrm{Cis}(X_1; X_2)$ is not closed in $C(X_1; X_2)$ and d is usually not complete on $\mathrm{Cis}(X_1; X_2)$. However, because complete metrics exist both $C(X_1; X_2)$ and $\mathrm{Cis}(X_1; X_2)$ are Baire spaces. □

21. EXERCISE. (a) *Let* $\{r_n : n = 1, \dots\}$ *be a counting of the rationals in* $I = [0, 1]$. *Let* $f_n \in C(I; I)$ *satisfy* $f_n(r_n) = 1$ *and* $f_n(x) = 0$ *for* $|x - r_n| \geq 1/n$. *Prove that* $\{f_n\}$ *converges to 0 in* $C\_(I \times I)$ *but not in* $C(I; I)$.

(b) *Let* $f_n(x) = nx$ *for* $0 \leq x \leq 1/n$ *and* $= 1$ *for* $1/n \leq x \leq 1$. *Prove that* $\{f_n\}$ *converges to a relation* f *in* $C(I \times I)$ *and so is* δ *Cauchy but* $f \notin C(I; I)$, *and so* $\{f_n\}$ *is not convergent in* $C(I; I)$.

(c) *Prove that* $*: C(X_1; X_2) \to C(C(X_1); C(X_2))$ *associating* f_* *to* f *(see* (b) *of Proposition* 16) *is an isometric map, i.e.,* $d(f_*, g_*) = d(f, g)$. □

At last we apply these results. First, if f is a fixed closed relation on X then $\mathcal{N}f$, Ωf, $\mathcal{G}f$, $\Omega\mathcal{G}f$, $\mathcal{C}f$, and $\Omega\mathcal{C}f$ are closed relations on X and so are usc as maps from X to $C(X)$. The pointwise closed relations ωf and $\mathcal{R}f$ are usually not closed, not usc. In general ωf is not lsc either. But we do have

22. PROPOSITION. *Let* $f: X \to X$ *be a continuous map. The pointwise closed relation* $\mathcal{R}f = \mathcal{O}f \cup \omega f$ *is* lsc *with closure* $\mathcal{N}f$. *In particular, the set of points* $\{x \in X : \omega f(x) = \Omega f(x)\}$ *is residual. If* f *is a homeomorphism* $\{x \in X : \omega f(x) = \Omega f(x)$ *and* $\alpha f(x) = \Omega f^{-1}(x)\}$ *is residual.*

PROOF. For each $n > 1$, f^n is a continuous map as so is continuous as a relation. $\mathcal{R}f = \underline{\mathrm{v}\{f^n : n = 1, \dots\}}$ is the lsc by Lemma 12. Since $\mathcal{O}f \subset \mathcal{R}f \subset \mathcal{N}f = \overline{\mathcal{O}f}$ it follows that $\mathcal{N}f = \overline{\mathcal{R}f}$. By Corollary 13, $\{x : \mathcal{R}f(x) = \mathcal{N}f(x)\}$ is the set of points at which $\mathcal{R}f$ is continuous and by Theorem 19 it is the countable intersection of dense open sets. For such a point $\mathcal{O}f(x) \cup \omega f(x) = \mathcal{O}f(x) \cup \Omega f(x)$ and so $\omega f(x) \subset \Omega f(x) \subset \mathcal{O}f(x) \cup \omega f(x)$. Now apply f^n using $f \circ \omega f = \omega f$ and $f \circ \Omega f = \Omega f$ (Proposition 1.12(a))

$$\omega f(x) \subset \Omega f(x) \subset \mathcal{O}f(f^n(x)) \cup \omega f(x).$$

But $\bigcap_n \{\mathcal{O}f(f^n(x)) \cup \omega f(x)\} = \omega f(x)$ because if the points $\{f(x), f^2(x), \dots\}$ are not all distinct they are eventually periodic and then $\omega f(x)$ in this periodic orbit as well. So we have $\omega f(x) = \Omega f(x)$ when $\mathcal{R}f(x) = \mathcal{N}f(x)$.

When f is a homeomorphism, we apply the result to f^{-1} to get $\{x : \alpha f(x) = \Omega f^{-1}(x)\}$ residual and the intersection of two residual sets is residual. □

In particular, Proposition 22 implies the Chapter 4 result that f topologically transitive ($\mathscr{N}f = X \times X$) implies $\{x: \mathscr{R}f(x) = X\} = \{x: \omega f(x) = X\}$ is residual. Also, if f is central ($1_X \subset \mathscr{N}f$), then the set of positive recurrent points $\{x: x \in \omega f(x)\}$, residual.

The really interesting applications are to the comparison of different functions. We first observe what can be said for closed relations in general.

23. THEOREM. *For a compact metric space X, the functions $\mathscr{C}$, $\Omega\mathscr{C}$, and $\mathscr{C} \wedge \mathscr{C}^{-1}$: $C(X \times X) \to C(X \times X)$ associating to f: $\mathscr{C}f$, $\Omega\mathscr{C}f$, and $\mathscr{C}f \cap \mathscr{C}f^{-1}$, respectively, are monotone* usc *maps, as is $|\mathscr{C}|: C(X \times X) \to C(X)$ associating to f the chain recurrent set $|\mathscr{C}f|$. If $j: D \to C(X \times X)$ is continuous* (*or even* usc) *for some metric space D there is a residual subset D_1 at every point of which all of these maps composed with j are continuous.*

The functions J^+, $J: C(X \times X) \times C(X) \to C(X)$ defined by $J^+(f, A) = \bigcap_{n=0}^{\infty} f^n(A)$ and $J(f, A) = \bigcap_{n=-\infty}^{\infty} f^n(A)$ are usc *maps monotone in each variable.*

For any fixed U in $C(X)$ the set $I_U = \{f: f(U) \subset \text{Interior}\, U\}$, the set of closed relations with respect to which U is inward, is an open subset of $C_+(X \times X)$. The usc *map $I_U \to C(X)$, $J^+(f, U) = \omega f[U] = \Omega\mathscr{C}f(U)$ gives, for each f in I_U, the f attractor corresponding to the inward set U. So if $j: D \to C(X \times X)$ is continuous* (*or even* usc) *then $J^+(j(d), U)$ is continuous at the points of some residual subset of the open set $j^{-1}(I_U)$ in D.*

PROOF. Recall from Proposition 1.8 and its proof that for positive ε_1 and ε_2 with $\varepsilon_1 < \varepsilon_2$ we have

$$\mathscr{C}(\overline{V}_{\varepsilon_1} \circ f \circ \overline{V}_{\varepsilon_1}) \subset \mathscr{O}(V_{\varepsilon_2} \circ f \circ V_{\varepsilon_2}) \subset \mathscr{N}(\overline{V}_{\varepsilon_2} \circ f \circ \overline{V}_{\varepsilon_2})$$

and $\mathscr{C}f$ is the intersection of the decreasing family $\{\mathscr{N}(\overline{V}_{\varepsilon_2} \circ f \circ \overline{V}_{\varepsilon_2}): \varepsilon_2 > 0\}$. So given $\varepsilon > 0$ there exists $\varepsilon_2 > 0$ so that $\varepsilon_1 < \varepsilon_2$ implies

$$\mathscr{C}(\overline{V}_{\varepsilon_1} \circ f \circ \overline{V}_{\varepsilon_1}) \subset \mathscr{N}(\overline{V}_{\varepsilon_2} \circ f \circ \overline{V}_{\varepsilon_2}) \subset V_\varepsilon \circ \mathscr{C}f \circ V_\varepsilon .$$

By monotonicity of $\mathscr{C}$ this means that $\rho(f/g) < \varepsilon_1$ implies $\rho(\mathscr{C}f/\mathscr{C}g) < \varepsilon$. So $\mathscr{C}$ is usc at f.

$\Omega\mathscr{C}f = \bigwedge_n\{(\mathscr{C}f)^n: n \geq 0\}$ is then usc by Lemma 12 and the upper-emicontinuity of composition. Because inversion is continuous and finite intersection is usc, $\mathscr{C} \wedge \mathscr{C}^{-1}$ is usc, see Proposition 16. $|\mathscr{C}|$ is the composition of monotone usc maps and so is usc. J^+ and J are usc because composition and image are usc maps and infinite intersection preserves usc by Lemma 12 again. By Exercise 10(b) $\{B: B \subset \text{Int}(U)\}$ is open in $C_+(X)$. Because $f \to f(U)$ is usc and monotone in f, I_U is open in $C_+(X \times X)$.

Continuity on residual sets follows from Theorem 19. □

For maps we define the subsets In^+ and In of $C(X; X) \times C(X)$,

$$\text{In}^+ = \{(f, A): f(A) \subset A\}, \qquad \text{In} = \{(f, A): f(A) = A\}. \tag{7.17}$$

The function on $C(X;X) \times C(X)$ defined by $(f, A) \to (A, f(A))$ is continuous by Proposition 20. As In^+ and In are the preimages of the closed sets C and $1_{C(X)}$ respectively, they are closed. Thus, In^+ and In can be regarded as usc maps from $C(X;X)$ to $C(C(X))$. In particular, for each f the collection $\mathrm{In}(f)$ of f invariant closed subsets of X is itself closed in $C(X)$.

24. THEOREM. *Let X be compact metric and let $j: D \to C(X;X)$ be a continuous map.*

Define the function $\widehat{\mathscr{C}}$, $\widehat{\mathscr{N}}$: $\mathrm{In}^+ \to C(X \times X)$ by $\widehat{\mathscr{C}}(f, A) = \mathscr{C}(f_A)$ and $\widehat{\mathscr{N}}(f, A) = \mathscr{N}(f_A)$. $\widehat{\mathscr{C}}$ is usc and $\widehat{\mathscr{N}}$ is lsc.

In particular, $\mathscr{N}: C(X;X) \to C(X \times X)$ associating $\mathscr{N}f$ to f is a lsc function. There is a residual subset D_1 of D at the points of which $\mathscr{N} \circ j$ is continuous.

$|\mathscr{N}| = |\ | \circ \mathscr{N}: C(X;X) \to C(X)$ associates to f the nonwandering set $|\mathscr{N}f| = |\Omega f|$. At the points of D_1 $|\mathscr{N}|$ is usc.

Define $\mathrm{Per}: C(X;X) \to C(X)$ *by*

$$\mathrm{Per}(f) = \overline{\bigcup_{n \geq 1} \{|f^n|\}}, \tag{7.18}$$

the closure of the set $|\mathscr{O}(f)|$ of periodic points of f. There is a residual subset D_2 of D at point of which $\mathrm{Per} \circ j$ is lsc.

PROOF. For each positive integer n, $e_n(f, A) = f_A^n$ defines a function from In^+ to $C(X \times X)$ which we will show is continuous. As composition is continuous and $e_n(f, A) = e_1(f^n, A)$ it suffices to treat the case $n = 1$.

$$\delta(f_A, g_B) \leq \delta(f_A, f_B) + \delta(f_B, g_B) \leq \delta(f_A, f_B) + d(f, g).$$

So for g close to f the second term is small. For the first, with f fixed and $\varepsilon > 0$ given choose $0 < \varepsilon_1 < \varepsilon$ an ε modulus of uniform continuity for f. Suppose $\rho(A/B) < \varepsilon_1$ then for $x \in B$ there exists $\tilde{x} \in A$ with $d(x, \tilde{x}) < \varepsilon_1$ and so $d(f(x), f(\tilde{x})) < \varepsilon$. This says $f_B \subset V_\varepsilon \circ f_A \circ V_{\varepsilon_1}$ and so $\rho(f_A/f_B) \leq \varepsilon$.

Now the map $\widehat{\mathscr{C}}$: $\mathrm{In}^+ \to C_+(X \times X)$ is just the composition

$$\mathrm{In}^+ \xrightarrow{e_1} C(X \times X) \xrightarrow{\mathscr{C}} C_+(X \times X),$$

which is continuous by Theorem 23. Thus, $\widehat{\mathscr{C}}$: $\mathrm{In}^+ \to C(X \times X)$ is usc.

$$\widehat{\mathscr{N}}(f, A) = \overline{\bigcup_{n \geq 1} \{f_A^n\}}, \qquad \text{i.e., } \widehat{\mathscr{N}} = \vee\{e_n : n \geq 1\},$$

and so $\widehat{\mathscr{N}}$ is lsc by Lemma 12. $\mathscr{N}f = \mathscr{N}(f, X)$ and so $\mathscr{N}$ is lsc as well. That D_1, the set of continuity points for $\mathscr{N} \circ j$, is residual follows from Theorem 19. By Theorem 23, $|\ |$ is usc, i.e., $|\ |: C(X \times X) \to C_+(X)$ is continuous and so at the points of D_1 the composition $|\ | \circ \mathscr{N} \circ j$ is usc.

Note $\operatorname{Per}(f) = \vee\{| \, | \circ \tilde{e}_n\}(f)$ where $\tilde{e}_n(f) = f^n$. Each $\tilde{e}_n$ is continuous but $| \, | \circ \tilde{e}_n(f) = |f^n|$ is only usc in f. By Theorem 23 there is, for each n, a residual subset D^n at points of which $| \, | \circ \tilde{e}_n \circ j$ is continuous. Let $D_2 = \bigcap_n\{D^n\}$ which is still residual. By Lemma 12, $\operatorname{Per} \circ j$ is lsc at each point of D_2. □

Let us pause here to compare our results for $|\mathscr{C} f|$ and for $|\Omega f|$. The chain recurrent set $|\mathscr{C} f|$ is usc in f and continuous at the points of a residual set. Thus, $|\mathscr{C} f|$ explosions never occur and implosions are rare. For $|\Omega f|$ we only have usc on a residual set. Nonwandering set implosions can both occur. But explosions are rare. The reverse is true for $\operatorname{Per}(f)$, the set of periodic points. But attempting to compare $\operatorname{Per}(f)$ and $|\Omega f|$ leads to interesting questions. Suppose for a particular $j \colon D \to C(X; X)$ that $\operatorname{Per}(j(d)) = |\Omega j(d)|$ for some point d in $D_1 \cap D_2$. In other words, for the map $f = j(d)$ suppose the periodic points are dense in the nonwandering set. It then follows that both $\operatorname{Per} \circ j$ and $|\Omega \circ j|$ are continuous at d. The argument came up in the proof of Corollary 13. Suppose $\{d_n\} \to d$ in D. Since $\operatorname{Per} \circ j$ is lsc at d and $|\mathscr{N}| \circ j$ is usc at d,

$$\begin{array}{ccccc} \operatorname{Per}(j(d)) \subset & \liminf\{\operatorname{Per}(j(d_n))\} & \subset & \limsup\{\operatorname{Per}(j(d_n))\} & \\ & \cap & & \cap & \\ & \liminf\{|\Omega(j(d_n))|\} & \subset & \limsup\{|\Omega(j(d_n))|\} & \subset |\Omega j(d)|. \end{array}$$

Because the extreme terms are equal by assumption, all six expressions are the same. Thus, if there is a residual set D_3 on which $\operatorname{Per} \circ j$ equals $|\Omega \circ j|$ then both functions are continuous at the points of the residual set $D_1 \cap D_2 \cap D_3$.

Recall that topological equivalence between homeomorphisms f and f_1 on X is expressed by a conjugacy, a homeomorphism h on X such that $h \circ f = f_1 \circ h$. h maps the f orbit of x to the f_1 orbit of $h(x)$. Zeeman's idea of tolerance stability is to replace this homeomorphism between orbits by the cruder association of closeness with respect to the metric δ. With this in mind we define (for $n = 1, 2, \ldots$),

$$o_n \colon \operatorname{Cis}(X; X) \times X \to C(X) \quad \text{by } o_n(f, x) = \{f^i(x) : |i| \le n\} \tag{7.19}$$

and

$$o \colon \operatorname{Cis}(X; X) \times X \to C(X) \tag{7.20}$$

by

$$o(f, x) = \overline{\{f^i(x) \colon n = 0, \pm 1, \pm 2, \ldots\}}.$$

$o(f, x)$ is called the *orbit closure* of x. It is just new notation for $\{x\} \cup \mathscr{R} f(x) \cup \mathscr{R} f^{-1}(x) = \alpha f(x) \cup \mathscr{O} f^{-1}(x) \cup x \cup \mathscr{O} f(x) \cup \omega f(x)$.

25. Lemma. *For each n, o_n is continuous. o is* lsc.

Proof. Because composition, image, and inversion are continuous for homeomorphisms, $u_i \colon \operatorname{Cis}(X; X) \times X \to X$ defined by $u_i(f, x) = f^i(x)$

is continuous for each integer i. If we include X into $C(X)$ by the singleton isometry we can regard u_i as mapping into $C(X)$. Then $o_n(f, x) = \bigvee\{u_i(f, x): |i| \le n\}$, which is continuous because finite union is continuous on $C(X)$ (c.f. Proposition 16 (d)). $o = \vee\{o_n : n = 1, 2, \ldots\}$ is then lsc by Lemma 12. □

26. LEMMA. *Let f be a homeomorphism on X.*

(a) *A closed subset A of X is an orbit closure, i.e., $A = o(f, x)$ for some $x \in X$, if and only if A is f invariant and any two distinct elements of A are $\mathcal{N}(f_A)$ comparable, i.e., $A \times A = \mathcal{N}(f_A)^{-1} \cup 1_A \cup \mathcal{N}(f_A)$.*

(b) *For a closed f invariant subset A of X the following conditions are equivalent. When they are satisfied we call A a* pseudo-orbit closure *for f.*

(1) *For every $\varepsilon > 0$ there exists $\{x_0, x_1, \ldots, x_n\} = F$ an ε chain for f in X such that $\delta(F, A) < \varepsilon$.*

(2) *For every $\varepsilon > 0$ there exists $\{\ldots, x_{-2}, x_{-1}, x_0, x_1, x_2, \ldots\} = G$ an infinite ε chain to f in X such that $\delta(\overline{G}, A) < \varepsilon$.*

(3) *For every $\varepsilon > 0$ there exists $\{\ldots, x_{-2}, x_{-1}, x_0, x_1, x_2, \ldots\} = G$ an infinite ε chain for f in A such that $\rho(\overline{G}/A) = \delta(\overline{G}, A) < \varepsilon$.*

(4) *Any two distinct elements of A are $\mathcal{C}(f_A)$ compariable, i.e., $A \times A = \mathcal{C}(f_A)^{-1} \cup 1_A \cup \mathcal{C}(f_A)$.*

PROOF. (a) $o(f, x)$ consists, in addition to the orbit of x, of $\alpha f(x)$ and $\omega f(x)$. Clearly, for y in the orbit or in $\alpha f(x)$, $\omega f(x) \subset \mathcal{N}(f_A)(y)$. By Proposition 1.12 the same is true for $y \in \omega f(x)$. Similarly, $\alpha f(x) \subset \mathcal{N}(f_A)^{-1}(y)$ for all y in $o(f, x)$. Of course, any two distinct elements of the orbit are $\mathcal{O}(f_A)$ and a fortiori $\mathcal{N}(f_A)$ comparable.

Conversely, if $A \times A = \mathcal{N}(f_A)^{-1} \cup 1_A \cup \mathcal{N}(f_A)$, then by Proposition 22 applied to f_A and f_A^{-1} there is a residual set of points x in A for which $\mathcal{N}(f_A)(x) = \mathcal{R}(f_A)(x)$ and $\mathcal{N}(f_A)^{-1}(x) = R(f_A^{-1})(x)$. For such x, $A = \mathcal{N}(f_A^{-1})(x) \cup x \cup \mathcal{N}(f_A)(x) = \mathcal{R}(f_A^{-1})(x) \cup x \cup \mathcal{R}(f_A)(x) = o(f, x)$.

(b) (4) $\Rightarrow$ (3). Let $E = \{y_1, \ldots, y_n\}$ be an ε set for A, i.e., $\rho(E/A) < \varepsilon$. Because any two distinct elements of A are $\mathcal{C}(f_A)$ comparable and $\mathcal{C}(f_A)$ is transitive we can order the points of E so that $y_j \in \mathcal{C}(f_A)(y_i)$ when $i < j$. By definition of $\mathcal{C}(f_A)$ we can insert between y_i and y_{i+1} an ε chain for f in A. Then extend the chain by $\{\ldots, f^{-2}(y_1), f^{-1}(y_1)\}$ on the left and by $\{f(y_n), f^2(y_n), \ldots\}$ on the right. The resulting infinite ε chain G lies in A and contains E. So $\rho(A/\overline{G}) = 0$ and $\rho(\overline{G}/A) \le \rho(E/A) < \varepsilon$. Thus, $\delta(\overline{G}, A) = \rho(\overline{G}/A) < \varepsilon$.

(3) $\Rightarrow$ (2) is obvious.

(2) $\Rightarrow$ (1). Truncate the infinite chain at n when $\{V_\varepsilon(x_i): |i| \le n\}$ covers the compact set A.

(1) $\Rightarrow$ (4). Given distinct points y, $\tilde{y}$ in A and $\varepsilon > 0$, choose a positive

$\varepsilon_1 < \min(\varepsilon/3, d(y, \tilde{y})/2)$ such that $d(x_1, x_2) \leq \varepsilon_1$ implies $d(f(x_1), f(x_2)) < \varepsilon/3$. Let $F = \{x_0, x_1, \dots, x_n\}$ be an ε_1 chain for f such that $\delta(F, A) < \varepsilon_1$. Each of y and $\tilde{y}$ are within ε_1 of points in F, different points because $2\varepsilon_1 < d(y, \tilde{y})$. Suppose the y point precedes the $\tilde{y}$ point and throw away the outliers. So after renumbering we have an ε_1 chain for f $\{x_0, \dots, x_n\}$ contained in $V_{\varepsilon_1}(A)$ and with $d(y, x_0)$, $d(\tilde{y}, x_n) < \varepsilon_1$. Let $y_0 = y$, $y_n = \tilde{y}$ and choose $y_i \in A$ with $d(y_i, x_i) < \varepsilon_1$ for $i = 1, 2, \dots, n-1$. Then for $i = 0, 1, \dots, n-1$,

$$d(f(y_i), y_{i+1}) \leq d(f(y_i), f(x_i)) + d(f(x_i), x_{i+1}) + d(x_{i+1}, y_{i+1}) < \varepsilon.$$

So $\{y_0, \dots, y_n\}$ is an ε chain in A connecting y to $\tilde{y}$.

With $\{\varepsilon_i\}$ a positive sequence tending to 0 choose such an ε_i chain for each i. Either they go from y to $\tilde{y}$ or from $\tilde{y}$ to y for infinitely many i. So either $y \in \mathscr{C}(f_A)(y)$ or $y \in \mathscr{C}(f_A)(\tilde{y})$. □

27. EXERCISE. *Prove that a pseudo-orbit-closure is indecomposible (Hint: if $\{A_1, A_2\}$ is a decomposition of A, prove $\mathscr{C}(f_A) \subset A_1 \times A_1 \cup A_2 \times A_2$).* □

Define the following subsets of $\mathrm{Cis}(X; X) \times C(X)$:

$$\begin{aligned} O_n &= \{(f, A) : A = o_n(f, x) \text{ for some } x \in X\}. \\ O &= \{(f, A) : \text{ For every } \varepsilon > 0 \text{ there exists } x \in X \\ &\qquad \text{such that } \delta(o(f, x), A) < \varepsilon\}. \\ R &= \{(f, A) : \text{ For every } \varepsilon > 0 \text{ there exists } x \in X \\ &\qquad \text{and } n \text{ such that } \delta(o_n(f, x), A) < \varepsilon\}. \\ T &= \{(f, A) : A \text{ is a pseudo-orbit-closure for } f\}. \\ \mathrm{In} &= \{(f, A) : A \text{ is } f \text{ invariant, i.e., } A = f(A)\}. \end{aligned} \tag{7.21}$$

Thus, $O(f)$ is the closure in $C(X)$ of the collection of all orbit closures for f. Since each $o(f, x)$ is f invariant and In is a closed subset of $C(X; X) \times C(X)$, it is clear that $O \subset \mathrm{In}$. On the other hand, the elements of $R(f)$ need not be f invariant, e.g., $O_n \subset R$, but by Lemma 26(b) if $A \in R(f)$ is f invariant, then A is a pseudo-orbit-closure. Finally, $\delta(o(f, x), A) < \varepsilon$ implies $\delta(o_n(f, x), A) < \varepsilon$ for large enough n and so $O(f) \subset R(f)$. In sum, we have the inclusions

$$O \subset R \wedge \mathrm{In} \subset T \subset \mathrm{In}. \tag{7.22}$$

28. THEOREM. *Let X be compact metric and let $j: D \to \mathrm{Cis}(X; X)$ be a continuous map.*

In, T, and O_n are closed subsets of $\mathrm{Cis}(X; X) \times C(X)$. So as maps from $\mathrm{Cis}(X; X)$ to $C(C(X))$ they are usc. In fact, O_n is continuous. There are residual subsets D_{In} and D_T at points of which $\mathrm{In} \circ j$ and $T \circ j$ are continuous, respectively.

R and O are pointwise closed relations and so define maps from $\mathrm{Cis}(X\,;X)$ *to* $C(C(X))$. *R is* lsc *and there is a residual subset D_R at points of which $R \circ j$ is continuous. There is a residual subset D_O at points of which $O \circ j$ is* lsc.

PROOF. We have already proved that In is closed for $C(X\,;X) \times X$. Intersect with $\mathrm{Cis}(X\,;X) \times X$.

By Lemma 26, $(f, A) \in T$ when $(f, A) \in \mathrm{In}$ and $(\widehat{\mathscr{C}}(f, A) \cup 1_A \cup \widehat{\mathscr{C}}(f, A)^{-1}, A \times A)$ is a point in $C \subset C(X \times X) \times C(X \times X)$. Because $\widehat{\mathscr{C}}$ is usc and C is a closed subset of $C_+(X \times X) \times C(X \times X)$ it follows that T is closed (cf. Theorem 24 and Exercise 10).

O_n is the projection to $\mathrm{Cis}(X\,;X) \times C(X)$ of $o_n \subset \mathrm{Cis}(X\,;X) \times X \times C(X)$ and so O_n is continuous by Lemmas 25 and 14.

$R(f) = \overline{\bigcup\{O_n(f)\}}$, i.e., $R = \mathrm{v}\{O_n\}$, and so R is pointwise closed and lsc by Lemma 12.

The existence of the residual sets D_{In}, D_T, and D_R now follow from Theorem 19.

$O(f)$ is the closure in $C(C(X))$ of $\{o(f, x)\colon x \in X\}$, and so O is pointwise closed and defines a map from $\mathrm{Cis}(X\,;X)$ to $C(C(X))$. Now define for positive integers m, n, $\widetilde{Q}_{mn} \subset \mathrm{Cis}(X\,;X) \times X \times C(X)$ by

$$\widetilde{Q}_{mn} = \{(f, x, A)\colon (f, A) \in \mathrm{In},\ x \subset A,\ \text{and } \rho(o_n(f, x)/A) \leq 1/m\}.$$

Because o_n is continuous $\widetilde{Q}_{mn}$ is closed and hence so is its projection $Q_{mn} \subset \mathrm{Cis}(X\,;X) \times C(X)$:

$$Q_{mn} = \{(f, A) \in \mathrm{In}\colon \rho(o_n(f, x)/A) \leq 1/m \text{ for some } x \in A\}$$

(see Lemma 14). Thus, Q_{mn} defines a usc map from $\mathrm{Cis}(X\,;X)$ to $C(X)$.

For each m and n there is a residual set in D at which $Q_{mn} \circ j$ is continuous. Intersecting them we get a residual set D_0 at points of which every $Q_{mn} \circ j$ is continuous. In particular, if we define $Q_m = \mathrm{v}\{Q_{mn}\colon n = 1, \ldots\}$ then by Lemma 12 each $Q_m \circ j$ is lsc at the points of D_0. We prove that O is lsc at the points of D_0.

To prove this it is sufficient to show that

$$\text{(7.23)} \qquad \rho(Q_m(f_1)/Q_m(f)) < 1/m \quad \text{implies} \quad \rho(O(f_1)/O(f)) < 4/m,$$

because with $d \in D_0$ there exists a neighborhood U_m of d in D such that $\rho(Q_m(j(d_1))/Q_m(j(d))) < 1/m$ by lower semicontinuity. By (7.23) $\rho(O(j(d_1))/O(j(d))) < 4/m$ for d_1 in U_m and this implies lower semicontinuity of O at d.

To prove (7.23), let $A \in O(f)$. Choose $x \in X$ such that $\delta(A, o(f, x)) < 1/m$ and let $B = o(f, x)$. Because B is an orbit closure for f, $B \in Q_m(f)$. In fact, $B \in \bigcup\{Q_{mn}(f)\colon n = 1, 2, \ldots\}$. So there exists B_1 in $Q_m(f_1)$ $1/m$ close to B, i.e., $\delta(B, B_1) < 1/m$. Choose $A_1 \in \bigcup\{Q_{mn}(f_1)\colon n = 1, 2, \ldots\}$

such that $\delta(B_1, A_1) < 1/m$. By definition of $Q_{mn}(f_1)$ there exists $x_1 \in A_1$ such that $\rho(o(f_1, x_1)/A_1) \leq \rho(o_n(f_1, x_1)/A_1) \leq 1/m$. Because A_1 is f_1 invariant $o(f_1, x_1) \subset A_1$ and so $\delta(o(f_1, x_1), A_1) = \rho(o(f_1, x_1)/A_1) \leq 1/m$. As $o(f_1, x_1)$ is an orbit closure for f_1 it lies in $O(f_1)$. Clearly, $\delta(o(f_1, x_1), A) < 4/m$.

Thus, for every A in $O(f)$ we have found an $o(f_1, x_1)$ which is $4/m$ close to A, i.e., $\rho(O(f_1)/O(f)) < 4/m$. □

29. Exercise. *Prove that* $\{(f, A): A = o(f, x)$ *for some* $x \in X\}$ *is a* G_δ *set in* $\mathrm{Cis}(X; X) \times C(X)$, *i.e., a countable intersection of open sets.* (*Hint*: *Show that* $\{(f, A) \in \mathrm{In}: (\mathcal{N}(f_A)^{-1} \cup 1_A \cup \mathcal{N}(f_A), A \times A) \subset U_\varepsilon^+\}$ *is open in* In). *Conclude that for* $f \in \mathrm{Cis}(X; X)$ $\{o(f, x): x \in X\}$ *is a residual subset of its closure* $O(f)$ *in* $C(C(X))$. □

We say that $j: D \to \mathrm{Cis}(X; X)$ satisfies *Zeeman's tolerance stability condition* if there is a residual subset of D at points of which $O \circ j$ is continuous. It is any easy exercise to check that $O \circ j$ is continuous at a point d of D is equivalent to each of the following conditions:

(1) For every $\varepsilon > 0$ there exists a neighborhood U of d in D such that for every d_1, $d_2 \in U$ and $x_1 \in X$ there exists $x_2 \in X$ so that $\delta(o(j(d_1), x_1), o(j(d_2), x_2)) < \varepsilon$.

(2) For every $\varepsilon > 0$ there exists a neighborhood U of d in D such that for every $d_1 \in U$ and $x \in X$ there exist x_1, $\tilde{x} \in X$ so that $\delta(o(j(d), x), o(j(d_1), x_1)) < \varepsilon$ and $\delta(o(j(d), \tilde{x}), o(j(d_1), x)) < \varepsilon$.

From Theorem 28 we have the following partial results due to Takens.

30. Corollary. *With* $j: D \to \mathrm{Cis}(X; X)$ *define the residual subsets* D_{In}, D_R, *and* D_O *as in Theorem* 28. *For* $d \in D_{\mathrm{In}} \cap D_R$ *we have*

(1′) *For every* $\varepsilon > 0$, *there exists a neighborhood* U *of* d *in* D *such that for every* d_1, $d_2 \in U$ *and* $x_1 \in X$ *there exist* x_2, $\tilde{x}_2 \in X$ *so that* $\rho(o(j(d_1), x_1)/o(j(d_2), x_2)) < \varepsilon$ *and* $\rho(o(j(d_2), \tilde{x}_2)/o(j(d_1), x_1)) < \varepsilon$.

For $d \in D_O$ *we have*

(2′) *For every* $\varepsilon > 0$, *there exists a neighborhood* U *of* d *in* D *such that for every* $d_1 \in U$ *and* $x \in X$ *there exists* $x_1 \in X$ *so that* $\delta(o(j(d), x), o(j(d_1), x_1)) < \varepsilon$.

Proof. For $d \in D_{\mathrm{In}} \cap D_R$ both $\mathrm{In} \circ j$ and $R \circ j$ are continuous at d. So given $\varepsilon > 0$ we can choose U a neighborhood of d so that for d_1, $d_2 \in U$ $\mathrm{In}(j(d_1))$ and $\mathrm{In}(j(d_2))$ are within ε of one another in $C(C(X))$, and $R(j(d_1))$ and $R(j(d_2))$ are within $\varepsilon/2$ of one another. As $o(j(d_1), x_1) \in \mathrm{In}(j(d_1))$ there exists A a closed set invariant with respect to $j(d_2)$, i.e., $A \in \mathrm{In}(j(d_2))$ such that $\delta(o(j(d_1), x_1), A) < \varepsilon$. Choose $x_2 \in A$. Clearly, $o(j(d_2), x_2) \subset A$ and so

$$\rho(o(j(d_1), x_1)/o(j(d_2), x_2)) \leq \delta(o(j(d_1), x_1), A) < \varepsilon.$$

Also $o(j(d_1), x_1) \in R(j(d_1))$ and so there exists $B \in R(j(d_2))$ such that $\delta(B, o(j(d_1), x_1)) < \varepsilon/2$. By definition of R there exist $\tilde{x}_2$ and n so that $\delta(o_n(j(d_2), \tilde{x}_2), B) < \varepsilon/2$ and so $\delta(o_n(j(d_2), \tilde{x}_2), o(j(d_1), x_1)) < \varepsilon$. Hence,

$$\begin{aligned}&\rho(o(j(d_2), \tilde{x}_2)/o(j(d_1), x_1))\\ &\quad\leq \delta(o_n(j(d_2), \tilde{x}_2), o(j(d_1), x_1))\\ &\quad< \varepsilon.\end{aligned}$$

For $d \in D_O$, $O \circ j$ is lsc. So given $\varepsilon > 0$ there exists U a neighborhood of d in D so that for $d_1 \in U$ every element of $O(j(d))$ is within $\varepsilon/2$ of some element of $O(j(d_1))$. $o(j(d), x) \in O(j(d))$ and so there exists $A \in O(j(d_1))$ with $\delta(o(j(d), x), A) < \varepsilon/2$. As $O(j(d_1))$ is the closure in $C(X)$ of the $j(d_1)$ orbit closures, there exists $x_1 \in X$ so that $\delta(A, o(j(d_1), x_1)) < \varepsilon/2$. Hence $\delta(o(j(d), x), o(j(d_1), x_1)) < \varepsilon$.

REMARK. By (1) continuity of $O \circ j$ at d means that in $(1')$ the points x_2 and $\tilde{x}_2$ can be chosen the same. $(2')$ says that every $j(d)$ orbit closure is δ close to some $j(d_1)$ orbit closure. (2) says, in addition, that the reverse is true. Of course, for d in the residual set $D_{\mathrm{In}} \cap D_R \cap D_O$ both $(1')$ and $(2')$ hold and for every $\varepsilon > 0$ the neighborhood U can be chosen so that both conditions apply. Even together they do not yield continuity. □

Just as $\mathrm{Per} \circ j = |\mathcal{N}| \circ j$ at the points of a residual set in D implies $\mathrm{Per} \circ j$ and $|\mathcal{N}| \circ j$ are both continuous at points of some residual set, so also a similar comparison between $O \circ j$ and $T \circ j$ would establish the tolerance stability condition.

31. EXERCISE. *With notation as in Theorem* 28, *prove that at the points of* D_R, $R \wedge \mathrm{In}$ *is usc. For* $d \in D_R \cap D_O$, $O(j(d)) = R(j(d)) \cap \mathrm{In}(j(d))$ *implies* $O \circ j$ *is continuous at* d. *Hence, if* $O \circ j = (R \wedge \mathrm{In}) \circ j$, *and a fortiori if* $O \circ j = T \circ j$ *on some residual subset of* D, *then* $O \circ j$ *satisfies the tolerance stability condition.* □

Supplementary exercises

32. Let D be metric and X compact metric.

(a) For $U \subset D \times X$ let f be the pointwise closure of U, i.e., $f(d) = \overline{U(d)}$. So $U \subset f \subset \overline{U} = \overline{f}$, where $\overline{U}$ is the closure of U in $D \times X$. If U is open prove that f is lsc. (Show that $\{d_n\} \to \{d\}$ implies $U(d) \subset \bigcup_n \bigcap_{k \geq n} U(d_k) \subset \liminf\{f(d_n)\}$.) Conclude that $f(d) = \overline{f}(d)$ if and only if $\overline{f}$ is continuous at d and that such points are residual in D. In particular, if U is dense as well as open in $D \times X$, show that the set $\{d\colon U(d)$ is dense in $X\}$ is residual in D.

(b) Prove the "Fubini theorem" for residual sets: If A is a residual subset of $D \times X$, then $\{d\colon A(d)$ is a residual subset of $X\}$ is residual in D.

33. (a) For $f \subset D \times X$, define $f^0 = \vee\{f_\alpha : f_\alpha \subset f$ and f_α is lsc$\}$. Prove f^0 is lsc and so f^0 is the largest lsc relation contained in f. Prove that $f^0(d) = f(d)$ implies f is lsc at d.

(b) With $I = [0, 1]$ let $A = I - \{1, 1/2, 1/4, \dots\}$, $A_u = \{(u + ut, u) : t \in A\}$ and $f = \bigcup\{A_u : u = 0, 1/2, 1/4, \dots\} \subset I \times I$. Draw a picture of f and prove $f: I \to C(I)$ is lsc at every point t of I except $t = 1/2, 1/4, \dots$. In particular, f is lsc at $t = 0$ with $f(0) = 0$. On the other hand, prove that $f^0(0) = \emptyset$ and so $f^0 \neq f$ at 0. ($f^0 = \bigcup\{A_u - (u, u) : u = 0, 1/2, \dots\}$).

(c) A more interesting example of this unpleasant phenomenon is an application of the usc map $|\;| : \mathrm{Cis}(X; X) \to C(X)$ associating to f the set of fixed points $|f|$. When X is the circle, $|\;|^0 = \emptyset$, i.e., there is no nonempty lsc map $k: \mathrm{Cis}(X; X) \to C(X)$ with $k \subset |\;|$. To see this, call $p \in |f|$ one-sided if on some oriented interval neighborhood of p, $f(x) \geq x$ for all x or $f(x) \leq x$ for all x. Prove:

(1) If p is a one-sided fixed point of f there is some neighborhood U of p such that for every $\varepsilon_1 > 0$ we can find $f_1 \in \mathrm{Cis}(X; X)$ with $d(f, f_1) < \varepsilon_1$ and $U \cap |f_1| = \emptyset$. (E.g., let $f_1 = q \circ f$ where q is an arbitrarily small positive or negative rotation.)

(2) Given $f \in \mathrm{Cis}(X; X)$ then for every $\varepsilon > 0$ there exists $f_1 \in \mathrm{Cis}(X; X)$ with $\bar{d}(f, f_1) < \varepsilon$ such that $|f_1|$ consists entirely of one-sided fixed points. (If $|f| = \emptyset$ let $f_1 = f$. Otherwise the complement of $|f|$ is a countable union of invariant open intervals on each of which either $f(x) > x$ or $f(x) < x$. First replace f by the identity on any intervals of length $< \varepsilon$. The endpoints of the remaining intervals are either now one sided or are isolated fixed points, attracting or repelling. Flatten f about each of these finitely many points to replace each p by an interval of fixed points containing p.)

(3) Suppose $k: \mathrm{Cis}(X; X) \to C(X)$ is lsc and contained in $|\;|$. If $\{f_n\} \to f$ then $k(f) \subset \liminf k(f_n) \subset \liminf |f_n|$. If $p \in |f|$ is one-sided, there exists $\{f_n\} \to f$ such that $|f_n| \cap U = \emptyset$ for U some neighborhood of p. So $p \notin k(f)$, i.e., $p \in k(f)$ implies p is not one-sided. Therefore there exists a sequence $\{f_n\} \to f$ such that $k(f_n) = \emptyset$ and so $k(f) = \emptyset$.

For the circle, prove that $|f^n| \neq \emptyset$ implies every periodic point has period dividing n. Hence, $\mathrm{Per}(f) = \bigcup\{|f^n| : n = 1, 2, \dots\}$. Adjust the proof to show that for X =the circle $\mathrm{Per}^0 = \emptyset$. (Compare this with a contrary claim in Palis, Pugh, Shub, and Sullivan (1975).)

34. Regarding X as a subset of $C(X)$. Prove that the original topology on X agrees with the topologies induced from $C(X)$, $C_+(X)$, and $C_-(X)$. Hence, if a relation $f \subset D \times X$ is a mapping from D to X, then $f: D \to C(X)$ is continuous if and only if it is usc if and only if it is lsc.

35. Develop the theory of real valued usc and lsc maps by using the following analogues on $\mathbb{R}$ of the corresponding pieces of structure for $C(X)$:

$$\begin{aligned} (7.24) \qquad C &= \{(a, b): b \leq a\} \subset \mathbb{R} \times \mathbb{R}, \\ \rho(a/b) &= \max(b - a, 0), \\ \delta(a, b) &= \max(\rho(b/a), \rho(a/b)) = |a - b|, \\ \bigvee\{a_d\} &= \sup\{a_d\}, \qquad \bigwedge\{a_d\} = \inf\{a_d\}. \end{aligned}$$

We obtain upper and lower topologies $_+\mathbb{R}$ and $_-\mathbb{R}$ on the reals and call $f: D \to \mathbb{R}$ usc if it is continuous to $_+\mathbb{R}$ and lsc if it is continuous to $_-\mathbb{R}$.

For the reals we have a duality lacking in $C(X)$: $\rho(a/b) = \rho(-b/-a)$ and so $i(x) = -x$ is a homeomorphism of $_+\mathbb{R}$ to $_-\mathbb{R}$ and $_-\mathbb{R}$ to $_+\mathbb{R}$. In particular, f is usc if and only if $-f$ is lsc.

For I a compact interval map $C_I \subset I \times I$ is a relation and thus a map $C_I: I \to C(I)$, $C_I(a) = \{b: a \leq b\} \cap I$. Prove that $\rho(C_I(a)/C_I(b)) = \rho(a/b)$ and so a function $f: D \to \mathbb{R}$ with $f(D) \subset I$ is continuous / usc / lsc if and only if $C_I \circ f: D \to C(I)$ is continuous / usc / lsc. By reducing to the bounded case and applying this transformation prove that if $f: D \to R$ is usc or lsc then it is continuous at the points of a residual set.

Using this result prove that if $\{f_n: D \to \mathbb{R}\}$ is a sequence of continuous functions converging pointwise to $f: D \to \mathbb{R}$, i.e., $f(x) = \lim\{f_n(x)\}$ for each $x \in D$, then f is continuous at the points of a residual set. (*Hint*: Show that $f = \limsup\{f_n\}$ and $-f = \limsup\{-f_n\}$ are each lsc at the points of a residual set.)

Prove that if $f: D \times X \to \mathbb{R}$ is continuous with X compact then $\bigvee_X f: D \to \mathbb{R}$ and $\bigwedge_X f: D \to \mathbb{R}$ are continuous with $\bigvee_X f(d) = \sup\{f(d, x): x \in X\}$ and $\bigwedge_X f(d) = \inf\{f(d_1, x): x \in X\} = -\bigvee_X(-f)(d)$ (use Lemma 4).

What is the definition of the real number analogue of $\overline{f}$ and of $f^0 = -(\overline{-f})$?

36. (a) The identity map $1_{C(X)}: C(X) \to C(X)$ can be regarded as a usc relation $C(X) \to X$ and so as the closed subset of $C(X) \times X$: $\in = \{(A, x): x \in A\}$. If H is a closed subset of $C(X)$, i.e., an element of $C(C(X))$ we denote by $\mathrm{v}(H)$ its image under this relation:

$$\mathrm{v}(H) = \{x: x \in A \text{ for some } a \in H\} = \bigcup\{A \in H\}.$$

In particular, by Proposition 1.1(d) $\mathrm{v}(H)$ is closed and so v defines a map $C(C(X)) \to C(X)$. v is monotone and distance decreasing. In fact, for $L, H \in C(C(X))$:

$$\rho(\mathrm{v}(L)/\mathrm{v}(H)) \leq \rho(L/H).$$

With $i_X: X \to C(X)$ defined in Proposition 16(a) by $i_X(x) = \{x\}$ we obtain two maps, $i_{C(X)}$ and $i_{X^*}: C(X) \to C(C(X))$. Show that v is a left inverse for each of them, i.e., $\mathrm{v} \circ i_{C(X)}(A) = \mathrm{v} \circ i_{X^*}(A) = A$.

(b) Show that with respect to the metrics defined by Proposition 20, $*: C(X_1; X_2) \to C(C(X_1); C(X_2))$ defined by $f \mapsto f_*$ is an isometric inclusion map. Also, $*$ preserves Lipschitz constants, i.e., if $d(f(x), f(y)) \leq L d(x, y)$ for all $x, y \in X_1$ then $d(f_*(A), f_*(B)) \leq L d(A, B)$ for all $A, B \in C(X_1)$.

(c) With X^n the n-fold product of copies of X, define $X^n \to C(X)$ by $(x_1, \dots, x_n) \mapsto \{x_1, \dots, x_n\}$. Show that this map is distance decreasing. (Hint: X^n is isometric to $C(\{1, \dots, n\}; X)$. Applying (b) and evaluate at the element $\{1, \dots, n\}$ of $C(\{1, \dots, n\})$.)

(d) With X_1 and X_2 compact show that $f: D \times X_1 \to X_2$ is continuous if and only if the adjoint map $f_{\#}: D \to C(X_1; X_2)$ is continuous where $f_{\#}(d)(x) = f(d, x)$. Prove that f continuous implies $f_*: D \times C(X_1) \to C(X_2)$ is continuous where $f_*(d, A) = f(\{d\} \times A)$.

(e) Use (d) to reprove the continuity case of Lemma 14: if $f: D \times X_1 \to C(X_2)$ is continuous, then $\mathrm{v}_{X_1} f: D \to C(X_2)$ is continuous. (Hint: Write $\mathrm{v}_{X_1} f$ as the composition of the inclusion $1_D \times c_{\{X_1\}}: D \to D \times C(X_1)$, $f_*: D \times C(X_1) \to C(C(X_2))$ and $\mathrm{v}: C(C(X_2)) \to C(X))$.

37. (a) By Theorem 23 the subset I_U of $\mathrm{Cis}(X; X)$ defined by $I_U = \{f: f(U) \subset \mathrm{Int}\, U\}$ is open. If $\{X = U_1, U_2, \dots, U_{n+1} = \emptyset\}$ is a sequence of sets with $\mathrm{Int}\, U_i \supset U_{i+1}$ $(i = 1, \dots, n)$ then $\bigcap\{I_{U_i}\}$ is the open set of homeomorphisms f for which $\{U_1, U_2, \dots, U_{n+1}\}$ is a filtration for f (see Exercise 5.23). Prove that the map which associated to each f in this open set the associated invariant decomposition is usc, i.e., it is a continuous map from $\bigcap\{I_{U_i}\} \to C_+(X)^n$.

(b) Prove that the set $\widehat{I}_U$ of $\mathrm{In} \subset \mathrm{Cis}(X; X) \times C(X)$ defined by $\widehat{I}_U = \{(f, A): f(A \cap U) \subset \mathrm{Int}\, U\}$ is an open subset of In and the map $\hat{\omega}_U: \widehat{I}_U \to C(X)$ defined by $\hat{\omega}_U(f, A) = \omega f[A \cap U] = \Omega\mathscr{C}(f_A)(A \cap U)$ is usc. $\widehat{I}_U$ consists of pairs such that $U \cap A$ is inward for the restriction f_A. Notice that for (f_1, A_1) near (f, A) we can regard f_{1A_1} as a perturbation of f_A obtained by changing not only the map $(f \to f_1)$ but even the space $(A \to A_1)$. But the correspondence between attractors $\omega f[A \cap U]$ for f_A and $\omega f_1[A_1 \cap U]$ for f_{A_1} is still usc.

(c) Prove that for every $\varepsilon > 0$, $\{(A, B): V_\varepsilon(A) \subset B\}$ is closed in $C_-(X) \times C_+(X)$ and a fortiori in $C(X) \times C(X)$. ($\{(A, B, C): {}_\varepsilon\bigcap(A, C) = \emptyset$ and $B \vee C = X\}$ is closed in $C_-(X) \times C_+(X) \times C(X)$. Project onto the first pair of factors.) Then prove $\{(f, U): V_\varepsilon(f(U)) \subset U\}$ is a closed subset of I_U for each $\varepsilon > 0$.

(d) Prove that $\mathrm{Ch} = \{(f, A) \in \mathrm{In}: A$ is a chain transitive subset of $X\}$ is closed $(= \{(f, A) \in \mathrm{In}: A \times A \subset \mathscr{C}(f_A)\})$ and is contained in T (see (7.21)).

(e) Let $\{r_n\}$ be a counting of the rationals in $I = [0, 1]$. Let $A_n = \{(r_i, 1/n): 1 \leq n\}$ and define $X = I \times 0 \cup \bigcup\{A_n\}$. With $f = 1_X$, prove that the set of basic sets of f (= components of X) is not closed in $C(X)$. In general, if E is a closed equivalence relation on X $E: X \to C(X)$ is usc, i.e., $E: X \to C_+(X)$ is continuous. Prove that the topology induced from $C_+(X)$ onto the image of E is Hausdorff and so coincides with the quotient topology induced from X on the set of equivalence classes. Show that this topology is in general different from the ones induced by $C\_(X)$ and $C(X)$.

(f) For $f \in \mathrm{Cis}(X; X)$ prove $y \in \mathscr{C}f(x)$ implies $x, y \in A$ for some $A \in T(f)$. Conclude:

$$\mathscr{C}f \cup 1_X \cup \mathscr{C}f^{-1} = \bigcup\{A \times A: A \in T(f)\},$$
$$\mathscr{C}f \cap \mathscr{C}f^{-1} = \bigcup\{A \times A: A \in \mathrm{Ch}(f)\}.$$

On the other hand, prove that

$$\mathscr{N}f \cup 1_X \cup \mathscr{N}f^{-1} = \bigcup\{A \times A: A \in O(f)\} = \bigcup\{A \times A: A \in R(f)\}.$$

38. Prove that for f a homeomorphism on X and $A \in R(f)$ there exists $B \in O(f)$ such that $A \subset B$. If $A \in R(f) \wedge \mathrm{In}(f)$ then A is also in $T(f)$ (see (7.22)). However, these two conditions are not sufficient to describe $R(f) \wedge \mathrm{In}(f)$. On $X = S_1 \times S_2 \times I$, circles S_1 and S_2 parametrized with 2π periodic coordinates θ_1 and θ_2, define the differential equation (with α irrational):

$$\frac{d\theta_1}{dt} = \sin^2(2\theta_2) + u^2, \quad \frac{d\theta_2}{dt} = (\sin^2(2\theta_2) + u^2)\alpha, \quad \frac{du}{dt} = -u.$$

So $A = S^1 \times 0 \times 0$ consists of the fixed points of the flow. With f the time-one homeomorphism A is, of course, f invariant and chain recurrent. Also $A \subset S^1 \times S^2 \times 0 = \omega f(x)$ for any $x = (\theta_1, \theta_2, u)$ with $u > 0$. However, $A \notin R(f)$.

39. In order to relate the results of this chapter to flows, we call $\varphi: X \times J \to X$, with J a closed interval containing 0, a *flow chunk* if φ is continuous, $\varphi(x, 0) = x$ for all $x \in X$, and $\varphi(\varphi(x, t_1), t_2) = \varphi(x, t_1 + t_2)$ for all $x \in X$ and for all $t_1, t_2 \in J$ such that $t_1 + t_2 \in J$. Prove:

(a) A flow chunk on $X \times [0, 1]$ is the restriction of a unique semiflow. A flow chunk on $X \times [-1, 1]$ is the restriction of a unique flow. We will use the same symbol φ for the flow chunk and its extensions.

(b) The set of flow chunks is a closed subset of $C(X \times J; X)$. We define $F(X)$ to be the set of semiflows on X with metric induced from $C(X \times [0, 1]; X)$ and $\mathrm{Fis}(X)$ to be the set of flows on X with metric induced form $C(X \times [-1, 1]; X)$. Show that for $\varphi_1, \varphi_2 \in F(X)$,

$$d(\varphi_1, \varphi_2) = \sup\{d(f_1^t, f_2^t): t \in [0, 1]\}$$

and for $\varphi_1, \varphi_2 \in \mathrm{Fis}(X)$ (here we denote the metric $\bar{d}$):

$$\begin{aligned}\bar{d}(\varphi_1, \varphi_2) &= \sup\{d(f_1^t, f_2^t): t \in [-1, 1]\}\\ &= \sup\{\bar{d}(f_1^t, f_2^t): t \in [0, 1]\}.\end{aligned}$$

(c) For any compact interval J the restriction of φ to $X \times J$ defines continuous maps

$$\begin{aligned}F(X) &\to C(X \times J; X), && (J \subset [0, \infty)),\\ \mathrm{Fis}(X) &\to C(X \times J; X), && (J \subset \mathbb{R}).\end{aligned}$$

(d) In_F^+ and In_F are closed subsets of $F(X) \times C(X)$ defined by

$$\begin{aligned}\mathrm{In}_F^+ &= \{(\varphi, A): \varphi(A \times [0, 1]) \subset A\},\\ \mathrm{In}_F &= \{(\varphi, A) \in \mathrm{In}_F^+: f(A) = A\}.\end{aligned}$$

$(\varphi, A) \in \mathrm{In}_F^+$ (or $\in \mathrm{In}_F$) if and only if A is a closed $\varphi +$ invariant subset (resp. a closed φ invariant subset). In particular, In_F^+ and In_F define usc maps.

(e) $\mathscr{C}, \mathscr{N}: F(X) \to C(X \times X)$ defined by $\mathscr{C}\varphi = \mathscr{C}f$ and $\mathscr{N}\varphi = \mathrm{v}\{f^t : t \in [0, \infty)\}$ are usc and lsc respectively.

(f) $T_F = T \cap \mathrm{In}_F^+ = T \cap \mathrm{In}_F$ is the set of pairs (φ, A) such that A is a pseudo-orbit closure with respect to f and A is φ invariant. We call such a set a pseudo-orbit closure for φ. If A is a pseudo-orbit closure for φ then A is indecomposible with respect to φ and hence is connected. T_F is a closed subset of In_F in $\mathrm{Fis}(X) \times C(X)$.

(g) Define o_t: $\mathrm{Fis}(X) \times X \to C(X)$ and o: $\mathrm{Fis}(X) \times X \to C(X)$ by $o_t(\varphi, x) = \{\varphi(x, s): |x| \le t\}$ and $o(\varphi, x) = \overline{\{\varphi(x, s): s \in \mathbb{R}\}}$. For each positive real t, o_t is continuous, o is lsc.

(h) $O_F = \{(\varphi, A)$: For every $\varepsilon > 0$ there exists $x \in X$ such that $\delta(A, o(\varphi, x)) < \varepsilon\}$ and $R_F = \{(\varphi, A)$: For every $\varepsilon > 0$ there exists $x \in X$ and $t > 0$ such that $\delta(A, o_t(\varphi, x)) < \varepsilon\}$. $O_F \subset R_F \cap \mathrm{In}_F \subset T_F$. O_F and R_F are pointwise closed. R_F defines an lsc map from $\mathrm{Fis}(X)$ to $C(C(X))$. If $j: D \to \mathrm{Fis}(X)$ is continuous then $O_F \circ j$ is lsc at the points of a residual subset of D.

40. Under certain conditions $\mathscr{C}: C(X; X) \to C(X \times X)$ is the closure of $\mathscr{N}: C(X; X) \to C(X \times X)$. As $\mathscr{N}$ is lsc this implies that on the residual set of points of $f \in C(X; X)$ at which $\mathscr{N}$ is continuous, $\mathscr{N}f = \mathscr{C}f$ and so $|\mathscr{N}f| = |\mathscr{C}f|$, i.e., the nonwandering set coincides with the chain recurrent set. However, this is a strictly C^0 result. For $j: D \to C(X; X)$ continuous $\mathscr{C} \circ j = \overline{\mathscr{N} \circ j}$ does not follow from $\mathscr{C} = \overline{\mathscr{N}}$.

(a) If $q \in \mathrm{Cis}(X; X)$ and $f \in C(X; X)$ prove that $d(q \circ f, f) \le d(q, 1_X)$ with equality when f is a homeomorphism. Using $f = q^{-1}$, show that $d(q, 1_X) = d(1_X, q^{-1}) = \overline{d}(q, 1_X)$. Show

$$\overline{d}(q_1 \circ q_2, 1_X) \leq d(q_1 \circ q_2, q_2) + d(q_2, 1_X)$$
$$= \overline{d}(q_1, 1_X) + \overline{d}(q_2, 1_X)$$

for $q_1, q_2 \in \mathrm{Cis}(X; X)$. Finally, if $f \in \mathrm{Cis}(X; X)$ and $f \circ V_{\varepsilon_1} \subset V_\varepsilon \circ f$ with $0 < \varepsilon_1 < \varepsilon$ show that $\overline{d}(q \circ f, f) < \varepsilon$ if $\overline{d}(q, 1_X) < \varepsilon_1$.

(b) A compact metric space X is called *generalized homogeneous* if for every $\varepsilon > 0$ there exists $\varepsilon_1 > 0$ such that for any positive n if $\{x_1, \dots, x_n\}$ and $\{y_1, \dots, y_n\}$ are two lists of n distinct points of X such that $d(y_i, x_i) < \varepsilon_1$ for $i = 1, \dots, n$, then there is a homeomorphism q on X such that $\overline{d}(q, 1_X) < \varepsilon$ and $q(y_i) = x_i$, $i = 1, \dots, n$. Prove that if X is generalized homogeneous and $y \in \mathscr{C}f(x)$ for $f \in C(X; X)$ then for any $\varepsilon > 0$ there exists $f_1 \in C(X; X)$ with $d(f, f_1) < \varepsilon$ such that $y = f_1^n(x)$ for some $n = 1, 2, \dots$. If $f \in \mathrm{Cis}(X; X)$ then f_1 can be chosen in $\mathrm{Cis}(X; X)$ so that $\overline{d}(f, f_1) < \varepsilon$. (Find an ε_1 chain $\{x_0, x_1, \dots, x_n\}$ with $x_0 = x$, $x_n = y$ such that $\{x_1, \dots, x_n\}$ and $\{f(x_0), \dots, f(x_{n-1})\}$ are lists of distinct points. Pick q so that $q(f(x_i)) = x_{i+1}$, $i = 0, \dots, n-1$ and let $f_1 = q \circ f$.) Then prove that for X generalized homogeneous, $\overline{\mathscr{N}} = \mathscr{C}$ and $\overline{\mathrm{Per}} = |\mathscr{C}|$ on $C(X; X)$ and on $\mathrm{Cis}(X; X)$. In particular, for f in a residual subset of $C(X; X)$, namely the continuity points of $\mathscr{N}$, $\mathscr{N}f = \mathscr{C}f$, and so $|\mathscr{N}f| = |\mathscr{C}f|$ (similarly, for $\mathrm{Cis}(X; X)$). Furthermore, $|\mathrm{Per}\, f| = |\mathscr{C}f|$ if and only if Per is usc at f. However, the set may not be residual in $C(X; X)$ or $\mathrm{Cis}(X; X)$.

(c) If X is a smooth compact manifold of dimension greater than 2 prove that X is generalized homogeneous. (Observe first that one can assume $\{x_1, \dots, x_n\} \cup \{y_1, \dots, y_n\}$ consists of $2n$ points. Otherwise perturb the x's to get points $\{z_1, \dots, z_n\}$ so that $\{x_1, \dots, x_n\} \cup \{z_1, \dots, z_n\}$ and $\{z_1, \dots, z_n\} \cup \{y_1, \dots, y_n\}$ both consist of $2n$ points and use $q = q_1 \circ q_2$. Then connect y_i to x_i by a short smooth arc path. Because dimension is greater than 2 the paths can be assumed disjoint and then thickened up to be the interior of n disjoint small balls. Let q be the identity outside the balls and move y_i to x_i inside ball i).

There is an alternative proof given by Nitecki and Shub (1975) which establishes the result for dimension equal to 2 as well.

(d) Show that the circle is not generalized homogeneous. (Let $x_1 < y_2 < y_1 < x_2$ with respect to an orientation of S^1.)

41. (a) Prove that if $f: X_1 \to X_2$ is a usc relation, then f^{-1} is usc. In particular, if $f: X_1 \to X_2$ is a continuous map $f^{-1}: X_2 \to X_1$ is a usc relation. Still assuming f is a map, prove that f^{-1} is an lsc, and hence continuous, relation exactly when f is an open map, i.e., O open in X_1 implies $f(O)$ is open in X_2.

(b) Consider the map $f\colon [0, 1] \to [0, 1]$ defined by $f(t) = 0$ for $0 \le t < 1/2$ and $f(t) = 1$ for $1/2 \le t \le 1$. Find the mistake in the following argument: A closed in $[0, 1]$ implies $f(A)$ is closed. Hence, by (7) of Proposition 11(a) applied to f^{-1}, f^{-1} is usc. By part (a), f is usc. But f is a map and so it is continuous.

42. For a point-closed relation f from X to Y prove the converse of Corollary 1.2: If for every $x \in X$ and $\varepsilon > 0$ there exists $\delta > 0$ such that $f(V_\delta(x)) \subset V_\varepsilon(f(x))$, then f is usc and so is a closed relation. Show that the point-closed hypothesis is needed. (*Hint*: Use $f = p \times A$.)

43. Extend Theorem 24 to relations, proving that $\widehat{\mathscr{C}}\colon C(X \times X) \times C(X) \to C(X \times X)$ defined by $\widehat{\mathscr{C}}(f, A) = \mathscr{C}(f_A) = \mathscr{C}(f \wedge A \times A)$ is usc. Use this result to reprove Theorem 4.5. (*Hint*: For U a neighborhood of B which is the intersection of an inward set and an outward set, $(\mathscr{C} f)_B \subset \mathscr{C}(f_V)$.)

Chapter 8. Invariant Measures for Mappings

For a topologically transitive map f on X, the orbits of most points are dense in X and so such orbits meander all over X. It may even happen that nearby points separate so that complete information about the orbit of one point may tell us little about the orbit of nearby points, e.g. when f is topologically mixing (see below).

These difficulties induce a retreat from dynamics to statistics. Rather than determine the location in X of every iterate $f^i(x)$ we ask instead how frequently the iterates enter different regions of the space. It is the hope of obtaining space-average answers to such time-average questions which motivates the introduction of invariant measures.

For a compact metric space X measure theory is introduced by using the Borel sets, the σ-algebra generated by the open sets. By "measurable" we will always mean "Borel measurable" and a "measure" μ is a Borel measure (with $\mu(X) = 1$ unless otherwise mentioned). We denote by $P(X)$ the set of such probability measure on X.

If u is any bounded, real valued measurable function and μ is any measure then the integral,

$$\langle u, \mu \rangle = \int u(x)\, \mu(dx) \tag{8.1}$$

is defined and

$$|\langle u, \mu \rangle| \leq \|u\|_0, \qquad \text{where } \|u\|_0 = \sup\{|u(x)| : x \in X\}. \tag{8.2}$$

In particular, the integral is defined on the vector space $C(X; \mathbb{R})$ of continuous real functions on X. Because uniform convergence preserves continuity, $C(X; \mathbb{R})$ is a Banach space when equipped with the sup norm $\| \ \|_0$.

For a real function u we define the *support of* u by

$$\text{supp}(u) = \{x : u(x) \neq 0\}, \tag{8.3}$$

which is a Borel set when u is measurable and an open set when u is continuous. We will call the closure of $\text{supp}(u)$ the *closed support of* u.

For μ in $P(X)$ we define the *support of* μ, $|\mu|$, to be the smallest closed set with probability 1. Thus

$$|\mu| = \bigcap\{A \in C(X) \colon \mu(A) = 1\}, \tag{8.4}$$

or

$$X - |\mu| = \bigcup\{U \colon U \text{ is open and } \mu(U) = 0\}.$$

Since $X-|\mu|$ is a union of countably many measure zero sets, $\mu(X-|\mu|) = 0$. So $\mu(|\mu|) = 1$. μ is called *dense* on X if $|\mu| = X$ or, equivalently, $\mu(U) > 0$ for every nonempty open subset.

For $x \in X$ we denote by δ_x the probability measure concentrated on the point x so that $\delta_x(x) = 1$. Clearly, $\langle u, \delta_x\rangle = u(x)$ for any measurable u and $|\delta_x| = x$.

On $P(X)$ we use the topology of weak convergence, the coarsest topology such that the function $\mu \to \langle u, \mu\rangle$ is continuous for each u in $C(X; \mathbb{R})$. Thus, a sequence $\{\mu_n\}$ converges to μ in $P(X)$ when $\lim\langle u, \mu_n\rangle = \langle u, \mu\rangle$ for every continuous real valued u. This integral limit equation will usually *not* be true when u is merely measurable.

There is a natural duality between $C(X; \mathbb{R})$ and the Borel measures on X described by the Riesz Representation Theorem which we apply to prove

1. PROPOSITION. *For X a compact metric space $C(X; \mathbb{R})$ is a separable Banach space and $P(X)$ is a compact metrizable space. The integral $\langle\ ,\ \rangle \colon C(X; \mathbb{R}) \times P(X) \to \mathbb{R}$ is continuous, the support function $|\ | \colon P(X) \to C(X)$ is lower semicontinuous and the point measure function $\delta \colon X \to P(X)$ is a homeomorphism onto a closed subset.*

PROOF. For separability we construct a sequence S which generates $C(X; \mathbb{R})$, i.e., $C(X; \mathbb{R})$ is the only closed subspace containing all of S. Then the rational linear combinations on S form a countable dense subset of $C(X; \mathbb{R})$.

For each positive integer n let $\{x_i^n \colon i = 1, \dots, m_n\}$ be a $1/n$ net for X, i.e., every point x lies in some $V_{1/n}(x_i^n)$ for $i = 1, \dots, m_n$. So if $\tilde{k}_i^n(x) = \max[(1/n) - d(x_i^n, x), 0]$, then $\tilde{k}^n = \sum_{i=1}^{m_n} \tilde{k}_i^n$ is a positive function and with $k_i^n = \tilde{k}_i^n/\tilde{k}^n$, $\{k_i^n \colon i = 1, \dots, m_n\}$ is a partition of unity for each n, i.e., $\sum_{i=1}^{m_n} k_i^n(x) = 1$ for all x. We now show that $S = \{k_i^n \colon i = 1, \dots, m_n$ and $n = 1, 2, \dots\}$ generates $C(X; \mathbb{R})$. Given $u \in C(X; \mathbb{R})$ and $\varepsilon > 0$ let $1/n$ be an ε modulus of uniform continuity for u. Then

$$u(x) = \sum_{i=1}^{m_n} u(x) k_i^n(x)$$

differs from linear combination

$$\sum_{i=1}^{m_n} u(x_i^n) k_i^n(x)$$

by at most ε because $k_i^n(x) > 0$ implies $d(x, x_i^n) < 1/n$.

Now if S generates $C(X;\mathbb{R})$ then the topology of weak convergence is the same as the coarsest topology such that the $\langle u, \ \rangle$'s with u in S are continuous. To see this define on $P(X)$ the pseudometric

$$d_u(\mu_1, \mu_2) = |\langle u, \mu_1\rangle - \langle u, \mu_2\rangle|. \tag{8.5}$$

If u differs from $\sum_{i=1}^n r_i u_i$ by at most ε then

$$d_{u_i}(\mu_1, \mu_2) < \varepsilon/|r_i|n \quad \text{for } i = 1, \dots, n$$

implies, by (8.2)

$$d_u(\mu_1, \mu_2) < 3\varepsilon.$$

In particular, if $S = \{k_i\}$ is a generating sequence for $C(X;\mathbb{R})$ with $\|k_i\|_0 \leq 1$ for $k = 1, 2, \dots$ then the pseudometric

$$d_S = \sup_{i=1}^{\infty} 2^{-i} d_{k_i} \tag{8.6}$$

determines the topology on $P(X)$.

That d_S is in fact a metric and compactness of the metric space $P(X)$ require some deeper results.

A real linear map l on $C(X;\mathbb{R})$ is called a linear functional. l is continuous when its norm $\|l\| = \sup\{|l(u)|: \|u\|_0 \leq 1\}$ is finite. l is called positive if $l(u) \geq 0$ when u is a nonnegative function. Observe that a positive functional is monotone ($u - v \geq 0$ implies $l(u) - l(v) \geq 0$) and so is continuous ($-\|u\|_0 \leq u \leq \|u\|_0$ implies $|l(u)| \leq \|u\|_0 l(1)$). For example, if μ is a Borel measure $\langle \ , \mu\rangle$ is a positive linear functional. The Riesz Theorem says, conversely, that for every positive linear functional l there is a unique Borel measure μ (with $\mu(X) = l(1)$) such that $l = \langle \ , \mu\rangle$ (see Halmos (1950) §56). In particular, if $\langle k_i, \mu_1\rangle = \langle k_i, \mu_2\rangle$ for all k_i in a generating sequence S then by uniqueness $\mu_1 = \mu_2$, and so d_S is a metric on $P(X)$. Furthermore, $P(X)$ is thus identified with the set of positive linear functionals l such that $l(1) = 1$. This is a closed, bounded, convex subset of the dual space of $C(X;\mathbb{R})$ and the topology on $P(X)$ is the weak* topology. Compactness is now an easy consequence.

The map $\mu \to \{\langle u, \mu\rangle : u \in C(X;\mathbb{R})\}$ embeds $P(X)$ in the compact product space $\prod\{[-\|u\|_0, \|u\|_0]: u \in C(X;\mathbb{R})\}$. Furthermore the condition that a point l of the product satisfy linearity, positivity and $l(1) = 1$ are closed conditions. So the homeomorphic image of $P(X)$ is closed and hence compact in the product.

Joint continuity of the integral follows from the inequality

$$\begin{aligned} |\langle u, \mu\rangle - \langle u_1, \mu_1\rangle| &\leq |\langle u, \mu\rangle - \langle u, \mu_1\rangle| + |\langle u, \mu_1\rangle - \langle u_1, \mu_1\rangle| \\ &\leq d_u(\mu, \mu_1) + \|u - u_1\|_0. \end{aligned}$$

So as (u_1, μ_1) approaches (u, μ), $\langle u_1, \mu_1\rangle$ approaches $\langle u, \mu\rangle$.

The point measure map δ is clearly continuous and one-to-one. So it is a homeomorphism onto its image by compactness.

Finally, let $\{\mu_n\}$ converge to μ in $P(X)$ and suppose $x \notin \liminf\{|\mu_n|\}$. We prove $x \notin |\mu|$. There exists $\varepsilon > 0$ and a subsequence $\{\mu_{n_i}\}$ such that $V_\varepsilon(x) \cap |\mu_{n_i}| = 0$. Let $u \in C(X; \mathbb{R})$ be nonnegative with support in $V_\varepsilon(x)$ and positive at x. $\langle u, \mu_{n_i} \rangle = 0$ for all i and so $\langle u, \mu \rangle = 0$. The support of u is thus an open set of μ measure zero containing x. Hence, the support map $| \ |$ is lsc. □

If μ is a measure on X_1 and $f: X_1 \to X_2$ is a measurable map, then the induced measure $f_* \mu$ on X_2 is defined by

$$f_* \mu(A) = \mu(f^{-1}(A)) \tag{8.7}$$

for all Borel subsets A of X_2. Equivalently, $f_* \mu$ is defined by the integral identity

$$\langle u, f_* \mu \rangle = \langle u \circ f, \mu \rangle \tag{8.8}$$

for all measurable real functions u on X_2.

Clearly, f_* preserves convex combinations, e.g.

$$f_*(t\mu_1 + (1-t)\mu_2) = t f_* \mu_1 + (1-t) f_* \mu_2, \qquad 0 \le t \le 1, \tag{8.9}$$

and relates point measures

$$f_* \delta_x = \delta_{f(x)}. \tag{8.10}$$

The construction is also functorial, i.e., $(f \circ g)_* \mu = f_*(g_* \mu)$ and $1_{X_1 *} \mu = \mu$. So if $X_1 = X_2$, $(f^n)_* = (f_*)^n$ for $n \ge 0$ and for all n if f is invertible.

Our main interest is in the continuous case.

2. Proposition. *$*: C(X_1; X_2) \times P(X_1) \to P(X_2)$ associating $f_* \mu$ to (f, μ) is continuous and the supports are related by*

$$|f_* \mu| = f(|\mu|). \tag{8.11}$$

In particular, if $X = X_1 = X_2$, then

$$\mathrm{In}_P = \{(f, \mu) : f_* \mu = \mu\}$$

is a closed subset of $C(X; X) \times P(X)$. For $f \in C(X; X)$ the set $\mathrm{In}_P(f)$ of f invariant measures is a closed, convex subset of $P(X)$. $|\mu|$ is a closed f invariant subset of X for each μ in $\mathrm{In}_P(f)$.

Proof. By a now familiar trick we have for $u \in C(X_2; \mathbb{R})$:

$$d_u(f_* \mu, f_{1*} \mu_1) \le d_{u \circ f}(\mu, \mu_1) + \|u \circ f - u \circ f_1\|_0.$$

By uniform continuity of u the right side approaches zero as (f_1, μ_1) approaches (f, μ).

An open set U of X_2 has $f_*\mu$ measure zero if and only if $f^{-1}(U)$ has μ measure zero, implying (8.11).

Hence, In_P is closed and for each f, $\mathrm{In}_P(f)$ is convex by (8.9). Finally, for $\mu \in \mathrm{In}_P(f)$ (8.11) says $|\mu| = f(|\mu|)$. □

By regarding the point measure map $\delta: X \to P(X)$ as an inclusion, (8.10) says that $f_*: P(X) \to P(X)$ is an extension of $f: X \to X$, continuous when f is. The invariant probability measures, the element of $\mathrm{In}_P(f)$, are just the fixed points of f_*. Now f need not have any fixed points. However, as was first shown by Krylov and Bougliabov, f_* always does, i.e., $\mathrm{In}_P(f)$ is nonempty. As we will now see, this is because, by (8.9), f_* preserves the linear structure on $P(X)$.

Given μ in $P(X)$ the limit point set of the orbit $\{f_*^n\mu\}$, $\omega(f_*)(\mu)$, need not contain invariant measures (e.g. $\mu = \delta_x$) but we can use the convex structure to replace this sequence by the sequence of Cesaro averages. Define for each positive integer n the continuous map:

$$(8.12) \qquad \sigma_n: C(X; X) \times P(X) \to P(X), \qquad \sigma_n(f, \mu) = \frac{1}{n}\sum_{i=0}^{n-1} f_*^i\mu.$$

Because f_* preserves convex combinations:

$$(8.13) \qquad \sigma_n(f, \mu) - f_*\sigma_n(f, \mu) = \sigma_n(f, \mu) - \sigma_n(f, f_*\mu) = \frac{1}{n}[\mu - f_*^n\mu].$$

Hence, if $\overline{\mu}$ is the limit of a subsequence $\{\sigma_{n_i}\}$ and so $f_*\overline{\mu}$ is the limit of $\{f_*\sigma_{n_i}\}$, then $\overline{\mu} = f_*\overline{\mu}$ because (8.13) implies $d_u(\sigma_n, f_*\sigma_n) \le 2\|u\|_0/n$. So every limit point of $\{\sigma_n\}$ is an invariant measure. Define $M(f, \mu)$ to be the set of such limit points:

$$(8.14) \qquad \begin{gathered} M: C(X; X) \times P(X) \to C(P(X)), \\ M(f, \mu) = \limsup_n \{\sigma_n(f, \mu)\} \subset P(X). \end{gathered}$$

3. Proposition. *For each $(f, \mu) \in C(X; X) \times P(X)$, $M(f, \mu)$ is a nonempty, closed, connected subset of $P(X)$ consisting entirely of f invariant measures. If $j: D \to C(X; X) \times P(X)$ is continuous, then there is a residual subset of D at points of which $M \circ j$ is usc.*

Proof. The limit point set of $\{\sigma_n\}$ is nonempty by compactness, is always closed and consists of invariant measures by the above remarks. Furthermore, while the sequence need not converge, we do have

$$(8.15) \qquad \sigma_{n+1}(f, \mu) - \sigma_n(f, \mu) = \frac{1}{n+1}[f_*^n\mu - \sigma_n(f, \mu)].$$

(Write $(n+1)(\sigma_{n+1} - \sigma_n)$ as $(n+1)\sigma_{n+1} - n\sigma_n - \sigma_n$.) With respect to the metric d_S of (8.6), (8.15) implies $d_S(\sigma_{n+1}, \sigma_n) \le 2/(n+1)$. That $M(f, \mu)$ is connected is an easy exercise (see exercise 4.27).

By Lemma 7.12, $\tilde{\sigma}_n: C(X; X) \times P(X) \to C(P(X))$ defined by $\tilde{\sigma}_n(f, \mu) = \bigvee_i \{\sigma_i(f, \mu): i \geq n\}$ is lsc for each n. By Theorem 7.19 there is a residual subset D_1 of D at points of which every $\tilde{\sigma}_n \circ j$ is continuous. At such points $M \circ j$ is usc by Lemma 7.12 again because $M \circ j = \bigwedge_n \{\tilde{\sigma}_n \circ j\}$. □

When $f \in C(X; X)$ is fixed, we will write $M(\mu)$ for $M(f, \mu)$, so that M is a map from $P(X)$ to $C(\text{In}_P(f))$. Because $\{f_* \sigma_n(f, \mu) = \sigma_n(f, f_* \mu)\}$ has the same limit points as $\{\sigma_n(f, \mu)\}$ (c.f. (8.13)) the map is f_* invariant:

$$M \circ f_* = M, \quad \text{i.e.,}$$
$$M(f, f_* \mu) = M(f, \mu) \quad \text{for } (f, \mu) \in C(X; X) \times P(X). \tag{8.16}$$

For $x \in X$ we will write $M(x)$ for $M(\delta_x)$. This point measure case is of special interest because if u is a bounded, real, measurable function then

$$\langle u, \sigma_n(f, \delta_x) \rangle = \frac{1}{n} \sum_{i=0}^{n-1} u(f^i x) \tag{8.17}$$

and so the limit, if it exists, is the time-average of u over the positive orbit of x. In any case we can define:

(8.18)
$$u^M(x) = \limsup_n \left\{ \frac{1}{n} \sum_{i=1}^{n-1} u(f^i x) \right\}, \qquad u^m(x) = \liminf_n \left\{ \frac{1}{n} \sum_{i=1}^{n-1} u(f^i x) \right\},$$

where we are using here the real number concept of lim sup and lim inf rather than the set one. That is, $u^M(x)$ and $u^m(x)$ are, respectively, the largest and smallest elements of the limit point set of the sequence $\{\langle u, \sigma_n(f, \delta_x) \rangle\}$ in $\mathbb{R}$. (This set is compact because the sequence remains in $[-\|u\|_0, \|u\|_0]$.) Thus $u^M(x) = u^m(x)$ exactly when the sequence of time averages of (8.17) converges.

We now describe a number of genericity results based on Birkhoff's Pointwise Ergodic Theorem. Recall that for topological results we are especially interested in the class of residual subsets because each such set is dense and the class is closed under countable intersections, unlike the class of merely dense sets. In the measure theoretic context we look for sets of full measure. A Borel subset A of X is called of *full measure* for f if $\mu(A) = 1$ for every f invariant μ in $P(X)$. This class of sets is again closed under countable intersection.

4. Proposition. *Let $f: X \to X$ be a continuous map.*

(a) *For any bounded, real valued, measurable function u on X, u^M and u^m are f invariant measurable functions (i.e., $u^M \circ f = u^M$ and $u^m \circ f = u^m$) and $\{x: u^M = u^m\}$ is an f^{-1} invariant Borel set of full measure. If $\mu \in \text{In}_P(f)$, then*

$$\int u^M(x) \mu(dx) = \int u(x)\, \mu(dx) = \int u^m(x)\, \mu(dx). \tag{8.19}$$

(b) *We will call* $x \in X$ *a* convergence point for f *if it satisfies the following equivalent conditions*:

(1) $M(x)$ *consists of a single measure.*

(2) $\{\sigma_n(f, \delta_x)\}$ *converges in* $P(X)$.

(3) $u^M(x) = u^m(x)$ *for all* u *in* $C(X; \mathbb{R})$.

(4) $u^M(x) = u^m(x)$ *for all* u *in some generating sequence* S *for* $C(X; \mathbb{R})$.

The set of convergence points for f, *denoted* $\mathrm{Con}(f)$, *is an* f^{-1} *invariant Borel set of full measure.*

Proof. $u_n = (1/n)\sum_{i=1}^{n-1} u \circ f^i$ is measurable for each n and for every f invariant measure μ, $\langle u_n, \mu\rangle = \langle u, \mu\rangle$. The lim sup and lim inf operators on sequences of real functions preserve measurability. So u^M and u^m are measurable with $u^M \geq u^m$. f invariance follows from (8.13) by analogy with (8.16). By the Lebesgue Dominated Convergence Theorem it follows that for any invariant measure μ,

$$\langle u^M, \mu\rangle \geq \langle u, \mu\rangle \geq \langle u^m, \mu\rangle$$

with equality equivalent to $u^M = u^m \bmod \mu$. That convergence takes place for almost every x (modulo μ) is the Birkhoff theorem (see Billingsley (1965) §1). Consequently, the Borel set $\{x : u^M(x) = u^m(x)\}$ has μ measure 1 for every invariant μ. It is f^{-1} invariant because $u^M(x) = u^m(x)$ if and only if $u^M(f(x)) = u^m(f(x))$.

For (b) the implications $(1) \Rightarrow (2) \Rightarrow (3) \Rightarrow (4)$ are obvious. Given (4) the sequence $\{\sigma_n(f, \delta_x)\}$ converges with respect to the metric d_S of (8.6) and this implies (1). So $\mathrm{Con}(f) = \bigcap_{u \in S}\{x : u^M(x) = u^m(x)\}$ and this is a Borel set of full measure by (a). □

For further results we require some measurability technicalities whose proofs we will only sketch:

5. **Exercise.** (a) *Prove that the* $C_+(X)$ *open sets and also the* $C_-(X)$ *open sets generate the Borel sets of* $C(X)$. (*Hint*: *A* C_- *open set is the countable union of* C_+ *closed sets.*) *Show that* $f_n : X_1 \to C(X)$ *measurable for* $n = 1, \ldots$ *implies* $\vee\{f_n\}$ *and* $\bigwedge\{f_n\}$ *are measurable.*

(b) *Prove that* $M : C(X; X) \times P(X) \to C(P(X))$ *is measurable. For fixed* $f \in C(X; X)$, *prove that* $M : \mathrm{Con}(f) \to \mathrm{In}_P(f)$ *is measurable.*

(c) *Suppose that* V *is a subspace of the vector space of all bounded real functions on* X. *Suppose that* $C(X; \mathbb{R}) \subset V$ *and that* V *is closed under countable, monotone, pointwise limits. Prove that* V *includes all bounded measurable functions.* (*Hint*: *Use the following steps.* (1) *Let* $\widetilde{V} = \{A : A \subset X$ *and the characteristic function of* A *lies in* $V\}$. *Because any bounded measurable function is a monotone limit of linear combinations of characteristic functions of Borel sets, it suffices to prove that* $\widetilde{V}$ *contains the Borel sets.*

(2) *$\tilde{V}$ is closed under countable, monotone set limits.* (3) *$\tilde{V}$ contains the class of G_δ sets which is closed under finite union and intersection and contains the closed sets.* (5) *$\tilde{V}$ contains the algebra generated by the open sets.* (6) *From* (2) *and* (5) *conclude that $\tilde{V}$ contains the σ algebra generated by the open sets, using, for example, Halmos* (1950) §6.)

(d) *If u is any bounded, real, measurable function on X, prove that $\langle u, \ \rangle$ is a bounded, real, measurable function on $P(X)$. In particular, for any Borel set A, $\mu \to \mu(A)$ is a measurable function of μ in $P(X)$.* □

For x in $\mathrm{Con}(f)$ we denote by μ_x the measure in $M(x)$ so that $\mu_x = \lim\{\sigma_n(f, \delta_x)\}$.

6. LEMMA. *For any bounded measurable u, the function $\langle u, \mu_x\rangle$ of x in* $\mathrm{Con}(f)$ *is measurable and equals $u^M(x)$ on a subset of* $\mathrm{Con}(f)$ *of full measure (depending on u). That is, on a set of full measure*

$$\int u(y)\,\mu_x(dy) = \lim\left\{\frac{1}{n}\sum_{i=0}^{n-1} u(f^i x)\right\} = u^M(x). \tag{8.20}$$

Furthermore, if u is continuous then the equations of (8.20) *hold for all x in* $\mathrm{Con}(f)$.

PROOF. Measurability follows from (b) and (d) of Exercise 5, and for u in $C(X;\mathbb{R})$, $\langle u, \mu_x\rangle = \lim\{\langle u, \sigma_n(f, \delta_x)\rangle\} = u^M(x)$ for all x in $\mathrm{Con}(f)$ by definition of the set.

Now let V be the set of bounded measurable functions u such that (8.20) holds for u on a subset of full measure. $C(X;\mathbb{R}) \subset V$ by the previous paragraph. Observe that for each u convergence to u^M occurs on a set of full measure by (a) of Proposition 4. By Exercise 5(c) it suffices to prove that V closed under linear combinations and countable, monotone, pointwise limits.

If $\langle u_i, \mu_x\rangle = \lim\{\langle u_i, \sigma_n(f, \delta_x)\rangle\}$ for x in E_i $(i = 1, 2)$, then the equation holds for $au_1 + bu_2$ on $E_1 \cap E_2$ which is of full measure if E_1 and E_2 are of full measure.

Now suppose $\{u_n\}$ is an increasing sequence of functions in V and u is a bounded measurable function with $u(x) = \lim\{u_n(x)\}$ for all x in X. $\{u_n^M\}$ is an increasing sequence with $u_n^M(x) \le u^M(x)$ for all n. Let $v(x) = \lim\{u_n^M(x)\}$. Clearly $v \le u^M$.

If E_n is a subset of $\mathrm{Con}(f)$ of full measure on which $\langle u_n, \mu_x\rangle = u_n^M(x)$ then $E_\infty = \bigcap\{E_n\}$ is a subset of full measure and for $x \in E_\infty$ the Dominated Convergence Theorem implies

$$\langle u, \mu_x\rangle = \lim\{\langle u_n, \mu_x\rangle\} = \lim\{u_n^M(x)\} = v(x).$$

So it suffices to prove that $\mu(\{x\colon v(x) = u^M(x)\}) = 1$ for any invariant measure μ. Because $v \le u^M$ it is enough to show that $\langle v, \mu\rangle = \langle u^M, \mu\rangle$.

By dominated convergence and (8.19)

$$\begin{aligned}\langle v, \mu\rangle &= \lim\{\langle u_n^M, \mu\rangle\} = \lim\{\langle u_n, \mu\rangle\} \\ &= \langle u, \mu\rangle = \langle u^M, \mu\rangle .\end{aligned}$$

REMARK. These results imply that every invariant measure μ can be described as an average of the measures μ_x. For any Borel set A we can integrate $\mu_x(A)$ over $x \in \operatorname{Con}(f)$ with respect to μ and define a measure on X. But integrating the extremes of (8.20) we see that we have just returned to μ. We write this formally as:

$$\mu = \int_{\operatorname{Con}(f)} \mu_x\, \mu(dx) \qquad (\mu \in \operatorname{In}_P(f)). \quad \square \tag{8.21}$$

The measure theoretic version of transitivity is ergodicity.

7. PROPOSITION. *Let f be a continuous map on X. An f invariant measure μ is called* ergodic *(with respect to f) if it satisfies the following equivalent conditions*:

(1) *If a bounded, real measurable function u is f invariant ($u \circ f = u$) then $\{x: u(x) = \langle u, \mu\rangle\}$ has μ measure 1. Thus, u is constant a.e. (μ).*
(2) *For each u in some generating sequence S for $C(X; \mathbb{R})$ $\{x: u^M(x) = \langle u, \mu\rangle\}$ has μ measure 1.*
(3) *$M^{-1}(\mu) = \{x \in \operatorname{Con}(f): \mu_x = \mu\}$ has μ measure 1.*
(4) *If A is an f^{-1} invariant Borel set then $\mu(A) = 0$ or 1.*
(5) *If A and B are Borel sets of positive μ measure, then $\limsup\{\mu(A \cap f^{-n}(B))\} > 0$.*
(6) *For Borel sets A and B $\lim\{(1/n)\sum_{i=0}^{n-1} \mu(A \cap f^{-i}(B))\}$ exists and equals $\mu(A)\mu(B)$.*

μ is called mixing *if it satisfies the stronger condition.*

(6′) *For Borel sets A and B, $\lim\{\mu(A \cap f^{-n}(B))\}$ exists and equals $\mu(A)\mu(B)$.*

Thus, every ergodic measure μ is equal to μ_x for some x in $\operatorname{Con}(f)$. Conversely, $\{x \in \operatorname{Con}(f): \mu_x$ is ergodic$\}$ is a Borel set of full measure.

PROOF. We will prove $(1) \Rightarrow (2) \Rightarrow (3) \Rightarrow (6) \Rightarrow (5) \Rightarrow (4) \Rightarrow (1)$.

$(1) \Rightarrow (2)$: u^M is f invariant and $\langle u^M, \mu\rangle = \langle u, \mu\rangle$.

$(2) \Rightarrow (3)$. Because $x \to \mu_x$ is measurable the set $M^{-1}(\mu)$ is a Borel set. Let $E = \{x \in \operatorname{Con}(f): u^M(x) = \langle u, \mu\rangle$ for all u in $S\}$. By (2) and Proposition 4 this is the countable intersection of sets of μ measure 1 and so $\mu(E) = 1$. Fix $x \in E$. For all u in S,

$$\langle u, \mu_x\rangle = u^M(x) = \langle u, \mu\rangle$$

by Lemma 6 and the definition of E. As S generates $C(X;\mathbb{R})$ this means $\mu_x = \mu$ and so $x \in M^{-1}(\mu)$. So $E \subset M^{-1}(\mu)$ and $\mu(M^{-1}(\mu)) = 1$.

(3) $\Rightarrow$ (6). Let c_A and c_B be the characteristic functions of A and B, respectively. By (3) and Lemma 6 applied to c_B there is a Borel set E with $\mu(E) = 1$ such that $x \in E$ implies

$$\mu(B) = \langle c_B, \mu\rangle = \langle c_B, \mu_x\rangle = \lim \left\{ \frac{1}{n} \sum_{i=0}^{n-1} c_B(f^i(x)) \right\}.$$

Multiply by $c_A(x)$ and integrate with respect to $\mu(dx)$ (because $\mu(E) = 1$ the integral over E is the same as the integral over X):

$$\begin{aligned}\mu(A)\mu(B) = \int c_A(x)\mu(B)\,\mu(dx) &= \lim \left\{ \frac{1}{n} \sum_{i=0}^{n-1} \int c_A(x) c_B(f^i(x))\,\mu(dx) \right\} \\ &= \lim \left\{ \frac{1}{n} \sum_{i=1}^{n-1} \mu(A \cap f^{-i}(B)) \right\}\end{aligned}$$

by dominated convergence.

(6) $\Rightarrow$ (5) (and (6$'$) $\Rightarrow$ (6)). If a sequence of real numbers converges then the sequence of Cesaro averages converges to the same limit. In particular, (6$'$) $\Rightarrow$ (6). Also the alternative to the conclusion of (5) is $\lim\{\mu(A \cap f^{-n}(B))\} = 0$ which contradicts (6) as $\mu(A)\mu(B) > 0$.

(5) $\Rightarrow$ (4). If A is f^{-1} invariant then so is its complement B. So $A \cap f^{-n}(B) = A \cap B = \varnothing$ for all n. By (5) this means either $\mu(A) = 0$ or $\mu(B) = 0$.

(4) $\Rightarrow$ (1). If u is f invariant and measurable then the three sets $A_+ = \{x : u(x) > \langle u, \mu\rangle\}$, $A_0 = \{x : u(x) = \langle u, \mu\rangle\}$ and $A_- = \{x : u(x) < \langle u, \mu\rangle\}$ are f^{-1} invariant Borel sets. If $\mu(A_+) = 1$, then

$$\langle u, \mu\rangle = \int_{A_+} u(x)\,\mu(dx) > \langle u, \mu\rangle,$$

which is impossible. So by (4) $\mu(A_+) = 0$. Similarly, $\mu(A_-) = 0$ and we have $\mu(A_0) = 1$.

(3) implies that any ergodic measure equals μ_x for some x in $\mathrm{Con}(f)$. Now for e in $\mathrm{Con}(f)$, (2) says that μ_e is ergodic if and only if for all u in S $\mu_e(\{x : u^M(x) = \langle u, \mu_e\rangle\}) = 1$. Since $\langle u, \mu_e\rangle = u^M(e)$ for u in $C(X;\mathbb{R})$ this is equivalent to $V_u(e) = 0$ where

$$V_u(e) = \int (u^M(x) - u^M(e))^2\,\mu_e(dx).$$

It thus suffices to show that for each $u \in C(X;\mathbb{R})$ V_u is measurable on $\mathrm{Con}(f)$ and $V_u = 0$ on a set of full measure. Squaring out we have

$$\begin{aligned}V_u(e) &= \langle (u^M)^2, \mu_e\rangle - 2u^M(e)\langle u^M, \mu_e\rangle + u^M(e)^2 \\ &= \langle (u^M)^2, \mu_e\rangle - (u^M)^2(e)\end{aligned}$$

since $\langle u^M, \mu_e\rangle = \langle u, \mu_e\rangle = u^M(e)$. Apply Lemma 6 to $v = (u^M)^2$ (which is invariant and so $v^M = v$). Conclude that V_u is measurable in e and vanishes on a set of full measure as required.

REMARK. Recall that when f is a homeomorphism, A is f^{-1} invariant if and only if it is f invariant. So by (4), μ is ergodic with respect to f if and only if it is ergodic for f^{-1}. Alternatively, observe that for a homeomorphism $f^n(A \cap f^{-n}(B)) = f^n(A) \cap B$ and use (6). This approach also shows that μ is mixing for f if and only if it is mixing for f^{-1}. □

We are now ready to deduce dynamic conclusions from measure theoretic hypotheses. The fundamental inspiration for all the applications is Poincaré's Recurrence Theorem. It comes from the observation that if U is any wandering open set for a continuous map f, i.e., the sequence $\{f^{-n}(U): n = 0, 1\dots\}$ is pairwise disjoint, then for any invariant measure μ the measures of the terms of this sequence are all the same and their sum is finite. So the common value must be zero. This means that the points of the support of μ must be nonwandering. In fact, if we restrict f to the f invariant set $|\mu|$ we have

$$\Omega(f_{|\mu|}) = |\mu| \tag{8.22}$$

which means that $f_{|\mu|}$ is central; $|\mu|$ is the closure of its recurrent points (cf. Proposition 4.17).

To sharpen these results we first describe the familiar relations ωf and Ωf by using the function θ defined on subsets of X by

$$\theta(A) = \begin{cases} 0, & A = \varnothing, \\ 1, & A \neq \varnothing. \end{cases} \tag{8.23}$$

Observe that for any μ in $P(X)$, $\mu \leq \theta$ and

$$y \in \omega f(x) \Leftrightarrow \limsup\{\theta(x \cap f^{-n}(U))\} > 0 \text{ for all neighborhoods } U \text{ of } y. \tag{8.24}$$

$$y \in \Omega f(x) \Leftrightarrow \limsup\{\theta(V \cap f^{-n}(U))\} > 0 \text{ for all neighborhoods } U \text{ of } y \text{ and } V \text{ of } x. \tag{8.25}$$

Now we refine these relations by analogy with the conditions for ergodicity. Define $\omega_{\#}f$, $\Omega_{\#}f$, and $\Omega_{\#\#}f$ by

$$y \in \omega_{\#}f(x) \Leftrightarrow \limsup\{(1/n)\textstyle\sum_{i=0}^{n-1}\theta(x \cap f^{-i}(U))\} > 0 \text{ for all neighborhoods } U \text{ of } y. \tag{8.26}$$

$$y \in \Omega_{\#} f(x) \quad \Leftrightarrow \quad \limsup\{(1/n)\sum_{i=0}^{n-1} \theta(V \cap f^{-i}(U))\} > 0 \text{ for all neighborhoods } U \text{ of } y \text{ and } V \text{ of } x. \tag{8.27}$$

$$y \in \Omega_{\#\#} f(x) \quad \Leftrightarrow \quad \lim\{\theta(V \cap f^{-n}(U))\} = 1 \text{ for all neighborhoods } U \text{ of } y \text{ and } V \text{ of } x. \tag{8.28}$$

The condition of (8.25) simply says that the sequence $\{f^{-n}(U)\}$ intersects V infinitely often. (8.27) further demands that the average frequency with which the intersections occur does not tend to zero. The relation between (8.24) and (8.26) is analogous. (8.28) demands that intersections occur for all but finitely many members of the sequence. Clearly, $\omega_{\#} f$ is a pointwise closed relation as is ωf, while $\Omega_{\#} f$ and $\Omega_{\#\#} f$ are closed relations. Because convergence of a sequence implies convergence of the Cesaro averages to the same limit we have

$$\begin{array}{ccccc} \Omega_{\#\#} f & \subset & \Omega_{\#} f & \subset & \Omega f \\ & & \cup & & \cup \\ & & \omega_{\#} f & \subset & \omega f. \end{array} \tag{8.29}$$

While it often happens that $\Omega_{\#\#} f(x) = \varnothing$, the following interpretation of $\omega_{\#} f$ shows in particular that $\omega_{\#} f(x) \neq \varnothing$ for any x and so, a fortiori, $\Omega_{\#} f(x) \neq \varnothing$.

8. Proposition. *Let f be a continuous map on X.*

(a) $\omega_{\#} f(x) = \bigvee\{|\mu|: \mu \in M(x)\}$, *the closure of the union of the supports of limit points of the sequence $\{\sigma_n(f, \delta_x)\}$. In particular, for $x \in \operatorname{Con}(f)$, $\omega_{\#} f(x) = |\mu_x|$.*

(b) $m[f] \subset |\omega_{\#} f| \subset |\omega f|$, *where $m[f]$ is the union of all minimal subsets of f.*

(c) $|\omega_{\#} f|$ *is a Borel set of full measure. In particular, if μ is any f invariant measure, $|\mu| \subset \overline{|\omega_{\#} f|}$. Furthermore, there exists an f invariant measure with support equal to $\overline{|\omega_{\#} f|}$.*

Proof. For u a nonnegative element of $C(X; \mathbb{R})$ let $U_\varepsilon = \{x: u(x) > \varepsilon\}$ and observe that

$$\varepsilon\theta(x \cap f^{-i}(U_\varepsilon)) \leq u(f^i(x)) \leq \|u\|_0 \theta(x \cap f^{-i}(U_0)).$$

By averaging these inequalities it is easy to deduce that

$$y \notin \omega_{\#} f(x) \quad \Leftrightarrow \quad u(y) > 0 \text{ and } u^M(x) = 0 \text{ for some nonnegative } u \text{ in } C(X; \mathbb{R}). \tag{8.30}$$

Furthermore, for u nonnegative in $C(X;\mathbb{R})$ we have

$$\begin{aligned} u^M(x)=0 &\Leftrightarrow \lim\{\langle u, \sigma_n(f, \delta_x)\rangle\}=0 \\ &\Leftrightarrow \langle u, \mu\rangle = 0 \quad \text{for all } \mu \text{ in } M(x) \\ &\Leftrightarrow \mu(U_0)=0 \quad \text{for all } \mu \text{ in } M(x). \end{aligned}$$

Since every open set is the support of some nonnegative function we see that the complement of $\omega_\# f(x)$ is the union of all open sets U such that $\mu(U)=0$ for all μ in $M(X)$, proving (a).

If A is a minimal invariant set, then for $x \in A$ and $\mu \in M(x)$ the support of μ is an invariant subset of A and so equals A. Hence, $x \in A = \omega_\# f(x)$. Thus, $m[f] \subset |\omega_\# f|$. $\omega_\# f \subset \omega f$ implies the second inclusion.

For u a nonnegative element of $C(X;\mathbb{R})$ let $Z_u = \{x\colon u(x)>0 \text{ and } u^M(x)=0\}$. By (8.20) $x \notin \omega_\# f(x)$ if and only if x lies in some Z_u in which case u can be chosen to lie in the generating set S of nonnegative functions constructed in the proof of Proposition 1. Thus, $X - |\omega_\# f| = \bigcup\{Z_u\} = \bigcup\{Z_u\colon u \in S\}$. As each Z_u is a Borel set, so is the countable union $X - |\omega_\# f|$. To see that $|\omega_\# f|$ has full measure it suffices to show $\mu(Z_u)=0$ for any continuous nonnegative u and any invariant measure μ.

Let $A = \{x\colon u^M(x)=0\}$ and c_A be its characteristic function. Since A is f^{-1} invariant we have

$$(c_A u)^M(x) = c_A(x)u^M(x) = 0$$

for all x. Hence, $\langle c_A u, \mu\rangle = \langle (c_A u)^M, \mu\rangle = 0$. Since $c_A u$ is nonnegative this means $\{x\colon c_A(x)u(x)>0\}$ has μ measure 0. But this set is Z_u.

Thus, any open set disjoint from $|\omega_\# f|$ has μ measure zero, for $\mu \in \mathrm{In}_p(f)$. So $|\mu| \subset \overline{|\omega_\# f|}$.

Finally, we can choose a sequence of invariant measures $\{\mu_i\}$ the union of whose supports is dense in $\overline{|\omega_\# f|}$. Then $\mu = \sum 2^{-i}\mu_i$ is an invariant measure with $|\mu| = \bigvee\{|\mu_i|\} = \overline{|\omega_\#(f)|}$. □

Notice that (a) and (c) together imply that $|\omega_\# f|$ is always dense in $\omega_\# f(X)$. However, we will later see that $|\Omega_\# f| = X$ need not imply $|\omega_\# f|$ is dense in X (compare Proposition 4.17).

9. Corollary. *For f a continuous map on X, suppose that X is the support of an invariant measure μ. Then $|\omega_\# f|$ is dense in X and $|\Omega_\# f| = X$. If μ is ergodic then $\Omega_\# f = X \times X$ and if μ is mixing $\Omega_{\#\#} f = X \times X$.*

Proof. That $|\omega_\# f|$ is dense in X follows (c) of the proposition. As $\omega_\# f \subset \Omega_\# f$ and the latter is a closed relation $|\Omega_\# f| = X$.

Because $|\mu| = X$, which is to say μ is a dense measure, $\mu(U)\mu(V) > 0$ for any pair of nonempty open sets. Furthermore, $\theta \geq \mu$ and so (6) and (6′) of Proposition 7 imply (8.27) and (8.28) respectively, for all pairs (x, y) in $X \times X$. □

Motivated by this result we will call f *topologically ergodic* (or *topologically mixing*) when $\Omega_{\#}f = X \times X$ (resp., $\Omega_{\#\#}f = X \times X$). Clearly, mixing implies ergodic implies topological transitivity for f. If f is a homeomorphism on X, then

$$(8.31) \qquad \Omega_{\#}(f^{-1}) = (\Omega_{\#}f)^{-1} \quad \text{and} \quad \Omega_{\#\#}(f^{-1}) = (\Omega_{\#\#}f)^{-1}$$

so that f^{-1} is ergodic (or mixing) when f is.

The mixing property is important because it is preserved under products.

10. LEMMA. *If $f_1 : X_1 \to X_1$ and $f_2 : X_2 \to X_2$ are topologically mixing, then so are the product $f_1 \times f_2 : X_1 \times X_2 \to X_1 \times X_2$ and the powers $f_1^n : X_1 \to X_1$ $(n = 1, 2, \ldots)$.*

PROOF. The power result is obvious and because $\{U_1 \times U_2 : U_1$ open in X_1 and U_2 open in $X_2\}$ is a base for the topology of $X_1 \times X_2$ it is easy to check that, in general,

$$\Omega_{\#\#}(f_1 \times f_2) = \Omega_{\#\#}f_1 \times \Omega_{\#\#}f_2$$

up to the obvious rearrangement of the factors. From this the lemma clearly follows.

REMARK. The analogous result is true for the measure theoretic version of mixing as well. □

In particular, if f is mixing then $f \times f$ on $X \times X$ is mixing and a fortiori is topologically transitive. This implies that for pairs of points (x, y) in some residual subset of $X \times X$, the orbit $\{(f^n(x), f^n(y)) : n = 0, 1 \ldots\}$ is dense in $X \times X$. In particular, there will be such pairs arbitrarily close to the diagonal 1_X and for such a pair the exact location of the $f^n(x)$'s provides no long run information about the location of the $f^n(y)$'s.

We earlier introduced sets of full measure as the measure theoretic analogue of residual sets. Unfortunately, the concepts do not overlap very well. It is a standard exercise to construct a residual subset of the unit interval of Lebesgue measure zero. In the dynamic context such sets arise naturally.

11. THEOREM. *Let f be a topologically transitive map on X. There exists a nonempty, closed, connected subset I^* of $\mathrm{In}_P(f)$ such that the Borel set $M^{-1}(I^*) = \{x : M(x) = I^*\}$ is residual; it contains a countable intersection of dense open sets. If $x \in X$ satisfies $\omega f(x) = X$, then $M(x) \subset I^*$. In particular, any dense ergodic measure lies in I^*. If I^* contains more than one element (the usual situation), then $\mathrm{Con}(f) \cap M^{-1}(I^*) = \varnothing$. So in that case, the residual set $M^{-1}(I^*)$ has measure zero with respect to every invariant measure μ, and the set of full measure $\mathrm{Con}(f)$ is of first category.*

PROOF. By Proposition 3 there is a residual subset R_1 at points of which $M : X \to C(\mathrm{In}_P(f))$ is usc. Because f is topologically transitive $R_2 = \{x : \omega f(x) = X\}$ is residual. Let $x_i \in R_i$ $i = 1, 2$. Because the orbit of

x_2 is dense in X there is a sequence $\{f^{n_k}(x_2)\}$ which converges to x_1. As M is usc at x_1, $M(x_1)$ contains the lim sup of the sequence $\{M(f^{n_k}(x_2))\}$. But M is an invariant function so this sequence is constantly $M(x_2)$. Thus, $M(x_2) \subset M(x_1)$. In particular, M is constant on the residual set $R_1 \cap R_2$ and I^* is the constant value. I^* is nonempty, closed and connected in $\text{In}_p(f)$ because this is true of every value of M. The previous argument also shows $M(x_2) \subset I^*$ for every x_2 in R_2.

If I^* contains more than one measure then $M(x) \neq I^*$ for $x \in \text{Con}(f)$. So $\text{Con}(f) \cap M^{-1}(I^*) = \emptyset$. Of course, it may happen that $M(x) \subset I^*$ for $x \in \text{Con}(f)$. In particular, if μ_x is a dense measure for some x in $\text{Con}(f)$ then by Proposition 8(a) $\omega_\# f(x) = |\mu_x| = X$. So $x \in R_2$ and $\mu_x \in I^*$. If μ is a dense ergodic measure then $\mu = \mu_x$ for some x and so $\mu \in I^*$.

REMARKS. (a) The above proof actually yields a sharper test for membership in I^*. If $\{x\colon \mu \in M(x)\}$ is dense in X, then $\mu \in I^*$, because μ is an element of the lim sup of $\{M(x_i)\}$ where $\{x_i\}$ is any sequence in this set. Also if $\bigvee\{|\mu|\colon \mu \in M(x)\} = X$ (e.g., if $M(x)$ contains a dense measure) then by Proposition 8(a) $x \in R_2$ and so $M(x) \subset I^*$. In particular, if f is minimal, $M(x) \subset I^*$ for all x.

(b) Observe that for $x \in M^{-1}(I^*)$, $\omega_\# f(x) = \bigvee\{|\mu|\colon \mu \in I^*\} \equiv |I^*|$. So if $|I^*| = X$, $M^{-1}(I^*) \subset R_2$.

(c) If I^* consists of a single measure, then $M^{-1}(I^*) \subset \text{Con}(f)$ and so $\text{Con}(f)$ is residual. Conversely, if $\text{Con}(f)$ is residual then $\text{Con}(f) \cap M^{-1}(I^*) \neq \emptyset$ and so I^* consists of a single measure. In addition, since $M(x_2) \subset I^*$ for x_2 in R_2, we have $R_2 \subset M^{-1}(I^*)$ when I^* is a singleton. □

12. COROLLARY. *For a continuous map f on X the following conditions are equivalent. When they hold we call f* strictly ergodic.

(1) *f admits a dense ergodic measure μ and $M^{-1}(I^*)$ has nonempty interior.*

(2) *f is minimal and $\text{Con}(f) = X$.*

(3) *f is minimal and I^* consists of a single measure.*

(4) *f admits a unique ergodic measure μ and $|\mu| = X$.*

(5) *f admits a unique invariant measure μ and $|\mu| = X$.*

PROOF. (1) $\Rightarrow$ (2). Since μ is a dense ergodic measure $\mu \in I^*$. Let A be the closure of the complement of $M^{-1}(I^*)$. A is f^{-1} invariant and as $M^{-1}(I^*)$ has interior points, A is a proper subset of X. Let U be a closed neighborhood of A which is a proper subset of X. If $x \in \bigcap_{n\geq 0} f^{-n}(U)$ then $\omega_\# f(x) \subset \omega f(x) \subset U$ and so by Proposition 8(a) $|\mu_1| \neq X$ for any μ_1 in $M(x)$. As $|\mu| = X$ and $\mu \in I^*$, $x \notin M^{-1}(I^*)$ and so $x \in A$. So A is a repellor by Theorem 3.6(b)(3). But by Corollary 9, f is topologically transitive. Hence A is empty and so $X = M^{-1}(I^*)$. If I^* contained more than one measure then $\text{Con}(f) \cap M^{-1}(I^*) = \emptyset$ and $\text{Con}(f) \neq \emptyset$ would contradict $X = M^{-1}(I^*)$. Hence, $I^* = \mu$ and every point converges to I^*, i.e., $\text{Con}(f) = X$. So $X = \omega_{\#}(f)(x) \subset \omega f(x)$ for every x and f is minimal as well.

(2) $\Rightarrow$ (3). As in Remark (c) the theorem following $\text{Con}(f) = X$ implies I^* consists of a single measure.

(3) $\Rightarrow$ (4). If f is minimal, then every invariant measure is dense and so every ergodic measure lies in I^*.

(4) $\Rightarrow$ (5). By (8.21) we can write any invariant measure as an integral of measures μ_x. By Proposition 7 we can restrict the integral to $\{x \in \text{Con}(f)\colon \mu_x \text{ is ergodic}\}$. For all x in this set $\mu_x = \mu$ and so μ is the only invariant measure.

(5) $\Rightarrow$ (1). If μ is the unique invariant measure, then $M(x) = \mu$ for all x in X. So $I^* = \mu$ and $M^{-1}(I^*) = X$. By Proposition 7 ergodic measures always exist. So μ is ergodic. It is dense by assumption in (5). □

Leaving the details to the reader, we now indicate how to obtain from these results their analogues for flows.

13. EXERCISE. *Let* $\varphi\colon X \times [0, \infty) \to X$ *be a semiflow with time-one map* f.

(a) *If* μ_1 *is a measure on* X_1 *and* $u \in C(X \times X_1; \mathbb{R})$, *then* $S(u) \in C(X; \mathbb{R})$ *where* $S(u)(x) = \int u(x, x_1)\mu_1(dx_1)$. *Furthermore, if* u *is bounded and measurable, then* $S(u)$ *is measurable* (*use* (c) *of Exercise* 5) *and* $\|S(u)\|_0 \le \|u\|_0$. *In particular, for* u *bounded and measurable on* X, $S_t(u) = (1/t)\int_0^t u(f^s(x))\,ds$ *is measurable on* X, *continuous if* u *is, and* $\|S_t(u)\|_0 \le \|u\|_0$ (*for all* $t > 0$).

(b) $\varphi_*(\mu, t) = f_*^t(\mu)$ *defines a semiflow* $\varphi_*\colon P(X) \times [0, \infty) \to P(X)$. *If* φ^t *is the restriction of* φ *to* $X \times [0, t]$ *and* λ^t *is Lebesgue measure on* $[0, t]$ *times* $(1/t)$, *then*

$$\sigma_t(\varphi, \mu) \equiv (\varphi^t)_*(\mu \times \lambda^t) = \frac{1}{t}\int f_*^s(\mu)\,ds, \tag{8.32}$$

where the latter formal equation follows from Fubini's Theorem which implies that for u *any bounded measurable function on* X *we have*

$$\begin{aligned} \langle u, \sigma_t(\varphi, \mu)\rangle &= \frac{1}{t}\int_0^t \left[\int u(f^s(x))\,\mu(dx)\right] ds \\ &= \int \left[\frac{1}{t}\int_0^t u(f^s(x))\,ds\right] \mu(dx) = \langle S_t(u), \mu\rangle. \end{aligned} \tag{8.33}$$

For any $t \ge 1$, *the analogue of* (8.15) *yields*

$$|\langle u, \sigma_t(\varphi, \mu)\rangle - \langle u, \sigma_{[t]}(\varphi, \mu)\rangle| \le 2\|u\|_0/t, \tag{8.34}$$

where $[t]$ *is the integer part of* t.

We will denote $\sigma_1(\varphi, \mu)$ *by* $\sigma(\mu)$. *So* $\sigma(\mu) = \int_0^1 f_*^s(\mu)\,ds$. *Observe that*

$$\sigma_n(\varphi, \mu) = \sigma_n(f, \sigma(\mu)) = \sigma(\sigma_n(f, \mu)). \tag{8.35}$$

$\sigma: P(X) \to P(X)$ is continuous and linear (cf. (8.33)).

(c) *$\mu \in P(X)$ is called φ invariant if $f_*^t \mu = \mu$ for all $t \geq 0$. Thus, the set of φ invariant measures, denoted $\mathrm{In}_P(\varphi)$ is the set of fixed points of the semiflow φ_* while $\mathrm{In}_P(f)$ contains, in addition, all periodic points of φ_* with period equal to $1/n$ for some positive integer n. Observe that if $\mu \in \mathrm{In}_P(f)$ then $f_*^t \mu$ remains in $\mathrm{In}_P(f)$ for all $t \geq 0$ and $\sigma(\mu) \in \mathrm{In}_P(\varphi)$ with $\sigma(\mu) = \mu$ for $\mu \in \mathrm{In}_P(\varphi)$. Thus, σ is a linear retraction of the closed convex set $\mathrm{In}_P(f)$ onto the closed convex subset $\mathrm{In}_P(\varphi)$. A Borel set A is called of full measure with respect to φ if $\mu(A) = 1$ for all $\mu \in \mathrm{In}_P(\varphi)$. Since $\mathrm{In}_P(\varphi) \subset \mathrm{In}_P(f)$ A is of full measure with respect to φ if i is full measure with respect to f. If A is φ^{-1} invariant, meaning $(f^t)^{-1}(A) = A$ for all $t \geq 0$, then the converse is true as well (for c_A the characteristic function of A, $S_t(c_A) = c_A$ when A is φ^{-1} invariant).*

(d) *$M(\varphi, \mu)$, the set of limit points as $t \to \infty$ of the function $\sigma_t(\varphi, \mu)$, is the same as the set of limit points of the sequence $\{\sigma_n(\varphi, \mu)\}$ by (8.34). From (8.35) we have:*

$$(8.36) \qquad M(\varphi, \mu) = M(f, \sigma(\mu)) = \sigma(M(f, \mu)) \subset \mathrm{In}_P(\varphi).$$

As before this function is invariant: $M(\varphi, \mu) = M(\varphi, f_*^t \mu)$ for all t. Measurability of $M(\varphi, \ )$ follows from (8.36). We will write $M(\varphi, x)$ and $M(f, x)$ when $\mu = \delta_x$. x is called a convergence point for φ when $\lim_{t\to\infty}\{\sigma_t(\varphi, \delta_x)\}$ exists or equivalently when $M(\varphi, x)$ consists of a single measure. By (8.36) the set of φ convergence points, $\mathrm{Con}(\varphi)$, is a φ^{-1} invariant Borel set containing $\mathrm{Con}(f)$ and so is of full measure. Observe that $\mathrm{Con}(f)$ is φ^{-1} invariant as well and for $x \in \mathrm{Con}(f)$, $\mu^f_{f^t(x)} = f_*^t(\mu_x^f)$ and $\mu_x^\varphi = \sigma(\mu_x^f)$ where we write μ_x^f and μ_x^φ for the single measure in $M(f, x)$ and $M(\varphi, x)$ respectively.

(e) *For u bounded and measurable we write $u_\varphi^M(x)$ and $u_\varphi^m(x)$ for the* lim sup *and* lim inf *of the real sequence $\{S_n(u)(x)\}$. Prove that*

$$(8.37) \qquad S_1(u_f^M) \geq S_1(u)_f^M = u_\varphi^M \geq u_\varphi^m = S_1(u)_f^m \geq S_1(u_f^m),$$

where we are now writing u_f^M and u_f^m for the functions defined by (8.18).

If $u_1 = u_2$ on a set of full measure with respect to φ, then the same is true of the equation $S_1(u_1) = S_1(u_2)$. (For a nonnegative function u, $u = 0$ a.e. $(\sigma(\mu))$ if and only if $S_1(u) = 0$ a.e. (μ) by (8.33).) In particular, $u_\varphi^M = u_\varphi^m$ on a φ^{-1} invariant Borel set of full measure. State and prove the analogue of Lemma 6.

(f) *Describe the analogues of the conditions of Proposition 7 defining ergodicity (with respect to φ) of $\mu \in \mathrm{In}_P(\varphi)$. Observe that μ ergodic for f implies $f_*^t \mu$ is ergodic for f for all $t \geq 0$ and $S(\mu)$ is ergodic with respect to φ. (If A is φ^{-1} invariant, then $f_*^s \mu(A) = \mu(A)$ is constantly 0 or 1 for all s.) Conversely, if μ is ergodic with respect to φ, then $\mu = \sigma(\mu_1)$*

for some μ_1 *ergodic with respect to* f. (*Choose* $\mu_1 = \mu_x^f$ *for* x *in the set* $\{x: \mu_x^\varphi = \mu\} \cap \{x \in \text{Con}(f): \mu_x^f$ *is ergodic*$\}$ *which has* μ *measure* 1.) *Hence,* $\{x \in \text{Con}(\varphi): \mu_x^\varphi$ *is ergodic with respect to* $\varphi\}$ *is a* φ^{-1} *invariant Borel set of full measure.*

(g) *If* μ *is* φ *invariant then the support* $|\mu|$ *is* φ *invariant. If* μ *is* f *invariant, then* $|\sigma(\mu)| = f^I(|\mu|)$. (*If* u *is continuous and nonnegative, then* $\{S_1(u) > 0\} = \bigcup\{f^{-s}(\{u > 0\}): 0 \le s \le 1\}$ *and* $\langle u \circ f^s, \mu\rangle$ *is continuous in* s.) *So if the map* f *admits a dense* (*ergodic*) *measure, the flow* φ *admits a dense* (*ergodic*) *measure.*

(h) *If* f *is topologically transitive then on a residual subset of* X, $M(\varphi, x)$ *is constantly* $\sigma(I^*)$.

(i) *If* $\text{In}_P(f) = \text{In}_P(\varphi)$, *i.e., every* f *invariant measure is* φ *invariant, then* $M(f, \mu) = M(\varphi, \mu)$ *for all* $\mu \in P(X)$ (*use* (8.36)). □

As usual, there are useful special results for the case where φ is the flow associated with a smooth vectorfield ξ on a compact manifold X. The tool we need is the Lie derivative of a C^1 function u with respect to ξ:

$$L_\xi(u)(x) = d_x u(\xi(x)) = \frac{d}{dt} u(f^t(x))|_{t=0}. \tag{8.38}$$

$L_\xi(u)$ is continuous on X and satisfies (for all x, t)

$$\frac{du(f^t(x))}{dt} = L_\xi(u)(f^t(x)) = L_\xi(u \circ f^t)(x). \tag{8.39}$$

14. Proposition. *Let* ξ *be a* C^r *vectorfield on a* C^{r+1} *compact manifold* X $(1 \le r \le \infty)$ *with associated* C^r *flow* φ. $\mu \in P(X)$ *is an invariant measure for* φ *if and only if*

$$\langle L_\xi(u), \mu\rangle = 0 \tag{8.40}$$

for all C^r *real functions* u *on* X.

Proof. The C^r functions generate $C(X; \mathbb{R})$. For example, the partitions of unity constructed for Proposition 1 could be made C^r (see, e.g. Lang (1972) Chapter II). So μ is invariant when $\langle u, f_*^t\mu\rangle$ is constant in t for all C^r functions u. But $\langle u, f_*^t\mu\rangle = \langle u \circ f^t, \mu\rangle$ and so from (8.39) one can derive (for u C^1):

$$\frac{d}{dt}\langle u, f_*^t\mu\rangle = \langle L_\xi(u \circ f^t), \mu\rangle. \tag{8.41}$$

(The convergence of the difference quotient to the derivative in (8.39) is uniform in x. Details to the reader.) When u is C^r, so is each $u \circ f^t$. Hence, (8.40) holds for all u C^r if and only if $\langle u, f_*^t\mu\rangle$ is constant in t for all u C^r. □

15. Corollary. *Let ξ be a C^r vector field on a C^{r+1} manifold X with flow φ $(1 \le r \le \infty)$. Let α be a nonnegative C^r real-valued function on X, and let ξ_1 be the C^r vector field defined by $\xi_1(x) = \alpha(x)\xi(x)$. Let φ_1 be the flow for ξ_1. Define $Z = \alpha^{-1}(0) \subset X$.*

If μ is a φ-invariant measure satisfying $\langle 1/\alpha, \mu \rangle < \infty$, then define the α-associated measure $\mu_1 = \langle 1/\alpha, \mu \rangle^{-1}(1/\alpha)\mu$, *i.e., for u bounded and measurable on X,*

$$\int u(x)\,\mu_1(dx) = \frac{\int (u(x)/\alpha(x))\,\mu(dx)}{\int (1/\alpha(x))\,\mu(dx)}. \tag{8.42}$$

μ_1 is a φ_1 invariant measure. A φ_1 invariant measure ν is the α-associate of some φ invariant μ if and only if $\nu(Z) = 0$. Every φ_1 invariant measure can be written at $t\mu_1 + (1-t)\nu$ where μ_1 is the α associate of μ, $|\nu| \subset Z$ and $0 \le t \le 1$.

If μ_1 is the α-associate of μ, then μ_1 and μ have the same sets of measure zero. In particular, $|\mu_1| = |\mu|$. Furthermore, μ_1 is ergodic with respect to φ_1 if and only if μ is ergodic with respect to φ.

Proof. If u is a C^r real function, then $L_{\xi_1}(u)(x) = \alpha(x)L_{\xi}(u)(x)$. So $\langle L_\xi(u), \mu \rangle = 0$ implies $\langle L_{\xi_1}(u), \mu_1 \rangle = 0$. The converse is true as well because $\langle 1/\alpha, \mu \rangle < \infty$ requires $\mu(Z) = 0$. So μ_1 is φ_1 invariant if and only if μ is φ invariant. Because $1/\alpha$ is a positive μ-integrable function μ_1 and μ are equivalent measures; they have the same sets of measure zero. In particular, $\mu_1(Z) = 0$.

If ν is φ_1 invariant and $\nu(Z) > 0$, then because Z consists of φ_1 fixed points ν_Z is a φ_1-invariant measure where

$$\nu_Z(A) = \nu(A \cap Z)/\nu(Z),$$

and $\mu_1(Z) = 0$, where

$$\mu_1 = (\nu - \nu(Z)\nu_Z)/(1 - \nu(Z))$$

is also φ_1 invariant. $\nu = t\mu_1 + (1-t)\nu_Z$ with $t = 1 - \nu(Z)$. If $t > 0$, then $\nu(Z) < 1$ and $\langle \alpha, \nu \rangle > 0$. So we can define $\mu = \alpha\nu/\langle \alpha, \nu \rangle$. It is easy to check that μ_1 is α-associated with μ (because $\mu_1(Z) = 0$).

Since μ_1 and μ have the same sets of measure 0 they have the same support. For ergodicity we must compare sets invariant φ with those invariant φ_1. Off the set Z the two systems are related by a change of time scale:

$$\begin{aligned} \frac{dx}{d\tau} &= \xi_1(x), \\ \frac{dt}{d\tau} &= \alpha(x), \end{aligned} \quad \Leftrightarrow \quad \frac{dx}{dt} = \xi(x). \tag{8.43}$$

So the solution path $\varphi_1(x, \tau)$ is just a reparametrization of the path $\varphi(x, t)$ unless the latter meets Z. In that case $\varphi_1(x, \tau)$ is a reparametrization of the piece of $\varphi(x, t)$ restricted to the 0 component of $\{s: \varphi(x, s) \notin Z\}$. So every φ invariant set is φ_1 invariant. Now assume μ_1 is ergodic φ_1 and that A is an φ invariant set. By φ_1 invariance and ergodicity either A or the complement has μ_1 measure 0 and so has μ measure 0. Hence, μ is ergodic. The same argument will prove the converse once we show that $\widetilde{Z} = \{\varphi(x, t): X \in Z \text{ and } t \in \mathbb{R}\}$ has measure zero with respect to μ and μ_1 because every φ_1 invariant set differs from a φ invariant set by a subset of $\widetilde{Z}$. We prove that $\mu(\widetilde{Z}_-) = 0$ where $\widetilde{Z}_- = \{\varphi(x, t): x \in Z \text{ and } t \leq 0\}$. The result then follows by using the inverse flows. For every point y of $\widetilde{Z}_-$ $\lim_{\tau\to\infty} \varphi(y, \tau)$ is an element of Z. So $(c_{\widetilde{Z}_-} \alpha)^M_{\varphi_1} = 0$, where $c_{\widetilde{Z}_-}$ is the characteristic function of $\widetilde{Z}_-$. Hence,

$$0 = \langle (c_{\widetilde{Z}_-} \alpha)^M_{\varphi_1}, \mu_1 \rangle = \langle c_{\widetilde{Z}_-} \alpha, \mu_1 \rangle = \frac{\langle c_{\widetilde{Z}_-}, \mu \rangle}{\langle 1/\alpha, \mu \rangle}$$

and so $\mu(\widetilde{Z}_-) = 0$.

REMARK. Of course, if α is positive $(Z = \emptyset)$ the two systems just differ by a time change and the result exhibits a correspondence between $\mathrm{In}_P(\varphi)$ and $\mathrm{In}_P(\varphi_1)$.

At the other extreme, if $1/\alpha$ is not μ integrable, let $\beta_N(x) = \min(1/\alpha(x), N)$. It is easy to check that every limit point of the sequence $\{\beta_N \mu / \langle \beta_N, \mu \rangle\}$ is a measure with support in Z. □

Supplementary exercises

16. (a) Prove that $d_u(\mu, \nu)$ defined by (8.5) is a continuous function of $(u, \mu, \nu) \in C(X; \mathbb{R}) \times P(X) \times P(X)$, by showing that

$$|d_u(\mu, \nu) - d_{u_1}(\mu_1, \nu_1)| \leq d_u(\mu, \mu_1) + d_u(\nu, \nu_1) + 2\|u - u_1\|_0.$$

Conclude that if S is a compact subset of $C(X; \mathbb{R})$ then $d_S(\mu, \nu) = \sup\{d_u(\mu, \nu): u \in S\}$ defines a continuous pseudometric on $P(X)$ (see Exercise 7.35). If S generates $C(X; \mathbb{R})$ as well (i.e., $C(X; \mathbb{R})$ is the smallest closed subspace containing S) then d_S is a metric and agrees with the topology on $P(X)$ by compactness.

(b) Define $H(X)$ or H to consist of the functions u in $C(X; \mathbb{R})$ with Lipschitz constant at most 1, i.e., $u \in H$ when $|u(x) - u(y)| \leq d(x, y)$ for all $x, y \in X$. With x_0 a point of X, prove that $H_{x_0} = \{u \in H: u(x_0) = 0\}$ is a compact, generating subset of $C(X; \mathbb{R})$. (For compactness, use Ascoli's theorem or directly embed H_{x_0} as a closed subset of product $\prod\{[-d(x, x_0), d(x, x_0)]: x \in X\}$. On the other hand, the generating sequence S of Proposition 1 consists of Lipschitz functions). Consequently, $d_H = d_{H_{x_0}}$ provides a metric for $P(X)$. We will follow Barnsley who calls it the *Hutchinson metric* (see Barnsley (1988) §9.6

and Hutchinson (1981)), Statisticians refer to it as the *bounded Lipschitz metric* (see Huber (1981) §2.4).

Using the Hutchinson metric, prove that the map $*: C(X_1; X_2) \to C(P(X_1); P(X_2))$ defined by $f \to f_*$ (see Proposition 2) is an isometric inclusion $(d_u(f_*\mu, f_{1*}\mu) = \langle |u \circ f - u \circ f_1|, \mu \rangle \le d(f, f_1)$ if $u \in H(X_2))$. Also, if L is a Lipschitz constant for f then it is for f_* as well. (If $u \in H(X_2)$ then $(1/L)u \circ f \in H(X_1))$.

(c) On $\{1, \dots, n\}$ impose the zero-one metric $(d(i, j) = 1$ where $i \ne j)$. $C(\{1, \dots, n\}; \mathbb{R})$ is $\mathbb{R}^n$ with norm $\|u\| = \sup\{|u_i|: i = 1, \dots, n\}$. $u \in H$ if it is a constant plus a member of $[0, 1]^n$. Identify $P(\{1, \dots, n\})$ with the unit simplex Δ^{n-1} in $\mathbb{R}^n$ and prove that the Hutchinson metric is given by $d(\mu, \nu) = (1/2)\sum_{i=1}^n |\mu_i - \nu_i|$, $(|\sum_{i=1}^n u_i(\mu_i - \nu_i)|$ is maximized for $u \in [0, 1]^n$ with $u_i = 1$ when $\mu_i \ge \nu_i$ and $= 0$ when $\mu_i < \nu_i$. Note that $\sum_{i=1}^n(\mu_i - \nu_i) = 0$.)

With X^n the n-fold product of copies of X define an isometric inclusion $X^n \to C(\Delta^{n-1}; P(X))$.

(d) With respect to the Hutchinson metric prove that $\delta_X: X \to P(X)$ of Proposition 1 is an isometric inclusion. So $\delta_{P(X)}: P(X) \to P(P(X))$ is an isometric inclusion. Define the map $a: P(P(X)) \to P(X)$ as follows: For $M \in P(P(X))$, i.e., a measure on $P(X)$, and $u \in C(X; \mathbb{R})$,

$$\langle u, a(M) \rangle = \int \left(\int u(x)\, \mu(dx) \right) m(d\mu).$$

(Compare v: $C(C(X)) \to C(X)$ of Exercise 7.36.)

Prove that a has Lipschitz constant 1 and is a left inverse for $\delta_{P(X)}$ and for δ_{X^*}. (*Hint*: If $u \in H(X)$ then $\mu \mapsto \langle u, \mu \rangle$ defines an element of $H(P(X))$.)

17. An *atom* for a measure μ is a point x such that $\mu(x) > 0$. μ has countably many atoms. In fact, the sum $\Sigma\{\mu(x): x \in X\} \le \mu(X) = 1$ and so has countably many nonzero terms. A measure is called *purely atomic* if the sum is 1. So a purely atomic measure is a countable sum of point measures $\mu = \sum\{\mu(x)\delta_x : x \in X\}$. A measure is called *nonatomic* if $\mu(x) = 0$ for every point x. Prove that μ is nonatomic if and only if the diagonal 1_X in $X \times X$ has measure zero with respect to the product measure $\mu \times \mu$ (Fubini's Theorem).

    If μ is an invariant measure for a map f then any atom of μ is a periodic point for f. Conversely, if x is a periodic point with period n $(f^n(x) = x)$ then $\sigma_n(f, x)$ is an invariant measure with support equal to the periodic orbit. Furthermore such a measure is ergodic. If μ is an ergodic measure and x is an atom for μ then $\mu = \sigma_n(f, x)$ (when $f^n(x) = x$).

    If f has only countably many periodic points prove that every invariant measure is purely atomic or ergodic nonatomic measures exist (use (8.21)).

18. If $\mu \in P(X)$ and $\mu(\mathrm{Con}(f)) = 1$, prove that $\lim\{\sigma_n(f, \mu)\} = \int_{\mathrm{Con}(f)} \mu_x \, \mu(dx)$. Conversely, if $\{\sigma_n(f, \mu)\}$ converges prove that $\mu(\mathrm{Con}(f)) = 1$ (use $u^M = u^m$ a.e. (μ) for u in a generating sequence).
19. A point c of a convex set C is called an extreme point if $e = tc_1 + (1-t)c_2$, $c_1, c_2 \in C$, and $0 < t < 1$ imply $c_1 = c_2 = e$. Thus e is extreme if it is not in the interior of any line segment in C. Prove that the ergodic measures are the extreme points of $\mathrm{In}_P(f)$. ((1) If A is f^{-1} invariant with complement B and $0 < \mu(A) < 1$ then μ is on the segment between μ_A and μ_B where $\mu_A(C) = \mu(C \cap A)/\mu(A)$. (2) If $\mu, \nu \in \mathrm{In}_P(f)$ and $\mu \gg \nu$, meaning $\mu(A) = 0$ implies $\nu(A) = 0$, then μ ergodic implies $\mu = \nu$. To prove this write $\nu = \alpha\mu$ with $\alpha \geq 0$ and $\langle \alpha, \mu \rangle = 1$ (Radon-Nikodym theorem). Prove that α is f invariant and so is constant a.e. (μ). (3) If $\mu = t\mu_1 + (1-t)\mu_2$ with $\mu_1, \mu_2 \in \mathrm{In}_P(f)$ and $0 < t < 1$ then $\mu \gg \mu_1, \mu_2$.) Prove that distinct ergodic measures μ_1, μ_2 are mutually singular (Use the Hahn decomposition of $\mu_1 - \mu_2$.)
20. Let $f: X \to X$ be a continuous map. Prove that the following conditions are equivalent: (1) f is topologically mixing. (2) For each $\varepsilon > 0$ and each nonempty open set U, the open sets $f^{-n}(U)$ are ε-dense for all but a finite number of positive integers n. (3) The sequence $\{f^n\}$ in $C(X \times X)$ converges to the element $X \times X$.

    Recall that f is topologically transitive precisely when $\limsup\{f^n\}(= \Omega f)$ is $X \times X$.
21. (a) A continuous map $f: X \to X$ is called *weak mixing* when $f \times f: X \times X \to X \times X$ is topologically transitive. It then follows that the n-fold product $f \times \cdots \times f: X^n \to X^n$ is topologically transitive for all positive integers n. Fill in the details of Furstenberg's proof (see his (1967) paper): (1) Define for U, V open in X the set of positive integers $N(U, V) = \{k: U \cap f^{-k}(V) \neq \varnothing\}$ and define the analogue $N(U_1 \times \cdots \times U_n, V_1 \times \cdots \times V_n)$ for the n-fold product. Show that the latter set is $\bigcap_{i=1}^n \{N(U_i, V_i)\}$. (2) For $k \in N(U_1 \times V_1, U_2 \times V_2)$ let $U_3 = U_1 \cap f^{-k}(U_2)$ and $V_3 = V_1 \cap f^{-k}(V_2)$. If $n \in N(U_3, V_3)$ then show that $U_1 \cap f^{-n}(V_1) \cap f^{-k}(U_2 \cap f^{-n}(V_2)) \neq \varnothing$, and conclude that $N(U_3, V_3) \subset N(U_1, V_1) \cap N(U_2, V_2)$. (3) Complete the proof by noting that f is topologically transitive if and only if $N(U, V) \neq \varnothing$ whenever U and V are nonempty open subsets of X.

    (b) Prove that f is weak mixing if and only if for every $\varepsilon > 0$ f^n is an ε-dense subset of $X \times X$ for infinitely many positive integers n. (*Hint*: $n \in \bigcap_{i,j}\{N(U_i, U_j)\}$ for some collection of open sets $\{U_i\}$ if and only if f^n intersects every $U_i \times U_j$.) Conclude that f is weak mixing if and only if some subsequence of $\{f^n\}$ converges in $C(X \times X)$ to the element $X \times X$. Prove that $\Omega(f \times f)(x, x) = X \times X$ for all $x \in X$

implies f is weak mixing. (*Hint*: with $U = U_1 \cap f)^{-k}(U_2)$, $N(U, U_3) \cap N(U, f^{-k}(U_4)) \subset N(U_1, U_3) \cap N(U_2, U_4))$. Finally, show that f is weak mixing if and only if for every $\varepsilon > 0$ and every nonempty open set U, the open sets $f^{-n}(U)$ are ε-dense for infinitely many positive integers n.

(c) Prove that if f is weak mixing then every power f^n is weak mixing. (Use Exercise 4.31 to show f^n is topologically transitive. Apply this result to $f \times f$ using (a).)

22. If $x_0, x_1, \dots, x_N$ is an ε chain for $f: X \to X$, then N is the *length of the chain.* f is called *chain mixing* if for every $\varepsilon > 0$ and every pair $(x, y) \in X \times X$ there exist N such that for all $n \geq N$ there exists an ε chain connecting x to y of length n. Prove that N_ε can be chosen to depend only on ε. (Let $\varepsilon_1 < \varepsilon/2$ be an $\varepsilon/2$ modulus of uniform continuity. Choose a finite ε_1 net for $X \times X$ and pick N large enough to work for all pairs in the net with ε_1.) Prove that f_1 and f_2 chain mixing imply $f_1 \times f_2$ and f_1^n are. If X is finite and consists of a single periodic orbit for f then f is chain transitive—in fact minimal—but f is not chain mixing. Prove conversely that if f is chain transitive but not chain mixing then there is a continuous map π of X onto a finite set F with at least two elements and a cyclic permutation $S: F \to F$ such that $S \circ \pi = \pi \circ f$. In particular, if f is chain transitive and X is connected, then f is chain mixing. Use the following steps:

(1) With $\mathbb{Z}_+$ the set of whole numbers and $T_1, T_2 \subset \mathbb{Z}_+$ let $T_1 + T_2 = \{t_1 + t_2 : t_1 \in T_1 \text{ and } t_2 \in T_2\}$. So $T \subset \mathbb{Z}_+$ is closed under addition if $T + T \subset T$. Prove the number theory result that if $\varnothing \neq T \subset \mathbb{Z}_+$ and $T + T \subset T$ then there exists N so that $nd \in T$ for $n \geq N$ where d is the greatest common divisor of the elements of T. ($d = \gcd(t_1, \dots, t_p)$. Use induction on p to reduce to the case $p = 2$ and $d = 1$. So $1 = n_1 t_1 - n_2 t_2$, $n_1, n_2 \in \mathbb{Z}_+$. If $n \geq N = n_2 t_2^2$ then $n = Qt_2 + R$ with $Q \geq n_2 t_2 - 1$ and $0 \leq R < t_2$. So $n = (Q - n_2 R)t_2 + Rn_1 t_1$.)

(2) With $\varepsilon > 0$ define $\varepsilon(x, y) = \{t \in \mathbb{Z}_+ :$ there exists an ε_1 chain from x to y of length t with $\varepsilon_1 < \varepsilon\}$. Because f is chain transitive $\varepsilon(x, y)$ is an infinite subset of $\mathbb{Z}_+$. Prove that $\varepsilon(x, y) + \varepsilon(y, z) \subset \varepsilon(x, z)$. Prove that $d = \gcd \varepsilon(x, x)$ is independent of x $(\varepsilon(x, x) + \varepsilon(x, y) + \varepsilon(y, x) + \varepsilon(y, y) = \varepsilon(x, x) + \varepsilon(x, y) + \varepsilon(y, y) + \varepsilon(y, x) \subset \varepsilon(x, x)$ and $\varepsilon(x, y) + \varepsilon(y, x) \subset \varepsilon(x, x)$. So $\gcd \varepsilon(x, x) | \gcd \varepsilon(y, y)$.) Prove that for $t_1, t_2 \in \varepsilon(x, y)$ $t_1 \equiv t_2 \bmod d$. (If $t_3 \in \varepsilon(y, x)$ then $t_1 + t_3 \equiv 0 \equiv t_2 + t_3 \bmod d$.) So there is a function $r: X \times X \to \mathbb{Z}/d\mathbb{Z}$ well defined by $\varepsilon(x, y) \subset r(x, y)$. Prove that r is locally constant and so is continuous. (Fix $t \in \varepsilon(x, y)$. Observe that $t \in \varepsilon(x', y')$ for (x', y') close to (x, y).) Prove that $r(x, y) + r(y, z) \equiv r(x, z)$ and $r(x, f(x)) \equiv 1$.

(3) If f is not chain mixing, prove there exists $\varepsilon > 0$ small enough that $d > 1$ (use (1)). Then fix $x_0 \in X$ and define $\pi: X \to \mathbb{Z}/d\mathbb{Z}$ by $\pi(x) = r(x_0, x)$. Check that $\pi(f(x)) = \pi(x) + 1 \pmod{d}$.

23. For T equal the set of integers, positive integers, reals or positive reals a subset U is called *relatively dense* in T if for some positive integer L_U, $U \cap [t, t + L_U] \neq \varnothing$ for every $t \in T$.

Let $f: X \to X$ be a topologically transitive homeomorphism.

(a) Prove that f is minimal if and only if it satisfies either of the following conditions (see Exercise 4.21):

(1) For every $\varepsilon > 0$ and every $x \in X$, $\{i: d(f^i(x), x) \leq \varepsilon\}$ is relatively dense in the integers.

(2) There exists x with $\omega f(x) = X$ such that for every $\varepsilon > 0$, $\{i: d(f^i(x), x) \leq \varepsilon\}$ is relatively dense in the positive integers.

(b) f is called *almost periodic* if it satisfies the following equivalent conditions:

(1) For every $\varepsilon > 0$ $\{i: d(f^i(x), x) \leq \varepsilon$ for all $x \in X\}$ is relatively dense in the integers.

(2) There exists x with $\omega f(x) = X$ such that for every $\varepsilon > 0$, $\{i: d(f^{i+j}(x), f^j(x)) \leq \varepsilon$ for all $j \geq 0\}$ is relatively dense in the positive integers.

Observe that (1) is equivalent to

(1′) For every $\varepsilon > 0$ $\{i: \overline{d}(f^i, 1_X) \leq \varepsilon\}$ is relatively dense, where $\overline{d}$ is the metric on $\mathrm{Cis}(X; X)$ of Proposition 7.20.

Prove that if $f: X \to X$ is an isometry of the metric d $(d(f(x_1), f(x_2)) = d(x_1, x_2))$ then f is minimal if and only if it is almost periodic (compare conditions (2) of (a) and (b)).

(c) A sequence $v = \{v_0, v_1, \dots\}$ in a Banach space E is bounded if $\|v\|_0 = \sup\{\|v_i\|: i = 0, \dots\}$ is finite, v is almost periodic if for every $\varepsilon > 0$ $U_\varepsilon = \{i: \|v_i - v_{i+j}\| \leq \varepsilon$ for $j = 0, 1, \dots\}$ is relatively dense in the positive integers. Let $s(v)$ be the shifted sequence so that $s^j(v)_i = v_{i+j}$ $(i, j = 0, 1, \dots)$ and let $M_n(v) = (1/n)\sum_{i=0}^{n-1} v_i$. Prove: If v is an almost periodic sequence in E then v is bounded and $\{M_n(v)\}$ converges in E. If v is an almost periodic sequence of nonnegative reals then $\lim\{M_n(v)\} > 0$ unless $v_i = 0$ for all i. (First, verify the following (compare (8.13) and (8.15)):

$$\|M_{n+i}(v) - M_n(v)\| \leq \frac{2i}{n}\|v\|_0, \qquad \|M_n(s^i(v)) - M_n(v)\| \leq \frac{2i}{n}\|v\|_0,$$
$$M_{nk}(v) = \frac{1}{k}[M_n(v) + M_n(s^n v) + \dots + M_n(s^{n(k-1)} v)].$$

Then choose $j_1, \dots, j_{k-1}$ between 0 and L_{U_ε} so that $n + j_1, \dots,$

$n(k-1)+j_{k-1}$ lie in U_ε and compare $M_n(v)$ and $M_{nk}(v)$ with

$$\widetilde{M} = \frac{1}{k}[M_n(v) + M_n(s^{n+j_1}v) + \cdots + M_n(s^{n(k-1)+j_{k-1}}v)].$$

With $t \geq n$ and k the integer part of t/n write

$$M_t(v) - M_n(v) = M_t(v) - M_{nk}(v) + M_{nk}(v) - \widetilde{M} + \widetilde{M} - M_n(v)$$

to prove that

$$\|M_t(v) - M_n(v)\| \leq 2\|v\|_0 \left(\frac{n}{t-n} + \frac{L_{U_\varepsilon}}{n}\right) + \varepsilon .$$

Then show that $\{M_t(v)\}$ is Cauchy. For a nonnegative sequence of reals with $v_0 > 0$ and $\varepsilon = v_0/2$ show that $M_{nk}(v) \geq v_0/2n$ with $n = L_{U_\varepsilon}$ and $k = 1, 2, \ldots$.)

(d) If f is almost periodic, then f is strictly ergodic. In fact, for $u \in C(X; \mathbb{R})$ $\{(1/n)\sum_{i=0}^{n-1} u \circ f^i\}$ converges to u^M uniformly on X. u^M is constant and equal to $\langle u, \mu\rangle$ where μ is the unique invariant measure on X.

In general, $d^M = \limsup\{(1/n)\sum_{i=0}^{n-1} d \circ (f \times f)^i\}$ is a pseudometric on X. If f is almost periodic, then d^M is continuous and is a metric. (Prove $\{d \circ (f^i \times f^i)\}$ is a.p. in $C(X \times X; \mathbb{R})$.) So X admits a metric with respect to which f is an isometry. Thus, f is almost periodic if and only if it is minimal and admits such an invariant metric.

(e) In general, the relation ωf is symmetric if and only if for every $x \in X$, $x \in \omega f(x)$ and $\omega f(x)$ is a minimal invariant subset. So ωf is then an equivalence relation. If f is an isometry with respect to d, prove that ωf is a closed equivalence relation and regarded as a map $X \to C(X)$ it is continuous. Furthermore, the restriction of f to each $\omega f(x)$ is almost periodic. (For symmetry and closure of ωf observe that $d(x, A) = \inf\{d(x, y): y \in A\}$ is an invariant function of x when A is an f invariant subset. Then ωf is an equivalence relation and is usc. But $\omega f = \mathscr{R} f$ is lsc.)

(f) Define almost periodicity for flows and extend the above results by analogy. Prove that if the time-one map f is almost periodic then the flow φ is almost periodic. Conversely, if φ is almost periodic then f^t is almost periodic if and only if it is minimal and this occurs for residual t (use the invariant metric and Exercise 6.15).

24. With $d(x, A) = \inf\{d(x, y): y \in A\}$ for $A \in C(X)$ prove that

$$\rho(A/\mu) = \int d(x, A)\, \mu(dx)$$

is a continuous, nonnegative real function on $C(X) \times P(X)$ and $\rho(A/\mu) = 0$ if and only if $|\mu| \subset A$. Observe that $\rho(A/\mu) \leq \rho(A/|\mu|)$ but the latter function is not continuous in μ. Prove that

$$\rho(A/\mu) < \varepsilon^2 \quad \text{implies} \quad \mu(X - V_\varepsilon(A)) = \mu(|\mu| - V_\varepsilon(A)) < \varepsilon .$$

Deduce that for $\varepsilon > 0$ $d_H(\nu, \mu) < \varepsilon^2$ implies $\nu(X - V_\varepsilon(|\mu|)) = \nu(|\nu| - V_\varepsilon(|\mu|)) < \varepsilon$, where d_H is the Hutchinson metric of Exercise 16.

Prove that for every $\varepsilon > 0$ and $f \in C(X; X)$ there exists $\varepsilon_1 > 0$ such that $d(f, g) < \varepsilon_1$ implies $\mu[X - V_\varepsilon(|\omega_\# f|)] < \varepsilon$ for every g invariant measure μ. (Use $\mathrm{In}_P: C(X; X) \to C(P(X))$ is usc.)

25. With $\mu \in \mathrm{In}_P(f)$, assume $(6')$ of Proposition 7 holds for all A and B in some ring generating the entire σ algebra of Borel sets. Prove that $(6')$ follows and so f is mixing with respect to μ (see, e.g. Billingsley (1965) Chapter 1).

Assume $f_1: X_1 \to X_1$ and $f_2: X_2 \to X_2$ are mixing with respect to μ_1 and μ_2. Prove that $f_1 \times f_2$ is mixing with respect to the product measure $\mu_1 \times \mu_2$. Prove that f_1^n is mixing for all positive integers n and for all nonzero integers n if f_1 is a homeomorphism.

Chapter 9. Examples—Circles, Simplex, and Symbols

For the examples of this chapter the reader should notice how extra algebraic structure on a space can simplify analysis of a dynamical system upon it.

The unit circle S in the complex plane is a group under multiplication. S can be identified with the quotient group $\mathbb{R}/\mathbb{Z}$ via the continuous homomorphism

$$e : \mathbb{R} \to S, \qquad e(t) = \exp(i2\pi t). \tag{9.1}$$

Observe that t is rational if and only if $e(t)$ is a root of unity.

Multiplication by $c = e(s)$ defines the rotation map $r^c : S \to S$ (also denoted r^s) and as s varies r^s is the time s map of the *rotation flow*. Each such rotation is an isometry of the angular distance d on S and preserves angular Lebesgue measure μ $(= e_*(\lambda)$ where λ is Lebesgue measure on $[0, 1])$, as well as orientation of S.

Most important for us is the homogeneity of S with respect to r^c. Any other rotation r^{c_1} commutes with r^c and so provides a conjugacy of r^c with itself. Hence r^{c_1} maps the r^c orbit of a point $z \in S$ to the r^c orbit of $c_1 z$. Thus, $\omega r^c(c_1 z) = r^{c_1}(\omega r^c(z))$. Since any two points of S can be related by some r^{c_1} it follows that all points of S behave the same with respect to r^c. For example, if c is a root of unity then every point of S is periodic with period that of c. If, on the other hand, c is not a root of unity then $\omega r^c(z) = S$ for all $z \in S$. In fact,

1. Proposition. *When c is not a root of unity the rotation $r^c : S \to S$ is uniquely ergodic, that is, it is a minimal map and μ is the unique r^c invariant measure.*

Proof. The orbit of 1 $\{1, c, c^2, \dots\}$ consists of distinct points. So given $\varepsilon > 0$ it contains distinct points of distance less than ε. If $d(c^j, c^{j+k}) < \varepsilon$ then $d(1, c^k) < \varepsilon$. Because r^c preserves orientation the intervals between successive points of $\{1, c^k, c^{2k}, \dots, c^{nk}\}$ cover S once $nd(1, c^k) > 2\pi$. As this piece of the orbit is ε dense we have $\omega r^c(1) = S$. By homogeneity $\omega r^c(z) = S$ for every $z \in S$ and r^c is minimal.

Similarly, $\mathrm{Con}(r^c) \neq \emptyset$ implies $z \in \mathrm{Con}(r^c)$ for some z in S and hence for every z in S by homogeneity. So r^c is strictly ergodic by Corollary 8.12.

REMARK. r^c is never topologically mixing. In fact, the product $r^c \times r^c$ is not topologically transitive because the metric d is invariant and so the open sets V_ε are $r^c \times r^c$ invariant. □

The torus $X = S \times S$ is the quotient of $\mathbb{R}^2$ by the integer lattice $\mathbb{Z}^2$ via the homomorphism $e \times e$. On $\mathbb{R}^2$ we define the constant vector field $\xi(x, y) = (1, a)$ with associated flow $\varphi(x, y, t) = (x + t, y + at)$. By translation invariance these define a vector field and flow on X which we will also denote by ξ and φ. When a is irrational, as we will now assume, φ is called an *irrational flow* on the torus. The time-s map f^s is the product $r^{e(s)} \times r^{e(as)}$. In particular, the time-one map $f = 1_S \times r^c$, where $c = e(a)$.

Thus, the flow preserves the product metric on X and the Lebesgue measure μ on the torus (which is the product of the Lebesgue measure on the circle factors). For $(x, y) \in \mathbb{R}^2$, $\omega f(e(x), e(y)) = e(x) \times \omega r^c(e(y)) = e(x) \times S$ and so $\omega\varphi(e(x), e(y)) = X$ by φ invariance. Hence the flow is minimal. As each f^s commutes with products $r^{c_1} \times r^{c_2}$ the above homogeneity arguments show that φ is strictly ergodic as well. Again φ is not topologically mixing.

Now let p be the point $(1,1)$ and let α be a C^1 nonnegative real function on X vanishing only at p. So $\alpha \circ (e \times e)$ (also denoted α) is a C^1 nonnegative real function on $\mathbb{R}^2$ vanishing precisely on the integer lattice. Let $\xi_1 = \alpha\xi$ with associated flow φ_1. That is, φ_1 is the solution flow of the system

$$\frac{dx}{d\tau} = \alpha(x, y), \qquad \frac{dy}{d\tau} = a\alpha(x, y). \tag{9.2}$$

2. THEOREM. *Case* 1. *Suppose* $1/\alpha$ *is not integrable with respect to* μ *on* X*—and this happens if* α *is at least* C^2*—then the point measure* δ_p *is the unique invariant measure for the flow* φ_1. *For every* $q \in X$ $M(q) = \delta_p$ *and so* $\mathrm{Con}(\varphi_1) = X$. $\omega_\#(\varphi_1)(q) = p$ *for every* q *in* X *and so* $p = m[\varphi_1] = |\omega_\#(\varphi_1)| = \omega_\#(\varphi_1)(X)$.

Case 2. *Suppose* $1/\alpha$ *is integrable with respect to* μ *on* X*—and this can happen—then* δ_p *and* $\mu_1 = (1/\alpha)\mu/\langle 1/\alpha, \mu\rangle$ *are the ergodic measures for* φ_1 *and* $\mathrm{In}_P(\varphi_1) = \{t\delta_p + (1-t)\mu_1 : 0 \le t \le 1\}$. *For* q *in a residual subset of* X $M(q) = I^* = \mathrm{In}_P(\varphi_1)$. *However this subset has Lebesgue measure zero. On the other hand,* $\{q : M(q) = \mu_1\}$ *is a set of first category of Lebesgue measure* 1.

In both cases the flow is topologically mixing. Thus, $\Omega_{\#\#}(\varphi_1) = X \times X$ *and a fortiori* $\Omega_\#(\varphi_1) = X \times X$.

PROOF. Integrability just depends on the behavior of α on $\mathbb{R}^2$ near $(0,0)$. If α is C^2, then for some $M > 0$ $\alpha(x, y) \le M(x^2 + y^2)$ and the reciprocal is not integrable. On the other hand, if $\alpha(x, y) = (x^2 + y^2)^b$ near $(0, 0)$ then α is C^1 if $b > 1/2$ and $1/\alpha$ is integrable if $b < 1$. It follows that both cases can occur.

Lebesgue measure μ is the only invariant measure for φ. So by Corollary 8.15 a φ_1 invariant measure μ_1 with $\mu_1(p) = 0$ exists only when $1/\alpha$ is μ integrable in which case μ_1 is the normalization of $(1/\alpha)\mu$. It is then φ_1 ergodic because μ is φ ergodic. Because p is the zero-set for α δ_p is the unique measure with support in this zero set. So in case 1, δ_p is the unique invariant measure while in case 2, $\mathrm{In}_P(\varphi_1)$ is the segment connecting δ_p and μ_1.

As indicated in (9.2) we will follow (8.43) using τ for the ξ_1 time variable and t for the ξ time variable. Denote by q^t the point $f^t(p) = (e(t), e(at))$ so that $\{q^t : t \in \mathbb{R}\}$ is the φ orbit of p. Observe that on the negative portion f_1^τ converges to p, i.e., $\omega(\varphi_1)(q^t) = p$ for $t \leq 0$.

Case 1. As $\varnothing \neq M(\varphi_1, \mu) \subset \mathrm{In}_P(\varphi_1) = \delta_p$, it follows that $M(\varphi_1, \mu) = \delta_p$ for every μ in $P(X)$. Thus, $\mathrm{Con}(\varphi_1) = X$ and by the flow analogue of Proposition 8.8(a) $\omega_{\#}(\varphi_1)(q) = |\delta_p| = p$ for all q in X.

Case 2. In the notation of Theorem 8.11, μ_1 lies in I^* because it is a dense ergodic measure. On the other hand, for q^t with $t \leq 0$ $\omega_{\#}(\varphi_1)(q^t) \subset \omega(\varphi_1)(q^t) = p$ and so $M(\varphi_1, q^t) = \delta_p$. But $\{q^t : t < 0\}$ is dense in X $(\omega(\varphi^{-1})(p) = X)$ and so $\delta_p \in I^*$ as well by the remarks following the proof of Theorem 8.11. Since I^* is connected it contains the entire segment between m_p and μ_1. By that Theorem 8.11 again the residual set $M^{-1}(I^*)$ has μ_1 measure 0 and so Lebesgue measure 0. On the other hand, μ_1 is ergodic and so $M(\varphi_1, q) = \mu_1$ except on a set of μ_1 measure 0 and so $\{q : M(\varphi_1, q) = \mu_1\}$ has Lebesgue measure 1.

To prove topological mixing we must show that for U and V open in X $f_1^\tau(U) \cap V \neq \varnothing$ for τ sufficiently large.

For $t \geq 0$ $F_t^+ = \{q^s : 0 \leq s \leq t\}$ and $F_t^- = \{q^s : -t \leq s \leq 0\}$ are compact pieces of the positive and negative φ orbit of p. Because p is a fixed point for φ_1 each F_t^+ is φ_1^{-1} +invariant and

$$f_1^{\tau_1}(F_t^+) \subset f_1^{\tau_2}(F_t^+), \qquad \tau_1 \leq \tau_2$$

and the monotone intersection as $\tau \to -\infty$ is p. The union as $\tau \to +\infty$ is the entire positive φ orbit of p. The reverse is true for the F_t^-'s.

Because the positive and negative φ orbits of p are dense in X they meet V and U. So we can replace U and V by $f^{\tau_1}(U)$ and $f^{-\tau_2}(V)$ for some fixed large τ_1 and τ_2 so as to assume $U \cap F_1^- \neq \varnothing$ and $V \cap F_1^+ \neq \varnothing$.

Now lift to $\mathbb{R}^2$ and draw little vertical line segments γ_U and γ_V of the same length, with γ_U contained in U and intersecting $U \cap F_1^-$ while γ_V in V intersects $V \cap F_1^+$. By shrinking the segments if necessary we can

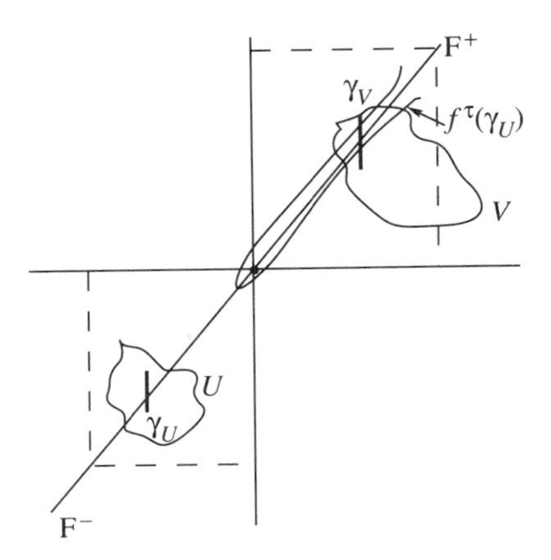

FIGURE 9.1

assume that the endpoints lie on rays associated with the positive φ orbit of p, i.e., on $\mathbb{Z} \times \mathbb{Z} + \mathbb{R}(1, a)$. Consequently, using the φ_1 flow on $\mathbb{R}^2$ the x coordinates of the endpoints of the curve $f^\tau(\gamma_U)$ tend to $+\infty$ with τ. But the intersection point with F_1^- tends to the origin remaining on the segment F_1^-. So as shown in Figure 9.1 $f^\tau(\gamma_U) \cap \gamma_V \neq \emptyset$ for τ sufficiently large. So back in X $f^\tau(U) \cap V \neq \emptyset$ for τ sufficiently large. □

The $n-1$ simplex Δ in $\mathbb{R}^n$ is the set of nonnegative vectors p such that $\sum_i p_i = 1$. Δ can be regarded as the space of probability measures on the finite metric space $\{1, 2, \ldots, n\}$.

Suppose that some population is subdivided into n different categories or types. A dynamical system on Δ gives a model for the change of distribution of relative frequencies among the types, provided population sizes are large enough that the use of real variables is appropriate.

For example, in evolutionary game dynamics the n types are strategies in a two player game with payoff a_{ij} to the i player when he meets a j player $(i, j = 1, 2, \ldots, n)$. We define on Δ the differential equation

$$\begin{aligned} &\frac{dp_i}{dt} = p_i(a_{ip} - a_{pp}), \qquad (i = 1, \ldots, n), \\ &a_{ip} = \sum_j p_j a_{ij}, \qquad a_{pp} = \sum_i p_i a_{ip} = \sum_{i,j} p_i p_j a_{ij}. \end{aligned} \tag{9.3}$$

Here a_{ip} is the average payoff that an i player receives against a random opponent when the population is in state p. (9.3) says that the relative rate of increase of the i subpopulation is the amount by which this average exceeds the mean payoff a_{pp}. Observe that $\sum_i dp_i/dt = 0$ and that $dp_i/dt = 0$ when $p_i = 0$. Thus, the flow associated with (9.3) preserves the simplex Δ and each face as well. In particular, the n vertices e^i are fixed points.

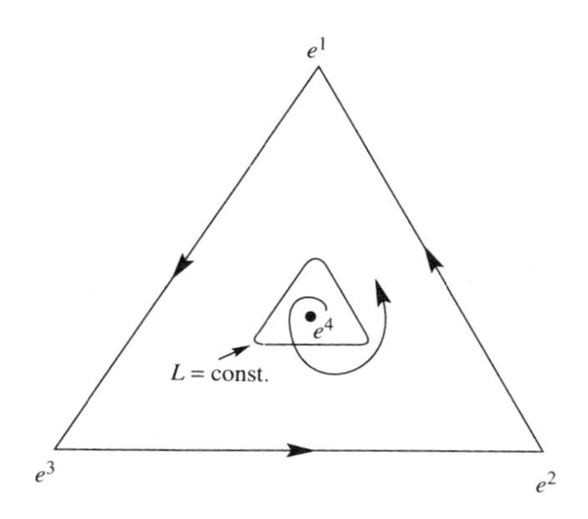

FIGURE 9.2

The discrete time analogue is obtained by replacing dp_i/dt by $\delta p_i/t$. In other words, we define the map $F: \Delta \to \Delta$ by regarding $\delta p = F(p) - p$ and so:

$$F(p)_i = p_i(1 + \tau(a_{ip} - a_{pp})). \tag{9.4}$$

(Here we assume that $\tau a_{ij} \geq -1$ to get $F(p) \in \Delta$.)

We will investigate some three strategy examples related to the paper-rock-scissors game with payoff matrix

$$E = \begin{pmatrix} 0 & 1 & -1 \\ -1 & 0 & 1 \\ 1 & -1 & 0 \end{pmatrix}. \tag{9.5}$$

Notice that because E is antisymmetric $E_{pp} = 0$ for all p in Δ.

First, let φ_ε be the solution flow for (9.3) with payoff matrix $\varepsilon I + E$ $(0 < \varepsilon < 1)$, i.e.,

$$\frac{dp_i}{dt} = p_i\left(\varepsilon p_i + E_{ip} - \varepsilon\left(\sum_j p_j^2\right)\right). \tag{9.6}$$

On the other hand, let F_τ be the map of (9.4) with payoff matrix E and $0 < \tau < 1$,

$$F_\tau(p)_i = p_i(1 + \tau E_{ip}). \tag{9.7}$$

The phase portraits for these pictures are all rather similar (see Figure 9.2).

3. PROPOSITION. *For the flows φ_ε of (9.6) and the maps F_τ of (9.7), with $0 < \varepsilon,\ \tau < 1$, the following hold:*

The set of recurrent points equals the set of fixed points which consists of the three vertices e^1, e^2, e^3 and the barycenter $e^4 = (1/3, 1/3, 1/3)$.

The chain recurrent set consists of two basic sets; the boundary of the simplex $\partial\Delta$ and the center e^4. Furthermore, $\{\partial\Delta, e^4\}$ form an attractor-repellor pair.

PROOF. It is easy to check that e^4 as well as e^1, e^2, and e^3 are fixed points. Because of the restrictions on ε and τ there are no other fixed points on the boundary. For example, on the edge between e^1 and e^2 where $p_3 = 0$, $dp_1/dt > 0$ and $F_\tau(p)_1 > p_1$.

Now define

$$L(p) = -\frac{1}{3}\sum_{i=1}^{3} \ln 3p_i = -\ln 3 - \frac{1}{3}\sum_i \ln p_i. \tag{9.8}$$

On the interior of the simplex concavity of the log implies that the smooth function L satisfies

$$L(p) \geq -\ln\left(\sum_{i=1}^{3} p_i\right) = 0$$

with equality only when $p_1 = p_2 = p_3$, that is, only at e^4.

As p approaches the boundary, $L(p)$ approaches ∞ and so we can regard L as a continuous, extended real valued function on all of Δ.

The restrictions to $\partial\Delta$ of these systems are chain transitive (see Figure 9.2 and Exercise 2.2) and so $\partial\Delta$ is contained in a basic set. For the remaining results we prove that L is a Lyapunov function with $|L| = \partial\Delta \cup e^4$.

From (9.6) we have

$$\frac{d\ln p_i}{dt} = \varepsilon p_i + E_{ip} - \varepsilon\left(\sum_j p_j^2\right), \tag{9.9}$$

while from (9.7)

$$\ln F_\tau(p)_i - \ln p_i = \ln(1 + \tau E_{ip}), \tag{9.10}$$

and so for the two systems, respectively,

$$\frac{dL}{dt} = \varepsilon\left(\sum_j p_j^2\right) - \frac{\varepsilon}{3} \geq 0,$$

$$\begin{aligned} L(F_\tau(p)) - L(p) &= -\sum_i \frac{1}{3}\ln(1 + \tau E_{ip}) \\ &\geq -\ln\left(\frac{1}{3}\sum_i (1 + \tau E_{ip})\right) = 0 \end{aligned}$$

for $p \in \Delta - \partial\Delta$ with equality only at $p = e^4$.

Hence L is a vector field Lyapunov function for (9.6) and a Lyapunov function for (9.7). It follows that for $p \in \Delta - \partial\Delta$ the α limit set of p is the fixed point e^4 while the ω limit set lies in $\partial\Delta$. □

Observe that the systems are symmetric with respect to the rotation map $\sigma(p_1, p_2, p_3) = (p_3, p_1, p_2)$ which maps $\sigma(e^i)$ to e^{i+1} (addition mod 3).

That is, the cyclic homeomorphism σ is a conjugacy of each of these systems with itself.

Figure 9.2 suggests that the interior orbits spiral outward near the boundary rather than converge to one of the boundary vertices. We will confirm this as part of our computation of the limit sets of invariant measures $M(\varphi_\varepsilon, p)$ and $M(F_\tau, p)$.

The point measure δ_i concentrated on the fixed point e^i $(i = 1, 2, 3, 4)$ is an invariant measure. By Proposition 3 these four fixed points constitute the entire set of recurrent points and so by Proposition 8.8 every invariant measure is a convex combination of these four point measures. So In_P is a simplex in $P(\Delta)$ with vertices δ_1, δ_2, δ_3 and δ_4, the ergodic measures.

Now define:

$$\mu_1^\varepsilon = \frac{1}{3+\varepsilon^2}[(1+\varepsilon)^2\delta_1 + (1+\varepsilon)(1-\varepsilon)\delta_2 + (1-\varepsilon)^2\delta_3], \quad \mu_2^\varepsilon = \sigma_*\mu_1^\varepsilon, \qquad \mu_3^\varepsilon = \sigma_*\mu_2^\varepsilon, \tag{9.11}$$

and similarly,

$$\begin{aligned} \nu_1^\tau &= \frac{1}{K_\tau}[(\ln(1+\tau))^2\delta_1 - \ln(1+\tau)\ln(1-\tau)\delta_2 + (\ln(1-\tau))^2\delta_3], \\ K_\tau &= (\ln(1+\tau))^2 - \ln(1+\tau)\ln(1-\tau) + (\ln(1-\tau))^2, \\ \nu_2^\tau &= \sigma_*\nu_1^\tau; \qquad \nu_3^\tau = \sigma_*\nu_2^\tau. \end{aligned} \tag{9.12}$$

Define M^ε and N^τ to be the two dimensional simplices in In_P with vertices $\{\mu_1^\varepsilon, \mu_2^\varepsilon, \mu_3^\varepsilon\}$ and $\{\nu_1^\tau, \nu_2^\tau, \nu_3^\tau\}$ respectively.

4. Lemma. *Let $f: X \to X$ be a continuous map and let $u \in C(X; \mathbb{R})$. If for $x \in X$ u is constant on the set $\omega_\# f(x)$ (e.g., if u is constant on $\omega f(x)$) then the sequence $\{(1/n)\sum_{i=0}^{n-1} u(f^i(x))\}$ converges to this constant value.*

Proof. Suppose the subsequence $\{a_{n_k}\}$ of $\{a_n = \frac{1}{n}\sum_{i=0}^{n-1} u(f^i(x))\}$ converges to a. By refining the subsequence we can assume that the subsequence $\{\sigma_{n_k}(f, \delta_x)\}$ converges to an invariant measure μ. By definition of weak convergence $\{a_{n_k}\} \to \langle u, \mu\rangle$ and so $a = \langle u, \mu\rangle$. But by Proposition 8.8(a) the support of μ lies in $\omega_\# f(x)$ upon which u is constant. So a is this constant value. A bounded sequence with a unique limit point converges.

Remark. With the same proof of the analogous result holds for flows as well. □

5. Theorem. *For the flow φ^ε associated with (9.6) the limit set $M(\varphi^\varepsilon, p)$ is the boundary of the simplex M^ε for any $p \in \Delta - (\partial\Delta \cup e^4)$. For the map F_τ associated with (9.7) the limit set $M(F_\tau, p)$ is the boundary of the simplex N^τ for any $p \in \Delta - (\partial\Delta \cup e^4)$.*

Proof. Because $\omega(\varphi^\varepsilon)(p)$ is a connected, chain transitive subset of $\partial\Delta$ it is either a single vertex or all of $\partial\Delta$. In any case, $\omega_\#(\varphi^\varepsilon)(p) \subset \omega(\varphi^\varepsilon)(p) \cap |\omega(\varphi^\varepsilon)| \subset \{e^1, e^2, e^3\}$. So the function $V(p) = \sum_{i=1}^3 p_i^2$ is constantly 1 on $\omega_\#(\varphi^\varepsilon)(p)$.

Now integrate (9.9) to get

$$\tag{9.13} \begin{gathered} \frac{1}{t}(\ln p(t)_i - \ln p_i) = \varepsilon x(t)_i - E_{ix(t)} - \varepsilon V_t, \\ p(t) = \varphi^\varepsilon(p, t), \quad x(t) = \frac{1}{t}\int_0^t p(s)\,ds, \quad V_t = \frac{1}{t}\int_0^t V(p(s))\,ds. \end{gathered}$$

Sum on i and take the limit, using Lemma 4 for V_t and observing that $\sum_i E_{ij} = 0$ for all j:

$$\tag{9.14} \lim_{t\to\infty} \sum_i \frac{1}{t} \ln p(t)_i = \varepsilon - 0 - 3\varepsilon = -2\varepsilon.$$

Suppose that $\{t_k\}$ is a sequence approaching ∞ such that $\{\sigma_{t_k}(\varphi^\varepsilon, p)\}$ converges to μ. By going to a subsequence we can assume that $\{p(t_k)\}$ converges to a point p^* which lies in $\omega(\varphi^\varepsilon)(p)$ and so in $\partial\Delta$. The average position sequence $\{x(t_k)\}$ converges to a point x^*. In fact, for the coordinate function $x_j(p) = p_j$:

$$\langle x_j, \mu\rangle = \lim_k\{\langle x_j, \sigma_{t_k}(\varphi^\varepsilon, p)\rangle\} = \lim_k\{x(t_k)_j\} = x_j^*.$$

On the other hand $\langle x_j, \delta_i\rangle = e_j^i$ $(i, j = 1, 2, 3)$ and μ is a linear combination of $\delta_1, \delta_2, \delta_3$ since $\omega_\#(\varphi^\varepsilon)(p) \subset \{e^1, e^2, e^3\}$. Therefore,

$$\tag{9.15} \mu = x_1^*\delta_1 + x_2^*\delta_2 + x_3^*\delta_3.$$

The determination of x^* depends upon p^*. Since $p^* \in \partial\Delta$ some $p_i^* = 0$. Suppose $p_3^* = 0$. We first deal with the case $p_1^*, p_2^* > 0$. So $p(t_k)_i$ remains bounded above 0 for $i = 1, 2$. Then (9.14) implies

$$\begin{aligned} &\lim_k \frac{1}{t_k} \ln p(t_k)_1 = \lim_k \frac{1}{t_k} \ln p(t_k)_2 = 0, \\ &\lim_k \frac{1}{t_k} \ln p(t_k)_3 = -2\varepsilon. \end{aligned}$$

Taking the limit in (9.13) we obtain that x^* is the solution of the system

$$\varepsilon x_i^* - E_{ix^*} = \varepsilon - 2\varepsilon e_i^3 \qquad (i = 1, 2, 3).$$

The coefficients of μ_1^ε are the unique solution and so $\mu = \mu_1^\varepsilon$ by (9.15).

Similarly, we get μ_2^ε and μ_3^ε under the assumptions $p_1^* = 0$ or $p_2^* = 0$ provided that the remaining other two coordinates are positive.

Now suppose that $p^* = e^1$ so that $p_3^* = 0$ and $p_2^* = 0$. By restricting further we can assume that both $\{(1/t_k)\ln p(t_k)_2\}$ and $= \{(1/t_k)\ln p(t_k)_3\}$

converge. A priori we know that the sum of these two negative sequences converges to -2ε. So there exist $u, v \geq 0$ with $u+v=1$ such that

$$\lim \frac{1}{t_k} \ln p(t_k)_1 = 0, \quad \lim \frac{1}{t_k} \ln p(t_k)_2 = -2\varepsilon u, \quad \lim \frac{1}{t_k} \ln p(t_k)_3 = -2\varepsilon v,$$

and so x^* in this case solves the system

$$\varepsilon x_i^* - E_{ix^*} = \varepsilon - 2\varepsilon e_i^2 u - 2\varepsilon e_i^3 v,$$

and so by linearity and (9.15), $\mu = u\mu_3^\varepsilon + v\mu_1^\varepsilon$, which lies on the edge of M^ε connecting μ_3^ε and μ_1^ε.

Hence $M(\varphi^\varepsilon, p)$ is contained in ∂M^ε. If the orbit $p(t)$ had converged to a vertex of Δ e^1 say, then $x(t)$ would converge to e^1 and $M(\varphi^\varepsilon, p) = \delta_1$. As $\delta_1 \notin M^\varepsilon$ it follows that instead $\omega(\varphi^\varepsilon)(p)$ is the entire boundary of Δ. In particular, we can choose subsequences $\{t_k\}$ such that $\{p(t_k)\}$ converges to any point of $\partial\Delta$. Even more, given $u, v > 0$ we can choose $\{t_k\}$ so that $\{p(t_k)_1\}$ converges to 1 and $\ln p(t_k)_2 / \ln p(t_k)_3$ converges to u/v. Hence, $M(\varphi^\varepsilon, p)$ is the entire boundary M^ε.

For the map F_τ the argument is similar but a bit more delicate. $\omega(F_\tau)(p)$ is still a chain transitive nonempty subset of $\partial\Delta$. So even if not connected it must consist of a single vertex or a subset which meets the interior of each of the three edges. Also, $\omega_{\#}(F_\tau)(p) \subset \{e^1, e^2, e^3\}$ as before.

Summing on (9.10) we have

$$\text{(9.16)} \qquad \begin{gathered} \frac{1}{n}(\ln F_\tau^n(p)_i - \ln p_i) = \frac{1}{n}\sum_{a=0}^{n-1} L_i^\tau(F_\tau^a(p)), \\ L_i^\tau(p) = \ln(1 + \tau E_{ip}), \end{gathered}$$

$$\begin{aligned} \sum_{i=1}^{3} L_i^\tau(p) &= \ln \prod_{i=1}^{3}(1 + \tau E_{ip}) \\ &= \ln[1 - \tau^2(p_1^2 + p_2^2 + p_3^2) + \cdots], \end{aligned}$$

where the remaining terms involve mixed powers like p_1p_2. So on $\omega_{\#}(F_\tau)(p)$ this sum is constantly $\ln(1-\tau^2)$ and by Lemma 4

$$\text{(9.17)} \qquad \lim \left\{ \sum_{i=1}^{3} \frac{1}{n} \sum_{a=0}^{n-1} L_i^\tau(F_\tau^a(p)) \right\} = \ln(1-\tau^2).$$

Now assume $\{\sigma_{n_k}(F_\tau, p)\}$ converges to $\mu = x_1^*\delta_1 + x_2^*\delta_2 + x_3^*\delta_3$ and as before assume that $\{F_\tau^{n_k}(p)\}$ converges to p^*.

$$\lim \left\{ \frac{1}{n_k} \sum_{a=0}^{n_k-1} L_i^\tau(F_\tau^a(p)) \right\} = \lim\{\langle \sigma_{n_k}(F_\tau, p), L_i^\tau \rangle\}$$

$$= \langle \mu, L_i^\tau \rangle = \sum_{j=1}^{3} \ln(1+\tau E_{ij}) x_j^* .$$

In the case where $p_3^* = 0$ but $p_1^*, p_2^* > 0$ then by (9.16) this limit is zero for $i = 1, 2$ and so by (9.17) the limit is $\ln(1-\tau^2)$ for $i = 3$. Thus, x^* is the solution of the system

$$\sum_j \ln(1+\tau E_{ij}) x_j^* = \ln(1-\tau^2) e_i^3 .$$

Solving this system we get that $\mu = \nu_1^\tau$. Similarly, if p^* lies in one of the other edges $\mu = \nu_2^\tau$ or ν_3^τ. When $p^* = e^1$ we go to a subsequence on which $\{(\ln F_\tau^{n_k}(p))_2 / (\ln F_\tau^{n_k}(p)_3)\}$ converges to u/v and get $\mu = u\nu_3^\tau + v\nu_1^\tau$ as in the flow case.

So we have that $M(F_\tau, p)$ is a connected subset of the boundary of the simplex N^τ in In_p. As before we see that $\omega(F_\tau)(p)$ cannot be a vertex. So the above arguments show that $M(F_\tau, p)$ contains ν_i^τ $(i = 1, 2, 3)$ and so by connectedness it contains at least two of the edges of ∂N^τ. Some extra work is needed to show that the entire ∂N^τ lies in each $M(F_\tau, p)$ but we defer the argument to the supplementary exercises.

REMARK. These examples show that the Cesaro limit sets $M(p)$, while necessarily connected, need not be convex. □

From Theorem 5 it follows that the maps for different ε's and different τ's are all nonconjugate.

6. COROLLARY. *For $s_1, s_2 > 0$ and $1 > \varepsilon_1, \varepsilon_2 > 0$ the time-s maps $f_{\varepsilon_1}^{s_1}$ and $f_{\varepsilon_2}^{s_2}$ are never topologically conjugate when $\varepsilon_1 \neq \varepsilon_2$. For $1 > \tau_1, \tau_2 > 0$ the maps F_{τ_1} and F_{τ_2} are never topologically conjugate when $\tau_1 \neq \tau_2$. f_ε^s and F_τ are not topologically conjugate except possibly when*

$$\frac{\ln(1+\tau)}{|\ln(1-\tau)|} = \frac{1+\varepsilon}{1-\varepsilon}. \tag{9.18}$$

PROOF. A topological conjugacy between $f_{\varepsilon_1}^{s_1}$ and $f_{\varepsilon_2}^{s_2}$ consists of a homeomorphism $h: \Delta \to \Delta$ such that $h \circ f_{\varepsilon_1}^{s_1} = f_{\varepsilon_2}^{s_2} \circ h$. Clearly, h permutes the vertices of Δ and maps the fixed point e^4 to itself. Also

$$h_* M(f_{\varepsilon_1}^{s_1}, p) = M(f_{\varepsilon_2}^{s_2}, h(p)).$$

But by part (i) of Exercise 8.13 we have (for $p \in \Delta - (\partial\Delta \cup e^4)$)

$$M(f_{\varepsilon_1}^{s_1}, p) = M(\varphi_{\varepsilon_1}, p) = M^{\varepsilon_1},$$
$$M(f_{\varepsilon_2}^{s_2}, h(p)) = M(\varphi_{\varepsilon_2}, h(p)) = M^{\varepsilon_2}.$$

Because h permutes the vertices of Δ, $h_* M^\varepsilon = M^\varepsilon$ for every ε. The result now follows because $M^{\varepsilon_1} \neq M^{\varepsilon_2}$ unless $\varepsilon_1 = \varepsilon_2$.

Thus, the limit set of measures M^ε is an invariant of topological conjugacy which distinguishes the different flows φ_ε.

Similarly, $N^{\tau_1} \neq N^{\tau_2}$ unless $\tau_1 = \tau_2$ and $M^\varepsilon \neq N^\tau$ except when ratio (9.18) holds. □

In contrast with this result there is a weaker notation of conjugacy for flows with respect to which all the φ_ε's are related. If φ_1 and φ_2 are flows on X_1 and X_2 then a *flow conjugacy* is a homeomorphism $h: X_1 \to X_2$ which maps the orbit of $x \in X_1$ onto the orbit of $h(x)$ in X_2 in an orientation preserving way.

7. EXERCISE. *Given* $\varepsilon_1, \varepsilon_2$ *in* $(0, 1)$ *construct a flow conjugacy* $h: \Delta \to \Delta$ *between* φ_{ε_1} *and* φ_{ε_2}. *(Hint: let* h *be the identity on* $\partial\Delta \cup e^4$ *and map each perpendicular from* e^4 *to an edge into itself.)* □

We turn now to a special class of maps called the shift maps. For X a compact metric space let X^∞ and $X^{\pm\infty}$ be the spaces of sequences and biinfinite sequences on X, respectively. Thus, a point of X^∞ (or $X^{\pm\infty}$) is a function from $\mathbb{Z}^+$ to X (resp. from $\mathbb{Z}$ to X). With the product topology these are compact metric spaces as well. The shift maps $s: X^\infty \to X^\infty$ and $s: X^{\pm\infty} \to X^{\pm\infty}$ are defined by:

$$s(x)_i = x_{i+1}. \tag{9.19}$$

On X^∞ s is an onto, continuous map and on $X^{\pm\infty}$ the shift is a homeomorphism.

If f is any continuous map on X, then the orbit map $o_f: X \to X^\infty$ defined by

$$o_f(x)_i = f^i(x) \tag{9.20}$$

is a homeomorphism of X onto a closed subset of X^∞. Because $s \circ o_f = o_f \circ f$ the restriction of s to the s +invariant subset $o_f(X)$ can be identified with f via the conjugacy o_f. Thus, the shift map on X^∞ "includes" all maps on X. When f is a homeomorphism then o_f can be regarded as a map to $X^{\pm\infty}$ by using (9.20) for negative i as well.

The term *symbolic dynamics* is applied to the shift maps when X is a finite set. The spaces of sequences are then totally disconnected as well as compact. We will examine in detail the case when $X = \mathbb{Z}_2 = \{0, 1\}$. This example is especially rich because $\mathbb{Z}_2^\infty$ has a number of different algebraic interpretations.

Regarded as the integers modulo 2, $\mathbb{Z}_2$ is a field and so $\mathbb{Z}_2^\infty$ is a ring (actually an algebra over $\mathbb{Z}_2$) with respect to coordinatewise addition and multiplication:

$$(x \oplus y)_i = x_i + y_i,$$
$$(x * y)_i = x_i y_i,$$

with addition and multiplication on the right mod 2.

We can also identify $x \in \mathbb{Z}_2^\infty$ with a subset of $\mathbb{Z}^+$ namely the support of $x = \{i : x_i = 1\}$. Then $*$ corresponds to the intersection and $\oplus$ to the symmetric difference of subsets. The *complement* of x, denoted $\overline{x}$, obtained by replacing 0's by 1's and vice versa, is given by

$$(\overline{x})_i = 1 + x_i.$$

Thus, for example, the complement of 0, the zero sequence, is the sequence of 1's. Clearly

$$x * \overline{x} = 0, \qquad x \oplus \overline{x} = \overline{0}.$$

BEWARE. We are saving the symbol 1 for another sequence. We will always denote by $\overline{0}$ the sequence of 1's.

Now on $\mathbb{Z}_2^\infty$ we define three real valued maps:

$$w_L(x) = \sum_{i=0}^{\infty} x_i 2^{-i-1}, \qquad w_C(x) = \sum_{i=0}^{\infty} x_i 2 \cdot 3^{-i-1}, \tag{9.21}$$
$$w_0(x) = \inf\{2^{-n} : x_i = 0 \text{ for } i < n\}.$$

8. EXERCISE. *Prove*:

(a) $w \geq 0$ *with equality only when* $x = 0$ *for* $w = w_L$, w_C, *and* w_0.

(b) $w(x \oplus y) \leq w(x) + w(y)$ *and* $w(x * y) \leq \min(w(x), w(y))$ *for* $w = w_L$ *and* w_C. *Observe that in these two cases the functions are measures on the subsets of* $\mathbb{Z}^+$.

(c) $w_0(x \oplus y) \leq \max(w_0(x), w_0(y))$ *and* $w_0(x * y) \leq \min(w_0(x), w_0(y))$.

(d) *All three maps are continuous,* w_L *associates to each sequence the point in* $[0, 1]$ *whose binary expansion is given by the sequence. So* w_L *maps onto the unit interval with point inverses unique except for the countable family of dyadic rationals which admit two binary expansions.* w_C *associates the point of* $[0, 1]$ *whose ternary expansion is the given sequence times* 2. *Thus,* w_C *is a homeomorphism onto the classical Cantor set.* □

The alternative algebraic structure on Z_2^∞ is that of the ring of 2-adic integers. Begin by including the natural numbers into $\mathbb{Z}_2^\infty$ by identifying each such number with its binary expansion written to the right. The natural numbers correspond to the sequences x with $x_i = 0$ for i large enough, and we identify such x with $\sum_{i=0}^{\infty} x_i 2^i$. Thus, $0 = 000\cdots$, $1 = 1000\cdots$, $2 = 0100\cdots$ etc.

Recall the usual algorithms for addition and multiplication for binary numbers but with "carrying" to the right instead of the left. These rules extend to pairs of sequences in $\mathbb{Z}_2^\infty$ and so define an addition and multiplication on $\mathbb{Z}_2^\infty$. At first glance more surprising is that we can always subtract as well because we can "borrow" from infinity. Alternatively, observe that 1 added to $\overline{0}$ is 0 and so $\overline{0} = -1$, the additive inverse of 1. In general, $x + \overline{x} = \overline{0}$ and so $x + (\overline{x} + 1) = 0$, proving that

$$\overline{x} + 1 = -x. \tag{9.22}$$

To derive such algebraic results as the associative law (which we just now used) as well as to contrast the two different algebraic structures, it is helpful to look at finite sequences. A list of n 0's and 1's is called a *word* of length n. Our original view of $\mathbb{Z}_2^\infty$ effectively regards the set of words of length n as the product $(\mathbb{Z}_2)^n$ of n copies of $\mathbb{Z}_2$ and regards $\mathbb{Z}_2^\infty$ as the inverse limit of the sequence of maps $\mathbb{Z}_2^{n+1} \to \mathbb{Z}_2^n$, projecting away from the last coordinate. Instead we can regard a word of length n as the binary expansion of an integer reduced $\bmod 2^n$. For arithmetic in $\mathbb{Z}_{2^n}$ consists of ordinary arithmetic written in binary with the information after the nth place discarded. $\mathbb{Z}_2^\infty$ with the 2-adic structure is the inverse limit of the sequence of maps $\mathbb{Z}_{2^{n+1}} \to \mathbb{Z}_{2^n}$, reduction $\bmod 2^n$.

Thus, to each $x \in \mathbb{Z}_2^\infty$ corresponds a sequence $\{x^{(n)}\}$ of congruence classes modulo 2^n with $x^{(n)}$ just the initial word of length n in x. The elements of the sequence are clearly compatible in that $x^{(n+1)} \equiv x^{(n)} \mod 2^n$. Conversely, to every such compatible sequence there corresponds a unique element of $\mathbb{Z}_2^\infty$. The projection to $\mathbb{Z}_2$ is of special interest and so we name it $\varepsilon_0 : \mathbb{Z}_2^\infty \to \mathbb{Z}_2$,

$$\varepsilon_0(x) = x_0 \in \mathbb{Z}_2, \tag{9.23}$$

and we call x *even* or *odd* according to whether $\varepsilon_0(x)$ is 0 or 1.

In general, we will write $x \equiv y \mod 2^n$ to mean that $x_i = y_i$ for $i = 0, \dots, n-1$ which is to say that x and y project to the same class in $\mathbb{Z}_{2^n}$.

If a is odd then it is a unit in the ring of 2-adic integers and we can divide by it. For then $ax = 1$ has a unique solution $x^{(n)} \mod 2^n$ and to the compatible sequence $\{x^{(n)}\}$ there corresponds the element a^{-1} in $\mathbb{Z}_2^\infty$. On the other hand, x is divisible by 2 if and only if it is even. In fact,

$$w_0(x) \le 2^{-n} \quad \text{if and only if} \quad x = 2^n a \text{ for some } a.$$

Notice that multiplication by 2 just attaches an initial zero, shifting the rest of the sequence to the right. In particular, we can describe the shift map algebraically

$$s(x) = \begin{cases} \frac{x-1}{2} & \text{if } x \text{ is odd}, \\ \frac{x}{2} & \text{if } x \text{ is even}. \end{cases} \tag{9.24}$$

9. EXERCISE. *Prove*:

(a) $w_0(x+y) \le \max(w_0(x), w_0(y))$ *and* $w_0(xy) = w_0(x)w_0(y)$. *In particular,* $w_0(2x) = w_0(x)/2$.

(b) *Define* $d(x, y) = w_0(x \oplus y) = w_0(x-y) = \inf\{2^{-n}: x_i = y_i$ *for* $i < n\} = \inf\{2^{-n}: x \equiv y \mod 2^n\}$. d *is a metric for* $\mathbb{Z}_2^\infty$ *which is translation invariant with respect to both* $\oplus$ *and* $+$. *For any* $\varepsilon > 0$ $\overline{V}_\varepsilon$ *is open and closed in* $\mathbb{Z}_2^\infty$ *and* $\overline{V}_{2^{-n}}(x) = \{y: y \equiv x \bmod 2^n\}$. *So the congruence classes* mod 2^n *form a base (of open and closed sets) for the topology of* $\mathbb{Z}_2^\infty$. *Observe that they are just cylindrical sets in the product.* □

We turn now to consider measures on $\mathbb{Z}_2^\infty$. Because the congruence classes mod 2^n form a base for the topology, any Borel measure μ on $\mathbb{Z}_2^\infty$ is determined by its values on these sets. If a is a word of length n, we denote by c_a the characteristic function of the set $\{x \in \mathbb{Z}_2^\infty: x \equiv a \bmod 2^n\}$. c_a is a continuous $\{0, 1\}$ valued function because the congruence classes are open and closed. $\langle c_a, \mu\rangle$ is the μ-probability that $x_i = a_i$ for $i = 0, \dots, n-1$ and these probabilities determine μ. Conversely, any nonnegative specification of these probabilities satisfying the obvious consistency conditions relating mod 2^n and mod 2^{n-1} classes yields a measure on $\mathbb{Z}_2^\infty$ (see, for example, Halmos (1950) §49).

In particular, there is a unique measure, which we denote by λ, such that all 2^n congruence classes have probability 2^{-n}.

10. EXERCISE. *Prove that* $w_L{}_*\lambda$ *is Lebesgue measure on the unit interval (compute the probabilities of the intervals* $[m/2^n, m+1/2^n]$ *for* $m = 0, \dots, 2^n-1$). $w_C{}_*\lambda$ *has support on the Cantor set. We met its distribution function* $L(t) = w_C{}_*\lambda([0, t])$, *the Cantor function, in Chapter* 3. *Prove that* $w_0{}_*\lambda$ *is the sum of point measures*: $w_0{}_*\lambda = \sum_{n=0}^\infty 2^{-n-1}\delta_{2^{-n}}$. □

With respect to both $\oplus$ and $+$ $\mathbb{Z}_2^\infty$ is a commutative topological group and λ is the associated Haar measure, that is, the unique translation invariant measure.

λ is also invariant with respect to the shift map but it is only one such invariant measure among many. If we regard each x of $\mathbb{Z}_2^\infty$ as a sample point for the experiment of a sequence of flips of a coin λ is the associated probability when we assume (1) the separate flips are independent events, and (2) the two outcomes 0 and 1 are equally likely. If we retain (1) but assume that the respective probabilities of 0 and 1 are p_0 and p_1, then we obtain a *Bernoulli measure* μ with $\langle c_a, \mu\rangle = p_0^i p_1^j$, where i and j are the number of occurrences of 0 and 1, respectively, in the word a.

In general, a measure μ is shift invariant if for a word a of length n $\langle c_a, \mu\rangle$ is the probability that $x_{i+j} = a_i$ for $i = 0, \dots, n-1$ when j is any whole number. It is easy to check each Bernoulli measure satisfies this property.

11. THEOREM. *For the shift map s on $\mathbb{Z}_2^\infty$ the periodic points are dense in $\mathbb{Z}_2^\infty$ and each Bernoulli measure is a mixing invariant measure. Furthermore, $I^* = \mathrm{In}_P(s)$. That is, for x in some residual subset of $\mathbb{Z}_2^\infty$ the limit set $M(x)$ consists of the entire set of invariant measures.*

PROOF. If a and b are words of length n and A and B are the congruence classes of a and b mod 2^n respectively, then for a Bernoulli measure μ it is easy to check that $\mu(A \cap s^{-k}(B)) = \mu(A)\mu(B)$ when $k \geq n$ because this is the probability that x agrees with a from 0 to $n-1$ and with b from k to $k+n-1$. It follows from this that each Bernoulli measure, for example λ, is mixing and a fortiori ergodic. (See exercise 8.25.)

x is periodic for the map s precisely when it is a periodic sequence of 0's and 1's. Clearly, for any n we can construct $\tilde{x}$ periodic and agreeing with x in the first n places and so $d(x, \tilde{x}) \leq 2^{-n}$. Hence, the periodic points are dense.

Also, given any word a of length n we can define ax by attaching a in front of x:

$$(ax)_i = \begin{cases} a_i, & i = 0, \dots, n-1, \\ x_{i-n}, & i \geq n. \end{cases} \tag{9.25}$$

So ax agrees with a in the first n places and $s^n(ax) = x$. In particular, $M(ax) = M(x)$. So if $\mu \in M(x)$ the set $\{y: \mu \in M(y)\}$ is dense in $\mathbb{Z}_2^\infty$. So from Theorem 8.11 and the remark that follows it, $\mu \in I^*$ whenever $\mu \in M(x)$ for some x. In particular, every s-ergodic measure lies in I^*.

$\mathrm{In}_P(s)$ is the closed convex hull of the ergodic measures and I^* is a closed set. So to show that I^* includes all of $\mathrm{In}_P(s)$ it suffices to show that $\widetilde{I} = \{\mu: \mu \in M(x)$ for some $x \in \mathbb{Z}_2^\infty\}$ has convex closure. For this it is enough to show that $\mu_0, \mu_1 \in \widetilde{I}$ imply $(\mu_0 + \mu_1)/2 \in \widetilde{I}$.

Assume $\mu_0 \in M(x)$ and $\mu_1 \in M(y)$. We can choose an increasing sequence of whole numbers $\{n_i\}$ such that $\{\sigma_{n_{2i}}(\delta_x)\}$ and $\{\sigma_{n_{2i+1}}(\delta_y)\}$ approach μ_0 and μ_1 respectively as $i \to \infty$. In other words, for a fixed word a of length n, $\langle c_a, \mu_0\rangle$ is the limit as $i \to \infty$ of $\langle c_a, \sigma_{n_{2i}}(\delta_x)\rangle$ which is the average number of times among the first n_{2i} shifts of x that the word a is the initial segment. Similarly, for $\langle c_a, \mu_1\rangle$. Now define

$$z = (x_{0-n_0})^{n_1}(y_{0-n_1})^{n_0}(x_{0-n_2})^{n_3}(y_{0-n_3})^{n_2}\cdots.$$

This means begin with n_1 repetitions of the initial length n_0 segment of x, then repeat the initial n_1 sequence of y n_0 times, the initial n_2 segment of x n_3 times etc. Now we claim that with $k_i = 2n_{2i}n_{2i+1}$ and $s_i = k_0 + k_1 + \cdots + k_i$,

$$\lim\{\sigma_{s_i}(\delta_z)\} = \frac{1}{2}(\mu_0 + \mu_1),$$

and it suffices to check for the arbitrary word a that

$$\lim\{\langle c_a, \sigma_{s_i}(\delta_z)\rangle\} = \frac{\lim\{\langle c_a, \sigma_{n_{2i}}(\delta_x)\rangle + \langle c_a, \sigma_{n_{2i+1}}(\delta_y)\rangle\}}{2}.$$

Now among the occurrences of a for the first s_i shifts of z most are inside one of the strings of x or y. The number of these is

$$2n_0n_1\left(\frac{\langle c_a, \sigma_{n_0}(x)\rangle + \langle c_a, \sigma_{n_1}(x)\rangle}{2}\right) + \cdots + 2n_{2i}n_{2i+1}\left(\frac{\langle c_a, \sigma_{n_{2i}}(x)\rangle + \langle c_a, \sigma_{n_{2i+1}}(x)\rangle}{2}\right).$$

Dividing this by $s_i = 2n_0n_1 + \cdots + 2n_{2i}n_{2i+1}$ we obtain the required limit (see Exercise 12 below). On the other hand, there are at most n times $n_0 + n_1 + \cdots + n_{2i+1}$ other occurrences of a and the ratio of this to s_i approaches 0 because

$$n_0 + n_1 + \cdots + n_{2i+1} = 2n_0n_1\left(\frac{n_0^{-1} + n_1^{-1}}{2}\right) + \cdots + 2n_{2i}n_{2i+1}\left(\frac{n_{2i}^{-1} + n_{2i+1}^{-1}}{2}\right)$$

(see Exercise 12 again). □

12. EXERCISE. *Let $\{m_i\}$ be a sequence of positive integers and let $\{a_i\}$ be a real sequence with limit a. Prove that the generalized Cesaro averages*

$$\left\{\frac{m_0a_0 + \cdots + m_ia_i}{m_0 + \cdots + m_i}\right\}$$

also approach a as i tends to infinity. □

The puzzling nature of Theorem 11 is exposed by using the map w_L which associates to $x \in \mathbb{Z}_2^{\infty}$ the real number in $[0, 1]$ whose binary expansion is given by x. By Exercise 10 the image $w_{L*}\lambda$ is Lebesgue measure on the unit interval. So the image under w_L of $\mathrm{Con}(\lambda)$ has Lebesgue measure equal one (w_L is one-to-one on this set). These are the so-called base 2 normal numbers in that the average frequency of any word a of length n in this binary expansion approaches 2^{-n}. However, this set of normal numbers is of first category because its complement contains the residual set which is the image under w_L of $\{x: M(x) = \mathrm{In}_P(s)\}$. For these points not only do the average word frequencies not converge to the λ values but the time average measures get arbitrarily close to every shift invariant measure.

Supplementary exercises

13. (a) Prove that an irrational rotation on the circle is an almost periodic map and that the irrational flow on the torus is an almost periodic flow (c.f. Exercise 8.23(d)).

(b) Let A be a closed subset of $\mathbb{Z}_2^\infty$ +invariant with respect to the shift map s. Assume that the restriction of s to A is almost periodic. Prove that A consists of a single periodic orbit (if $x \in \mathbb{Z}_2^\infty$ and the positive integer n satisfies $d(s^{n+j}(x), s^j(x)) \leq 1/2$ for all j then x is periodic of degree n).

14. Complete the proof of Proposition 5 by showing that $M(F_\tau, p)$ is the entire boundary of N^τ. We already know that it is a connected subset containing the three vertices. To complete the proof choose an increasing sequence of positive integers so that $\{F^{n_{2i}}(p)\}$ tends to a point p^* with $p_3^* = 0$ and $p_1^*, p_2^* > 0$ while $\{F^{n_{2i+1}}(p)\}$ tends to p^{**} with $p_2^{**} = 0$ and $p_1^{**}, p_3^{**} > 0$ such that for all i $n_{2i} \leq k \leq n_{2i+1}$ implies $F^k(p)_1 \geq 1/3$. Prove that $\lim n_{2i+1} - n_{2k} = \infty$ and then adjust the proof of connectedness of the set $M(F, p)$ to show that the set of limit points of all sequences $\{\sigma_{k_i}(F, p)\}$ with $n_{2i} \leq k_i \leq n_{2i+1}$ forms a connected set. Conclude that $M(F_\tau, p)$ contains the interval between ν_1^τ and ν_2^τ.

15. Let ξ_1 and ξ_2 be C^1 vector fields on manifolds X_1 and X_2, respectively. If $h: X_1 \to X_2$ is a C^1 map such that at every point x of X_1 $T_x h(\xi_1(x)) = \xi_2(h(x))$ then the two vector fields are called *h-related* and this implies h maps each f_1^t to f_2^t, i.e., h provides a semiconjugacy of the ξ_1 flow with the ξ_2 flow. In particular, if h is a diffeomorphism then h is a C^1 conjugacy between the two flows.

If h is a diffeomorphism such that $T_x h(\xi_1(x)) = \alpha(h(x))\xi_2(h(x))$ for some positive C^1 function α on X_2, prove that h provides a flow conjugacy between φ_1 and φ_2 (cf. the discussion before Exercise 7).

Unfortunately the most important examples of conjugacies are merely continuous (see Chapter 11) and so cannot be so described using the vector fields directly.

16. (a) Prove that any compact, totally disconnected metric space X with no isolated points is homeomorphic to $\mathbb{Z}_2^\infty$. Conclude that any open and closed subset of $\mathbb{Z}_2^\infty$ is homeomorphic to $\mathbb{Z}_2^\infty$. (Construct inductively a sequence $\{\mathcal{F}_i\}$ of covers of X with $\mathcal{F}_0 = \{X\}$ and for $i \geq 1$: (1) $\mathcal{F}_i$ is a pair-wise disjoint family of open and closed sets. (2) For some positive integer k_i each member of $\mathcal{F}_{i-1}$ contains exactly 2^{k_i} members of $\mathcal{F}_i$. (3) The diameter of each member of $\mathcal{F}_i$ is at most $1/i$. Compare X with the inverse limit space of the sequence $\{\mathcal{F}_i\}$ and compare the inverse limit with $\mathbb{Z}_2^\infty$.)

(b) Prove that any compact, totally disconnected metric space X with no isolated points is generalized homogeneous, cf. exercise 7.40(b). (In $\mathbb{Z}_2^\infty$, let $\{x_1, \ldots, x_n\}$ and $\{y_1, \ldots, y_n\}$ be two lists of distinct points with $d(x_i, y_i) \leq 1/2^k$, i.e., $x_i \equiv y_i \bmod 2^k$. Choose K so that the distinct points among the x's and y's have distinct 2^K congruence classes. For each 2^k congruence class let q permute the mod 2^K congruence classes it contains so that $q(x_i) \equiv y_i \bmod 2^K$ for $i = 1, \ldots, n$. Then continue inductively on the 2^l congruence classes for $l > k$ lifting q so that $q(x_i) \equiv y_i \bmod 2^l$. Observe that $\overline{d}(q, 1) \leq 2^{-k}$. Conclude that on the spaces $C(X; X)$ and $\text{Cis}(X; X)$ $\overline{\mathscr{N}} = \mathscr{C}$ and $\overline{\text{Per}} = |\mathscr{C}|$.)

17. (a) Let $f\colon \mathbb{Z}_2^\infty \to \mathbb{Z}_2^\infty$ be a continuous map. Define the associated *orbit parity* map $h_f\colon \mathbb{Z}_2^\infty \to \mathbb{Z}_2^\infty$ by

$$h_f(x)_i = \varepsilon_0(f^i(x)) = f^i(x)_0, \qquad i = 0, 1, \ldots. \tag{9.26}$$

Thus, the sequence of $h_f(x)$ is the sequence of 0's and 1's indicating the parities of the successive points of the orbit of f. With s the shift map, show that:

$$\varepsilon_0 \circ h_f = \varepsilon_0, \qquad h_s = 1, \qquad h_f \circ f = s \circ h_f. \tag{9.27}$$

Thus, h_f maps f to s (see Definition 1.15), providing a semiconjugacy from f to s. The term "semiconjugacy" is used because h_f is not in general invertible. Prove that it is continuous by showing

$$d_0(h_f(x), h_f(y)) \leq 2^{-n}, \qquad n = 1, 2, \ldots$$

if and only if

$$d_0(f^i(x), f^i(y)) \leq 1/2, \qquad i = 0, 1, \ldots, n-1.$$

In particular, $h_f(x) = h_f(y)$ if and only if $d_0(f^i(x), f^i(y)) < 1/2$ for all $i \geq 0$. So h_f is one-to-one precisely when the latter condition implies $x = y$.

(b) Define the map $t\colon \mathbb{Z}_2^\infty \to \mathbb{Z}_2^\infty$ by

$$t(x) = \begin{cases} \frac{3x+1}{2} & \text{if } x \text{ is odd}, \\ \frac{x}{2} & \text{if } x \text{ is even}, \end{cases} \tag{9.28}$$

where $3 = 11000\cdots$ (compare (9.24)). Show that t is an onto map and the preimage of each point is a pair consisting of one even and one odd point. Prove for $x, y \in \mathbb{Z}_2^\infty$ that

$$x \equiv y \bmod 2^n \Leftrightarrow x_0 = y_0 \text{ and } t(x) \equiv t(y) \bmod 2^{n-1}. \tag{9.29}$$

Use (9.29) to show that for the orbit parity map h_t $\{x\colon h_t(x) \equiv y \bmod 2^n\}$ consists of a single mod 2^n conjugacy class for each y in $\mathbb{Z}_2^\infty$. Then prove that h_t is homeomorphism of $\mathbb{Z}_2^\infty$ preserving distance d_0 and the measure λ. Conclude that h_t is a conjugacy of t with the shift map and that λ is an invariant measure for t.

The *Collatz Problem* (also known as the $3x+1$ *Problem*) is the (open) question whether the t orbit of every positive integer terminates eventually in the period 2 cycle $1, 2, 1, 2\ldots$. The above discussion certainly does not solve the problem. Instead, it suggests a possible explanation why this apparently simple problem has proved so difficult. λ is mixing for t because it is so for s. So $\{x: M(x, \tau) = \lambda\}$ has λ measure 1 in $\mathbb{Z}_2^\infty$. For the Collatz Problem we scrutinize a measure zero subset, the positive integers in $\mathbb{Z}_2^\infty$, and try to derive upon it approach to a periodic orbit. (For a lovely survey of the Collatz Problem see Lagarias (1986).)

18. (a) On $X = Z_2^\infty$ regarded as the 2-adic integers define $f: X \to X$ by $f(x) = x + 1$. Prove that the map f is almost periodic ($f^{2^n}(x) \equiv$ mod 2^n) and so f is uniquely ergodic with invariant measure the Haar measure λ. However, f^n is not even chain transitive if n is even. In particular, for no value of n is f^n chain mixing.

(b) More generally, let $\{d_k\}$ be an increasing sequence of positive integers such that d_k divides d_{k+1} and let X be the inverse limit of the sequence $p_k: Z_{d_{k+1}} \to Z_{d_k}$ (reduction mod d_k). Define $f: X \to X$ by $f(\{x_k\}) = \{x_k + 1 \pmod{d_k}\}$ and prove that f is almost periodic but no iterate of f is chain mixing.

(c) Return to exercise 8.22 and suppose that $f: X \to X$ is chain transitive. For $\varepsilon > 0$ define $d(\varepsilon)$ to be the greatest common divisor of the $\varepsilon(x, x)$'s, defined there. Prove $\varepsilon_1 < \varepsilon_2$ implies $d(\varepsilon_2)$ divides $d(\varepsilon_1)$ and that f^n is chain mixing on each f^n basic set if and only if $d(\varepsilon)$ divides n for all $\varepsilon > 0$. If, on the other hand, $d(\varepsilon) \to \infty$ as $\varepsilon \to 0$, prove that there is a semiconjugacy of f onto a translation map of part (b).

Chapter 10. Fixed Points

In extending our study of dynamical systems on metric spaces we now revoke our standing assumption of compactness. Instead, we sharpen our demands on the maps beginning first with Lipschitz conditions.

For a map, f, between metric spaces X_1 and X_2 the Lipschitz constant $L(f)$ is the smallest bound on the rate at which f expands distances, that is, $L(f)$ is the smallest nonnegative, extended real number such that for all $x_1, x_2 \in X_1$:

$$d_2(f(x_1), f(x_2)) \leq L(f) d_1(x_1, x_2). \tag{10.1}$$

f is called a *Lipschitz* map, or equivalently, f satisfies a Lipschitz condition, if $L(f) < \infty$. Notice that this condition implies uniform continuity which is, in turn, strictly stronger than continuity now that compactness is no longer assumed.

The important Banach Fixed Point Theorem illustrates how Lipschitz conditions can be used to get convergence results when compactness is replaced by the weaker hypothesis of completeness.

1. THEOREM. *Let X be a complete metric space and let $f: X \to X$ have Lipschitz constant $\lambda < 1$. (Such a map is called a contraction.) f has a unique fixed point e and for every $x \in X$*

$$e = \lim_{n \to \infty} \{f^n(x)\}, \qquad d(x, e) \leq d(x, f(x))/(1 - \lambda). \tag{10.2}$$

PROOF. By induction $d(f^n(x), f^{n+1}(x)) \leq \lambda^n d(x, f(x))$ for $n = 0, 1, \dots$. So the series $\sum_{n=0}^{\infty} d(f^n(x), f^{n+1}(x))$ converges and by the triangle inequality

$$d(x, f^{k+1}(x)) \leq \sum_{n=0}^{k} d(f^n(x), f^{n+1}(x)) \leq d(x, f(x))/(1 - \lambda). \tag{10.3}$$

Replacing x by $f^l(x)$ we have

$$\begin{aligned} d(f^l(x), f^{l+k+1}(x)) &\leq d(f^l(x), f^{l+1}(x))/(1 - \lambda) \\ &\leq \lambda^l d(x, f(x))/(l - \lambda). \end{aligned}$$

Thus, $\{f^n(x)\}$ is a Cauchy sequence and by completeness, it converges to a point e. As usual $e = \lim\{f^{n+1}(x)\} = f(e)$ and so e is a fixed point. If e_1 is also a fixed point then

$$d(e, e_1) = d(f(e), f(e_1)) \le \lambda d(e, e_1)$$

which implies $d(e, e_1) = 0$ since $\lambda < 1$. Hence, e is the unique fixed point and so is independent of the choice of initial point x. Finally, letting k tend to ∞ in (10.3) we get the inequality in (10.2). □

This fundamental result generalizes in several different ways. A point e is called an *attractive fixed point* of a map $f: X \to X$ if $e = \lim\{f^n(x)\}$ for every $x \in X$. The dynamics of a contraction or any map with an attractive fixed point are so simple that the concept is not much needed for the study of a particular system. It is, however, valuable for the building of theorems. The following results of Hirsch and Pugh stretch the range of the arguments to which the Banach Fixed Point Theorem applies.

2. PROPOSITION. *Let f be a map on a metric space X admitting an attractive fixed point e.*

(a) *If A is a nonempty, closed, f $+$invariant subset of X, then e lies in A.*

(b) *With Y a complete metric space, assume $F: X \times Y \to X \times Y$ is a continuous map fibering over f, that is, $F(x, y) = (f(x), g(x, y))$ $((x, y) \in X \times Y)$ where $g: X \times Y \to Y$. For $x \in X$ let $g_x: Y \to Y$ be defined by $g_x(y) = g(x, y)$. Assume that for some $\lambda < 1$, $L(g_x) \le \lambda$ for all $x \in X$, i.e., the family of maps $\{g_x\}$ is uniformly contracting. Then F has an attractive fixed point.*

PROOF. (a) Begin with $x \in A$. As the limit of the sequence $\{f^n(x)\}$ in A, e lies in A.

(b) Let q be the fixed point of the contraction g_e. Given $(x, y) \in X \times Y$ let $(x_n, y_n) = F^n(x, y)$ so that $x_{n+1} = f(x_n)$ and $y_{n+1} = g_{x_n}(y_n)$. Because e is attractive for f, $\lim\{x_n\} = e$ and so by continuity of F $\lim\{\varepsilon_n\} = 0$ where $\varepsilon_n \equiv d(g_{x_n}(q), q)$. To complete the proof that (e, q) is attractive for F, we must show $\lim\{d(y_n, q)\} = 0$. This follows from

$$\begin{aligned} 0 \le d(y_{n+1}, q) &\le d(g_{x_n}(y_n), g_{x_n}(q)) + d(g_{x_n}(q), q) \\ &\le \lambda d(y_n, q) + \varepsilon_n, \end{aligned}$$

and the exercise below. □

3. EXERCISE. *On the real line the fixed point of the contraction $f(x) = \lambda x + \varepsilon$ is $\varepsilon/(1 - \lambda)$ (where $|\lambda| < 1$). If $0 \le \lambda < 1$ and $\{\varepsilon_n\}$ is a nonnegative sequence with $\lim\{\varepsilon_n\} = 0$ prove that $0 \le x_{n+1} \le \lambda x_n + \varepsilon_n$ for all n implies $\lim\{x_n\} = 0$. (Hint: prove the lim sup is between 0 and $\varepsilon_n/(1 - \lambda)$ for all n.)* □

The reader should observe the contrast between an attractive fixed point and a fixed point which is an attractor. Of course, the former condition is global and the latter local so that an attractor need not be attractive. But the converse is not true either.

4. EXERCISE. (a) *Construct a homeomorphism* $f: X \to X$ *with* X *compact such that* X *has an attractive fixed point* e *but* X *is the only attractor (and* $X \neq e$ *), (see Exercise* 3.7(a)*). Construct a map* f *of a compact space* X *with a unique fixed point* e *lying in every nonempty, closed, invariant set and yet* e *is not an attractive fixed point (spiral in to the previous example).*

(b) *Let* f *be a map on a compact space and* e *be a fixed point of* f *which is an attractor. Prove that* e *is attractive if and only if* e *lies in every closed, nonempty* f $+$ *invariant set. Prove that the repellor dual to* e *is then empty and conclude that* $e = \bigcap_n \{f^n(X)\}$. □

We get a different view of the Banach Fixed Point Theorem by concentrating on the inequality of (10.2). Think of a point x as almost fixed if $f(x)$ is near x, i.e., if $d(x, f(x))$ is small. For a contraction, Theorem 1 says that near an almost fixed point x lies a fixed point e and the distance between them, $d(x, e)$, is bounded by $d(x, f(x))$ times $1/(1 - L(f))$. This sort of result, useful in itself, is also valuable for perturbations of the map. We will now extend this aspect of the theorem to Lipschitz perturbations of various sorts of linear mappings.

Given a continuous linear map T between Banach spaces (written B-spaces) E_1 and E_2, the norm of T:

$$\|T\| \equiv \sup\{\|T(x)\|_2 : \|x\|_1 \leq 1\}$$

is the smallest bound on the rate at which T expands the length of a vector. By linearity the norm of T is the same as its Lipschitz constant:

$$\|T(x_1) - T(x_2)\| = \|T(x_1 - x_2)\| \leq \|T\| \, \|x_1 - x_2\| .$$

In particular, a linear map $T: E \to E$ is a contraction when $\|T\| < 1$. More generally, we will be interested when T is a contracting map with respect to some equivalent norm on E.

A norm $\| \ \|_1$ on E is called *equivalent* to $\| \ \|$ if there exists $C > 0$ such that for all x in E:

$$C^{-1}\|x\| \leq \|x\|_1 \leq C\|x\| . \tag{10.4}$$

The smallest such constant C is the maximum of the norms of the "identity" maps between E equipped with $\| \ \|$ and with $\| \ \|_1$. We recall some results on norms and spectra that we will need.

5. EXERCISE. (a) *If a one-to-one, onto linear map* T *between* B *spaces is continuous then its inverse is continuous as well (cf. The Closed Graph Theorem).*

(b) *If* $E_1, \ldots, E_n$ *are B-spaces, then on the product* $E_1 \times \cdots \times E_n$ *define the norms*:

$$\|(x_1, \ldots, x_n)\|_0 = \max(\|x_1\|, \ldots, \|x_n\|)$$
$$\|(x_1, \ldots, x_n)\|_p = (\|x_1\|^p + \cdots + \|x_n\|^p)^{1/p}, \qquad 1 \le p < \infty.$$

Show that $\| \ \|_0 \le \| \ \|_p \le n^{1/p} \| \ \|_0$ *implying that all of these norms are equivalent. Unless otherwise mentioned we will always use the max norm* $\| \ \|_0$ *on the product.*

(c) *A* splitting *of a* B *space* E *is a pair of closed subspaces* E_1, E_2 *such that* $E_1 \cap E_2 = 0$ *and* $E_1 + E_2 = E$. *Equivalently, a pair* E_1, E_2 *of closed subspaces splits* E *if the map of* $E_1 \times E_2$ *into* E *by* $(x_1, x_2) \to x_1 + x_2$ *is one-to-one and onto. The map is then a linear isomorphism* (*cf.* (a)) *and we write* $E = E_1 \oplus E_2$.

(d) *If* $T \colon E \to E$ *is a continuous linear map, then the spectrum of* T, *written* $\sigma(T)$, *is the set of complex numbers* λ *such that* $\lambda 1 - T$ *is not a linear isomorphism* (*making sense of this requires complexification of the Banach space*). *For example,* T *is an isomorphism if and only if* $0 \notin \sigma(T)$. *Recall that* $\sigma(T)$ *is a compact subset of the complex plane the maximum modulus of which* $\rho(T)$ *is called the spectral radius of* T. *The spectral radius is computable by the formula*

$$\rho(T) = \limsup\{\|T^n\|^{1/n}\}.$$

So in particular, $\rho(T) \le \|T\|$. (*See, e.g., Yosida* (1965) §VIII.2).

(e) *If* $\|T\| < 1$ *prove that* $1 - T$ *is an isomorphism with inverse*

$$(1 - T)^{-1} = \sum_{n=0}^{\infty} T^n. \tag{10.5}$$

(*Show that the series converges to a limit* S *which satisfies* $1 + TS = S = 1 + ST$.) *Use this formula to prove directly that* $\rho(T) \le \|T\|$.

(f) *For a linear operator* T *on a real* B *space* E *the spectrum is closed under conjugation* ($\lambda 1 - T$ *invertible implies* $\bar{\lambda} 1 - T$ *is*). *Now suppose* $\sigma(T)$ *is written as the disjoint union of two closed, conjugation invariant subsets* σ_1 *and* σ_2. *There exists a splitting of* E *by* T *+invariant subspaces* E_1 *and* E_2 *such that* $g(T|E_1) = \sigma_1$ *and* $\sigma(T|E_2) = \sigma_2$. (*The proof of this result is best accomplished by using Dunford's operational calculus* (*see Yosida* §VIII.5) *the projection of* E *to* E_1 *is obtained by integrating the resolvent function about a loop or family of loops containing* σ_1 *in its interior and* σ_2 *in its exterior.*)
□

We call a linear map T on a B space E *contractive* if it is a contraction with respect to some equivalent norm on E. The following lemma of Holmes shows that T is contractive precisely when its spectrum is contained in the open unit disk of the complex plane.

6. LEMMA. *Let T be a continuous linear map on a B space E. For a positive number λ the following conditions are equivalent*:

(1) *For some constants $C > 0$ and $\lambda_1 < \lambda$, $\|T^n\| \leq C\lambda_1^n$ for $n = 0, 1, \dots$.*

(2) *The spectral radius of T is less than λ.*

(3) *With respect to some equivalent norm on E, the norm of T is less than λ.*

PROOF. (1) and (2) both say that $\limsup\{\|T^n\|^{1/n}\} < \lambda$ which is implied by $\|T\| < \lambda$ where the norm of T can be computed with respect to any equivalent norm on E. Hence, (3) implies (1) and (2) and the latter two are equivalent.

On the other hand, given (1) choose λ_2 between λ_1 and λ and define on E the norm

$$\|x\|_1 = \sum_{n=0}^{\infty} \lambda_2^{-n} \|T^n x\|. \tag{10.6}$$

Clearly, $\|x\|_1 \geq \|x\|$ and

$$\|x\|_1 \leq \sum_{n=0}^{\infty} \lambda_2^{-n} C\lambda_1^n \|x\| = \frac{C}{1 - (\lambda_1/\lambda_2)} \|x\|.$$

Thus, $\| \ \|_1$ is a norm equivalent to $\| \ \|$. But

$$\|Tx\|_1 = \sum_{n=1}^{\infty} \lambda_2^{-n+1} \|T^n x\| \leq \lambda_2 \|x\|_1.$$

So with respect to $\| \ \|_1$, the norm of T is at most λ_2.

REMARK. It is convenient for some applications to observe that we can replace the infinite sum in (10.6) by a finite one. Just choose λ_3 between λ_2 and λ and choose a whole number N large enough that $C\lambda_2^{-N}\lambda_1^{N+1} < \lambda_3 - \lambda_2$. If we then define $\|x\|_1 = \sum_{n=0}^{N} \lambda_2^{-n} \|T^n x\|$ we obtain an equivalent norm and $\|Tx\|_1 \leq \lambda_2 \|x\|_1 + (\lambda_3 - \lambda_2)\|x\| \leq \lambda_3 \|x\|_1$. □

A map T is called *expansive* if it is a linear isomorphism and its inverse is a linear contraction. The identity $\lambda 1 - T^{-1} = -\lambda T^{-1}(\lambda^{-1} 1 - T)$ implies $\sigma(T^{-1}) = \{1/\lambda : \lambda \in \sigma(T)\}$. It follows from the lemma that T is expansive if and only if its spectrum is disjoint from the closed unit disk. Observe that a contractive linear map is not necessarily an isomorphism.

Now for the prototype of our perturbation results.

7. PROPOSITION. *A sufficiently small Lipschitz perturbation of a linear contraction has an attractive fixed point. More precisely, if $f(x) = T(x) + g(x)$ with T linear and g Lipschitz satisfying $\|T\| + L(g) < 1$, then f is a contraction and so has an attractive fixed point.*

While trivial $(L(f) \leq \|T\| + L(g))$ this proposition illustrates a repeating pattern. The perturbation result can be stated without reference to a particular norm but the estimates require that we begin with a norm adapted to

T. We see the same pattern in the following Lipschitz version of the Inverse Function Theorem.

8. PROPOSITION. *Let T be a linear isomorphism of a B space E and let $g: E \to E$ be a Lipschitz function. Assume that $\|T^{-1}\| L(g) < 1$. Then the function $f: E \to E$ defined by $f(x) = T(x) + g(x)$ is a homeomorphism whose inverse function is Lipschitz. Writing $f^{-1}(y) = T^{-1}(y) + \tilde{g}(y)$ we have*

$$L(\tilde{g}) \leq \frac{\|T^{-1}\|^2 L(g)}{1 - \|T^{-1}\| L(g)}. \tag{10.7}$$

PROOF. Because $y = T(x) + g(x)$ if and only if $x = T^{-1}(y) - T^{-1}(g(x))$ the value of the inverse function f^{-1} at y is well defined as the unique fixed point of the contracting map k_y defined by

$$k_y(x) = T^{-1}(y) - T^{-1}(g(x)).$$

Observe that $L(k_y) \leq \|T^{-1}\| L(g) < 1$.

Now if $x_1 = f^{-1}(y_1)$ and $x_2 = f^{-1}(y_2)$, then

$$\begin{aligned}\|T^{-1}\| \, \|y_1 - y_2\| &\geq \|T^{-1}(y_1) - T^{-1}(y_2)\| \\ &= \|(x_1 - x_2) + (T^{-1} g(x_1) - T^{-1} g(x_2))\| \\ &\geq (1 - \|T^{-1}\| L(g)) \|x_1 - x_2\|.\end{aligned}$$

Consequently,

$$L(f^{-1}) \leq \frac{\|T^{-1}\|}{1 - \|T^{-1}\| L(g)}. \tag{10.8}$$

Since $\tilde{g}(y) = x - T^{-1}(y) = -T^{-1}(g(x))$ it follows that $L(\tilde{g}) \leq \|T^{-1}\| L(g) L(f^{-1})$ and so (10.7) follows from (10.8).

REMARK. Applying inequality (10.2) with initial point $x = 0$ to the contracting map k_y with $y = 0$ we get

$$\|\tilde{g}(0)\| \leq \frac{\|T^{-1}\|}{1 - \|T^{-1}\| L(g)} \|g(0)\|. \tag{10.9}$$

Observe that $f(0) = g(0)$ and $f^{-1}(0) = \tilde{g}(0)$. We focus attention upon 0 because it is fixed by T and T^{-1}. □

Of course, for a linear map T, 0 is always a fixed point. It is the only fixed point unless 1 is an eigenvalue of T, i.e., $1 - T$ has a nonzero kernel. On the other hand, if $1 - T$ is not onto, if, for example, $v \in E$ is not in the image of $1 - T$, then the affine perturbation $f(x) = T(x) + v$ has no fixed point. Avoiding these cases we call T a *nondegenerate* linear map if $1 \notin \sigma(T)$ or, equivalently, if $1 - T$ is a linear isomorphism.

9. PROPOSITION. *Let T be a nondegenerate linear map of a B-space E and let $g: E \to E$ be a Lipschitz function. Assume that $\|(1-T)^{-1}\|L(g) < 1$. Then the function $f: E \to E$ defined by $f(x) = T(x) + g(x)$ has a unique fixed point e and*

$$\|e\| \le \frac{\|(1-T)^{-1}\|}{1 - \|(1-T)^{-1}\|L(g)}\|g(0)\|. \tag{10.10}$$

PROOF. Let $k(x) = (1-T)(x) - g(x)$. By Proposition 8 k is a homeomorphism and the fixed point e of f is clearly $k^{-1}(0)$. (10.10) then follows from (10.9). □

It often happens that the perturbing map is defined only on a ball about 0. The general procedure for applying these results is to extend the map via retraction onto the ball. We observe that the natural retraction is Lipschitz.

10. LEMMA. *Let E be a B-space and ε a positive number. The function $r: E \to E$ defined by*

$$r(x) = \begin{cases} x, & \|x\| \le \varepsilon, \\ \dfrac{\varepsilon x}{\|x\|}, & \|x\| \ge \varepsilon \end{cases}$$

maps E onto the ball $\overline{V}_\varepsilon(0)$. r is a Lipschitz map with $L(r) \le 2$.

PROOF. Given $x_1, x_2 \in E$ we look at three cases: First,

$$\|r(x_1) - r(x_2)\| = \|x_1 - x_2\|, \qquad \|x_1\|, \|x_2\| \le \varepsilon. \tag{10.11}$$

Next, suppose $\varepsilon \le \|x_1\| \le \|x_2\|$, then

$$\begin{aligned} \|r(x_1) - r(x_2)\| &\le \left\| \frac{\varepsilon x_1}{\|x_1\|} - \frac{\varepsilon x_1}{\|x_2\|} \right\| + \left\| \frac{\varepsilon x_1}{\|x_2\|} - \frac{\varepsilon x_2}{\|x_2\|} \right\| \\ &= \varepsilon\|x_1\| \left(\frac{1}{\|x_1\|} - \frac{1}{\|x_2\|} \right) + \frac{\varepsilon}{\|x_2\|}\|x_1 - x_2\|. \end{aligned}$$

For the real-valued function $q(t) = 1/t$ the maximum value of $|q'(t)|$ on the interval between $\|x_1\|$ and $\|x_2\|$ is $1/\|x_1\|^2$. So by the Mean Value Theorem,

$$\varepsilon\|x_1\| \left(\frac{1}{\|x_1\|} - \frac{1}{\|x_2\|} \right) \le \frac{\varepsilon}{\|x_1\|}(\|x_2\| - \|x_1\|) \le \frac{\varepsilon}{\|x_1\|}\|x_1 - x_2\|.$$

Substituting, we get the second estimate

$$\|r(x_1) - r(x_2)\| \le \varepsilon \left(\frac{1}{\|x_1\|} + \frac{1}{\|x_2\|} \right) \|x_1 - x_2\|, \qquad \|x_1\|, \|x_2\| \ge \varepsilon. \tag{10.12}$$

Finally, suppose $\|x_1\| < \varepsilon < \|x_2\|$ and proceed as above with $\varepsilon x_1/\|x_1\|$ replaced by $\varepsilon x_1/\varepsilon$

$$\|r(x_1) - r(x_2)\| \le \varepsilon\|x_1\| \left(\frac{1}{\varepsilon} - \frac{1}{\|x_2\|} \right) + \frac{\varepsilon}{\|x_2\|}\|x_1 - x_2\|,$$

and by the Mean Value Theorem applied to q on $[\varepsilon, \|x_2\|]$ we have

$$\varepsilon\|x_1\|\left(\frac{1}{\varepsilon}-\frac{1}{\|x_2\|}\right) \le \frac{\|x_1\|}{\varepsilon}(\|x_2\|-\varepsilon)$$
$$\le \frac{\|x_1\|}{\varepsilon}(\|x_2\|-\|x_1\|) \le \frac{\|x_1\|}{\varepsilon}\|x_1-x_2\|.$$

So, for the third case

$$(10.13)\quad \|r(x_1)-r(x_2)\| \le \left(\frac{\|x_1\|}{\varepsilon}+\frac{\varepsilon}{\|x_2\|}\right)\|x_1-x_2\|, \qquad \|x_1\| \le \varepsilon \le \|x_2\|.$$

Putting the three cases together we have $L(r) \le 2$. □

The use of (and slight problem with) this approach is illustrated by the following local version of Proposition 9.

11. EXERCISE. *Let T be nondegenerate linear map on a B space E and let λ, δ be positive numbers with $\lambda < 1$. Assume g is a Lipschitz function defined on the closed ball $\overline{V}_\delta(0)$, mapping to E, and satisfying $\|g(0)\| \le (1-\lambda)\delta/\|(1-T)^{-1}\|$. Define $f\colon \overline{V}_\delta(0) \to E$ by $f(x) = T(x) + g(x)$. Assuming $2\|(1-T)^{-1}\|L(g) < \lambda$, apply Lemma 10 and Proposition 9 to $T + g \circ r$ to prove that f has a unique fixed point e in $\overline{V}_\delta(0)$. Assuming only $\|(1-T)^{-1}\|L(g) < \lambda$ prove the same result directly by showing that $(1-T)^{-1} \circ g$ is a contraction of $\overline{V}_\delta(0)$.* □

If the entire unit circle is disjoint from $\sigma(T)$ then T is clearly nondegenerate. This stronger condition, hyperbolicity, satisfies special properties because removal of the circle disconnects the plane.

12. LEMMA. *A linear map T on a B space is called* hyperbolic *when it satisfies the following equivalent conditions:*

(1) *The unit circle of the complex plane is disjoint from the spectrum $\sigma(T)$.*

(2) *There exists a splitting E^+, E^- by T + invariant subspaces such that the restriction of T to E^+ is contractive and to E^- is expansive.*

(3) *There exists an equivalent norm on E and a splitting E^+, E^- by T + invariant subspaces so that the restrictions of T satisfy $\|T|E^+\| < 1$, $T|E^-$ is a linear isomorphism with $\|(T|E^-)^{-1}\| < 1$, and the identification of E with the product $E^+ \times E^-$, identifies $\|\ \|$ with the max norm on the product.*

A norm $\|\ \|$ satisfying condition (3) *is called* adapted *to the hyperbolic linear map T.*

PROOF. (1) $\Leftrightarrow$ (2) By Lemma 6 the spectrum of a linear map is interior (or exterior) to the unit circle if and only if the map is contractive (resp. expansive). So given (2), the spectra of the two restrictions $T|E^+$ and $T|E^-$ are disjoint from the circle and the spectrum of the product map T is the union of the two pieces, proving (1). On the other hand, given (1), $\sigma(T)$ is separated into two pieces σ^+, σ^- interior and exterior to the circle. By the functional analysis result recalled as part (f) of Exercise 5 we obtain a T +invariant splitting E^+, E^- with $\sigma(T|E^{\pm}) = \sigma^{\pm}$. So the restrictions are contractive and expansive.

(2) $\Leftrightarrow$ (3) Obviously (3) implies (2). For the other direction apply Lemma 6 to get norms on $E^{\pm}$ so that $\|T|E^+\|$ and $\|(T|E^-)^{-1}\|$ are less than 1. Then use the isomorphism of $E^+ \times E^-$ with E (cf. Exercise 5(c)) to transfer the max norm on the product to E. □

E^+ and E^- are called, respectively, the *stable* and *unstable* subspaces for the hyperbolic linear map T. The points of these subspaces are characterized by their dynamic behavior. If $v \in E^+$ then the positive orbit $\{T^n(v)\}$ converges to 0. In fact, the sequence of norms $\{\|T^n(v)\|\}$ tends to 0 at an exponential rate. On the other hand, if $v \notin E^+$ then the sequence $\{T^n(v) \colon n \geq 0\}$ is not even bounded in norm. The analogue for E^- is a bit less elegant because we have not assumed T is an isomorphism. If $v \in E^-$ then the sequence $\{(T|E^-)^{-n}(v)\}$ converges to 0, but if $v \notin E^-$ there does not exist a bounded chain, i.e., a sequence $\{\dots, z_{-2}, z_{-1}, z_0\}$ with $z_0 = v$ and $T(z_{-n}) = z_{-n+1}$. While useful exercises, these characterizations are special cases of the following perturbation result. This multifaceted theorem is the rock upon which hyperbolicity theory is built. It illustrates how the splitting property of a hyperbolic linear map has much stronger consequences than mere nondegeneracy.

13. THEOREM. *Let T be a hyperbolic linear map on a B space E with stable/unstable subspaces $E^{\pm}$. Assume that the norm on E is adapted to T so that $\lambda = \max(\|T|E^+\|, \|(T|E^-)^{-1}\|) < 1$. Let $g \colon E \to E$ be a Lipschitz map satisfying $\lambda_g \equiv \lambda + L(g) < 1$. The perturbed map $f \colon E \to E$ defined by $f(x) = T(x) + g(x)$ satisfies the following properties*:

(a) (characterization of unique fixed point). *f has a unique fixed point e in E and*

$$\|e\| \leq \frac{1}{1 - \lambda_g} \|g(0)\|. \tag{10.14}$$

Furthermore, if $\{\dots z_{-2}, z_{-1}, z_0, z_1, z_2, \dots\}$ is a bounded, bi-infinite 0 chain in E, i.e., for some constant C and for all n $z_{n+1} = f(z_n)$ and $\|z_n\| \leq C$, then $z_n = e$ for all n.

(b) (stable/unstable manifold theorem). *There exists a Lipschitz function $w^+ \colon E^+ \to E^-$ whose graph $W^+ \equiv \{(x, w^+(x)) \colon x \in E^+\}$ is an f + invariant subset of E (where we identify E with $E^+ \times E^-$) and the restriction of f to W^+ is a contraction*

$$L(w^+) \leq \lambda_g, \qquad L(f|W^+) \leq \lambda_g. \tag{10.15}$$

There exists a Lipschitz function $w^- \colon E^- \to E^+$ whose graph $W^- \equiv \{(w^-(y), y) \colon y \in E^-\}$ is an f invariant subset of E. The restriction of

f to W^- is a homeomorphism whose inverse is a contraction

$$L(w^-) \leq \lambda_g, \qquad L((f|W^-)^{-1}) \leq \lambda_g. \tag{10.16}$$

The intersection $W^+ \cap W^-$ consists of e, the attractive fixed point of $f|W^+$ and $(f|W^-)^{-1}$. $W^\pm$ are called the stable *and* unstable manifolds *for the fixed point e. For a point $z \in E$ the positive orbit sequence $\{z, f(z), f^2(z), \dots\}$ is bounded if and only if $z \in W^+$, in which case the sequence converges to e. On the other hand, there exists a bounded 0 chain $\{\dots, z_{-2}, z_{-1}, z_0\}$ (i.e., $f(z_n) = z_{n+1}$ for all $n < 0$) with $z_0 = z$ if and only if $z \in W^-$, in which case the sequence is the $(f|W^-)^{-1}$ orbit sequence of z and converges to e.*

(e) (shadowing lemma). *For $\{\dots, z_{-2}, z_{-1}, z_0, z_1, \dots\}$ a bi-infinite ε chain ($\|z_{n+1} - f(z_n)\| \leq \varepsilon$ for all n), there exists a unique bi-infinite 0 chain $\{\dots, w_{-2}, w_{-1}, w_0, w_1, \dots\}$ ($w_{n+1} = f(w_n)$ for all n) such that the distance sequence $\{\|w_n - z_n\|\}$ is uniformly bounded. Furthermore, an explicit bound is given by*

$$\|w_n - z_n\| \leq \frac{1}{1-\lambda_g}\varepsilon. \tag{10.17}$$

If, in addition, the sequence $\{\|z_{n+1} - f(z_n)\|\}$ tends to 0 as n approaches $+\infty$ (or, alternatively, as n approaches $-\infty$), then the sequence $\{\|w_n - z_n\|\}$ tends to 0 as well, as n approaches $+\infty$ (respectively, $-\infty$).

(d) (conjugacy theorem). *In addition to the previous conditions, assume that T is a linear isomorphism with*

$$\|(T|E^+)^{-1}\| L(g) < 1$$

and that g is a bounded function so that

$$\|g\|_0 = \sup\{\|g(x)\| : x \in E\} < \infty.$$

Then there exists a unique homeomorphism $h: E \to E$ providing a conjugacy between the linear map T and the perturbed map f, i.e., $f \circ h = h \circ T$. While h is usually not Lipschitz, its deviation from the identity map can be estimated by:

$$\|z - h(z)\| \leq \frac{1}{1-\lambda_g}\|g\|_0 \quad \text{for all } z \in E. \tag{10.18}$$

In parts (a)–(c) g can be linear (then $\lambda_g = \lambda + \|g\|$) and the results show that the linear map f is still hyperbolic. In that case the orbit sequence characterizations show that the maps $w^\pm$ are linear so that $W^\pm$ are the stable/unstable subspaces for f. In particular when $g = 0$ the characterizations show that $w^\pm = 0$ and specialize to the descriptions of $E^\pm$ given earlier. Returning to the nonlinear case we remark that, although we will not prove it here, additional smoothness assumptions on g yield corresponding smoothness results for the maps $w^\pm$ and so for the manifolds $W^\pm$.

The reader should observe how the noncompactness of the vector space and the global definitions of the mappings are vital for these results. In a compact space every sequence has compact closure. Here we use for certain sequences the weaker condition of boundedness to characterize the fixed point e and the subsets W^+ and W^-.

The proof of Theorem 13 is based on the following local and time dependent version due to Conley. The streamlined proof we give is Irwin's.

14. THEOREM. *For each integer n a B-space is given as $E_n^+ \times E_n^-$ with the max norm on the product. With $\infty > \varepsilon > 0$ fixed, we denote by B_n the closed ε ball about 0 in E_n which is the product $B_n^+ \times B_n^-$ of the ε balls in $E_n^{\pm}$. We are given Lipschitz maps $(F_n, G_n)\colon B_n = B_n^+ \times B_n^- \to E_{n+1} = E_{n+1}^+ \times E_{n+1}^-$. For $(u_n, v_n) \in B_n^+ \times B_n^-$*

$$F_n(u_n, v_n) \in E_{n+1}^+,$$
$$G_n(u_n, v_n) \equiv R_n(v_n) + S(u_n, v_n) \in E_{n+1}^-$$

with $R_n\colon E_n^- \to E_{n+1}^-$ a linear isomorphism for each n.

We are given a positive $\lambda < 1$ and assume that the Lipschitz constants satisfy

$$L(F_n) \quad \text{and} \quad \|R_n^{-1}\|(1 + L(S_n)) \le \lambda \quad \text{for all } n. \tag{10.19}$$

Furthermore, the origin images satisfy

$$\frac{\|F_n(0,0)\|}{1-\lambda}, \qquad \frac{\|R_n^{-1}\|\,\|S_n(0,0)\|}{1-\lambda} \le \varepsilon \quad \text{for all } n. \tag{10.20}$$

For $T = \mathbb{Z} \cap [k, \infty)$, $\mathbb{Z} \cap (-\infty, k]$ or $\mathbb{Z}$ we call $\{(u_n, v_n) \in E_n \text{ for } n \in T\}$ a positive, negative, or bi-infinite orbit sequence *if $(u_n, v_n) \in B_n$ for all $n \in T$ and when $n, n+1 \in T$*

$$u_{n+1} = F_n(u_n, v_n) \qquad v_{n+1} = G_n(u_n, v_n). \tag{10.21}$$

For the positive or negative case we call (u_k, v_k) (where k is the endpoint value of T) the base point of the sequence.

(a) *There exists a unique bi-infinite orbit sequence $\{(u_n^*, v_n^*)\colon n \in \mathbb{Z}\}$.*

(b) *For each n there exist Lipschitz maps*

$$w_n^+\colon B_n^+ \to B_n^- \quad \text{and} \quad w_n^-\colon B_n^- \to B_n^+$$

with graphs

$$W_n^+ = \{(u_n, w_n^+(u_n))\colon u_n \in B_n^+\} \subset B_n \subset E_n,$$
$$W_n^- = \{(w_n^-(v_n), v_n);\ v_n \in B_n^-\} \subset B_n \subset E_n$$

satisfying

$$L(w_n^+) \quad \text{and} \quad L(w_n^-) \le \lambda. \tag{10.22}$$

$W_n^+ \cap W_n^- = (u_n^*, v_n^*)$ and (u_n, v_n) is the base point of a positive (or negative) orbit sequence if and only if $(u_n, v_n) \in W_n^+$ (resp. $\in W_n^-$).

$$(10.23) \qquad (F_n, G_n)^{-1}(W_{n+1}^+) = W_n^+, \qquad B_{n+1} \cap (F_n, G_n)(W_n^-) = W_{n+1}^-.$$

The relation $(F_n, G_n)^{-1} \cap (W_{n+1}^- \times W_n^-)$ defines a function denoted $(F_n, G_n)^{-1}|W_{n+1}^-$, and the restrictions satisfy

$$(10.24) \qquad L((F_n, G_n)|W_n^+) \quad and \quad L((F_n, G_n)^{-1}|W_{n+1}^-) \leq \lambda.$$

PROOF. The proof is based on the observation that $\{(u_n, v_n): n \in T\}$ is an orbit sequence exactly when for $n, n+1 \in T$

$$u_{n+1} = F_n(u_n, v_n), \qquad v_{n+1} = R_n(v_n) + S(u_n, v_n)$$

or equivalently

$$u_{n+1} = F_n(u_n, v_n), \qquad v_n = R_n^{-1}(v_{n+1} - S_n(u_n, v_n)).$$

Now for T an infinite subset of $\mathbb{Z}$ we define $E_T \equiv$ to be the set of bounded sequences (u, v) indexed by T, i.e., $(u_n, v_n) \in E_n$ for $n \in T$ and $\|(u, v)\| = \sup_{n\in T} \|(u_n, v_n)\| < \infty$. Clearly E_T can be identified with $E_T^+ \times E_T^-$ and so the ε ball B_T in E_T becomes the product of ε balls $B_T^+ \times B_T^-$. Since E_T with the sup norm is a B space, B_T is a complete metric space. We look at three cases $T = \mathbb{Z}^+$ $(= \mathbb{Z} \cap [0, \infty))$, $\mathbb{Z}^-$ $(= \mathbb{Z} \cap [-\infty, 0])$, and $\mathbb{Z}$. Define for $(x, y) \in B_0 = B_0^+ \times B_0^-$ maps $K: B_{\mathbb{Z}} \to B_{\mathbb{Z}}$, $K_x^+: B_{\mathbb{Z}^+} \to B_{\mathbb{Z}^+}$, and $K_y^-: B_{\mathbb{Z}^-} \to B_{\mathbb{Z}^-}$. For $(u, v) \in B_T$ we let (U, V) be the associated value of the appropriate function K.

(1) For $(u, v) \in B_{\mathbb{Z}}$, $(U, V) = K(u, v)$ is defined by

$$(10.25) \qquad \begin{aligned} U_n &= F_{n-1}(u_{n-1}, v_{n-1}), \\ V_n &= R_n^{-1}(v_{n+1} - S_n(u_n, v_n)), \end{aligned} \qquad n \in \mathbb{Z}.$$

(2) For $x \in B_0^+$ and $(u, v) \in B_{\mathbb{Z}^+}$, $(U, V) = K_x^+(u, v)$ is defined by

$$(10.26) \qquad \begin{aligned} U_0 &= x, \\ U_n &= F_{n-1}(u_{n-1}, v_{n-1}), \qquad n \geq 1, \\ V_n &= R_n^{-1}(v_{n+1} - S_n(u_n, v_n)), \qquad n \geq 0. \end{aligned}$$

(3) For $y \in B_0^-$ and $(u, v) \in B_{\mathbb{Z}^-}$, $(U, V) = K_y^-(u, v)$ is defined by

$$(10.27) \qquad \begin{aligned} U_n &= F_{n-1}(u_{n-1}, v_{n-1}), \qquad n \leq 0, \\ V_0 &= y, \\ V_n &= R_n^{-1}(v_{n+1} - S_n(u_n, v_n)), \qquad n \leq -1. \end{aligned}$$

Each of these is a contracting map of the corresponding B_T. We focus on (1). The proofs for (2) and (3) are identical.

$$L(K) \leq \max_n(L(F_{n-1}), \|R_n^{-1}\|(1 + L(S_n))) \leq \lambda < 1.$$

To show that the image $K(B_{\mathbb{Z}})$ in $E_{\mathbb{Z}}$ actually lies in $B_{\mathbb{Z}}$ we note that

$$\begin{aligned}\|U_n\| &\leq \|F_{n-1}(0, 0)\| + L(F_{n-1})\|(u_{n-1}, v_{n-1})\| \\ &\leq (1 - \lambda)\varepsilon + \lambda\varepsilon = \varepsilon.\end{aligned}$$

$$\begin{aligned}\|V_n\| &\leq \|R_n^{-1}\| \, \|v_{n+1}\| + \|R_n^{-1}\|L(S_n)\|(u_n, v_n)\| + \|R_n^{-1}\| \, \|S_n(0, 0)\| \\ &\leq \|R_n^{-1}\|(1 + L(S_n))\varepsilon + \|R_n^{-1}\| \, \|S_n(0, 0)\| \\ &\leq \lambda\varepsilon + (1 - \lambda)\varepsilon = \varepsilon.\end{aligned}$$

Because K is a contraction it has a unique fixed point. This is the unique bi-infinite orbit sequence (u^*, v^*), proving (a).

To each x in B_0^+ associate the fixed point of K_x^+, denoted $(U^+(x), V^+(x))$, in $B_{\mathbb{Z}^+}$. This defines a function $(U^+, V^+)\colon B_0^+ \to B_{\mathbb{Z}^+}$. With $n = 0$, $U_0^+(x) = x$ and $w_0^+\colon B_0^+ \to B_0^-$ is defined to be the function V_0^+ with graph W_0^+. If $(\tilde{u}, \tilde{v})$ is a positive orbit sequence starting at 0 then for $x = \tilde{u}_0$ we see that $(\tilde{u}, \tilde{v})$ is a fixed point of K_x^+ and so $(\tilde{u}, \tilde{v}) = (U^+(x), V^+(x))$. In particular, $\tilde{v}_0 = w^+(x)$ and the base point $(\tilde{u}_0, \tilde{v}_0) \in W_0^+$.

Similarly, we define $(U^-(y), V^-(y))$ to be the fixed point of K_y^- in $B_{\mathbb{Z}}$, with $V_0^-(y) = y$ and $U_0^- \equiv w_0^-\colon B_0^- \to B_0^+$ with graph W_0^-. Again if $(\tilde{u}, \tilde{v})$ is a negative orbit sequence starting at 0 with $y = \tilde{v}_0$ then $(\tilde{u}, \tilde{v}) = (U^-(y), V^-(y))$ and $(\tilde{u}_0, \tilde{v}_0) \in W_0^-$.

A point lies in the intersection of W_0^+ and W_0^- if and only if it is the base of a positive and a negative orbit sequence. Putting them together we get a bi-infinite orbit sequence of which there is only one. So (u_0^*, v_0^*) is the unique intersection point of W_0^+ and W_0^-.

Thus, we have defined and characterized by orbit sequences the stable and unstable manifolds for $n = 0$. By the same argument or by simply renumbering we obtain them for all n. From the orbit characterizations we now derive the invariance results (10.23).

$(u_n, v_n) \in W_n^+$ if and only if $(u_n, v_n) \in B_n$ and (u_n, v_n) is the base of a positive orbit sequence. The next term is then $(F_n, G_n)(u_n, v_n)$ and there follows a positive orbit sequence based at $(F_n, G_n)(u_n, v_n)$. As this argument is reversible $(u_n, v_n) \in W_n^+$ if and only if $(u_n, v_n) \in B_n$ and $(F_n, G_n)(u_n, v_n) \in W_{n+1}^+$. Similarly, $(u_n, v_n) \in W_n^-$ if and only if it lies in B_n and there exists $(u_{n-1}, v_{n-1}) \in W_{n-1}^-$ with $(F_{n-1}, G_{n-1})(u_{n-1}, v_{n-1}) = (u_n, v_n)$.

There remain the Lipschitz estimates which are a typical application of Proposition 2(a).

For a fixed pair of points x, $\tilde{x}$ in B_0^+ we consider the product map $K_x^+ \times K_{\tilde{x}}^+$ on $B_{\mathbb{Z}^+} \times B_{\mathbb{Z}^+}$. Let $\Delta x = x - \tilde{x}$ and for $((u, v), (\tilde{u}, \tilde{v})) \in B_{\mathbb{Z}^+} \times B_{\mathbb{Z}^+}$, let $\Delta u = u - \tilde{u}$ and $\Delta v = v - \tilde{v}$. Define A to be the subset of $B_{\mathbb{Z}^+} \times B_{\mathbb{Z}^+}$ consisting of all pairs such that

$$\|\Delta u_n\| \le \lambda^n \|\Delta x\| \quad \text{and} \quad \|\Delta v_n\| \le \lambda^{n+1} \|\Delta x\|, \qquad n \ge 0.$$

A is a closed subset of $B_{\mathbb{Z}^+} \times B_{\mathbb{Z}^+}$ because uniform convergence preserves closed pointwise inequalities. We prove that A is $+$invariant for $K_x^+ \times K_{\tilde{x}}^+$.

Let $((U, V), (\tilde{U}, \tilde{V})) = (K_x^+(u, v), K_{\tilde{x}}^+(\tilde{u}, \tilde{v}))$ with $((u, v), (\tilde{u}, \tilde{v})) \in A$, then

$$\begin{aligned} \|\Delta U_0\| = \|\Delta x\| = \lambda^0 \|\Delta x\|, \ \|\Delta U_n\| &\le L(F_n) \max(\|\Delta u_{n-1}\|, \|\Delta v_{n-1}\|) \\ &\le \lambda \cdot \lambda^{n-1} \|\Delta x\| = \lambda^n \|\Delta x\|, \qquad n \ge 1, \end{aligned}$$

and

$$\begin{aligned} \|\Delta V_n\| &\le \|R_n^{-1}\| \, \|\Delta v_{n+1}\| + \|R_n^{-1}\| L(S_n) \max(\|\Delta u_n\|, \|\Delta v_n\|) \\ &\le (\|R_n^{-1}\| \lambda^{n+2} + \|R_n^{-1}\| L(S_n) \lambda^n) \|\Delta x\| \le \lambda^{n+1} \|\Delta x\|, \qquad n \ge 0. \end{aligned}$$

So by Proposition 2(a) the fixed point $((U^+(x), V^+(x)), (U^+(\tilde{x}), V^+(\tilde{x})))$ lies in A. In particular,

$$\|w_0^+(x) - w_0^+(\tilde{x})\| = \|\Delta V_0^+\| \le \lambda \|\Delta x\| = \lambda \|x - \tilde{x}\|.$$

Because $\lambda < 1$ and we are using the max norm of E_0, it follows that $\|(x, w_0^+(x)) - (\tilde{x}, w_0^+(\tilde{x}))\| = \|x - \tilde{x}\|$. When we apply (F_1, G_1) to the two points in W^+ we get $(U_1^+(x), V_1^+(x))$ and $(U_1^+(\tilde{x}), V_1^+(\tilde{x}))$. So

$$\begin{aligned} &\|(F_1, G_1)(x, w_0^+(x)) - (F_1, G_1)(\tilde{x}, w_0^+(\tilde{x}))\| \\ &\quad = \max(\|\Delta U_1^+\|, \|\Delta V_1^+\|) \le \lambda \|\Delta x\| = \lambda \|(x, w^+(x)) - (\tilde{x}, w^+(\tilde{x}))\|. \end{aligned}$$

This proves the W^+ estimates of (10.22) and (10.24). For the W^- estimates we start with $y, \tilde{y} \in B_0^-$ and let A' be the subset of $B_{\mathbb{Z}^-} \times B_{\mathbb{Z}^-}$ consisting of pairs $((u, v), (\tilde{u}, \tilde{v}))$ such that

$$\|\Delta u_n\| \le \lambda^{1-n} \|\Delta y\| \quad \text{and} \quad \|\Delta v_n\| \le \lambda^{-n} \|\Delta y\|, \qquad n \le 0.$$

We argue similarly that A' is closed and $K_y^- \times K_{\tilde{y}}^-$ +invariant. Then apply Proposition 2(a) again and argue as before to complete the proof of (10.22) and (10.24). □

15. Exercise. *Prove the following addendum to Theorem* 14: *If, as* n *approaches* $+\infty$ *(or alternatively as* n *approaches* $-\infty$*), the sequences* $\{\|F_n(0, 0)\|\}$ *and* $\{\|S_n(0, 0)\|\}$ $(= \{\|G_n(0, 0)\|\})$ *approach* 0, *then so does the sequence* $\{\|u_n^*, v_n^*)\|\}$. *(Hint: Show that the subset* A *consisting of*

$(u, v) \in B_{\mathbb{Z}}$ *with* $\lim\{\|u_n, v_n)\|\} = 0$ *is nonempty, closed and* K *+invariant.)* □

Notice that the proof showed that (U_o^+, V_o^+), associating to x not only the point $(x, w_0^+(x))$, but its entire positive orbit sequence, is a Lipschitz map of B_0^+ to $B_{\mathbb{Z}^+}$. Similarly, $(U_o^-, V_o^-): B_0^- \to B_{\mathbb{Z}^-}$ is Lipschitz.

Now observe that for $\lambda_1 < 1$ we can replace the norm distance on B_T by

$$d_{\lambda_1}((u, v), (\tilde{u}, \tilde{v})) = \sup\{\lambda_1^{|n|}\|(u_n, v_n) - (\tilde{u}_n, \tilde{v}_n)\|: n \in T\}. \tag{10.28}$$

It is easy to check that the topology given by this metric on B_T is that of pointwise convergence, that is, the restriction of the product topology on $\prod\{E_n: n \in T\}$. B_T is a closed subset of the product and so is a complete metric space with respect to d_{λ_1}. Provided that $\lambda_1 > \lambda$ the maps K, K_x^+ and K_y^- are all contractions with respect to this metric as well. The Lipschitz constants are bounded by λ/λ_1. From this we prove the following:

16. Addendum. *For each* $(x, y) \in B_0$, *the orbit sequences* (u^*, v^*), $(U_o^+(x), V_o^+(x))$ *and* $(U_o^-(y), V_o^-(y))$ *vary continuously with respect to the initial data* $\{R_n\}$, $\{S_n\}$, *and* $\{F_n\}$ *provided that the estimates* (10.19) *and* (10.20) *hold with fixed* λ *and* ε.

Proof. The statement has been left vague to accommodate two different sorts of continuous variation. Assume that for each whole number M we are given $\{R_n^M\}$, $\{S_n^M\}$, and $\{F_n^M\}$ such that as M tends to infinity $\{R_n^M\}$ converges to R_n in norm and $\{S_n^M\}$, $\{F_n^M\}$ converge to S_n and F_n respectively for each n. Alternatively, we make the stronger assumption that convergence holds uniformly in n. Throughout, the analogues of (10.19) and (10.20) hold with the same fixed λ and ε. We conclude that for each n $\{(u_n^{*^M}, v_n^{*^M})\}$ converges to (u_n^*, v_n^*) and we obtain uniform convergence in n under the alternative uniform hypothesis. A similar conclusion holds for $(U_o^+(x), V_o^+(x))$ and $(U_o^-(y), V_o^-(y))$. Since the latter proofs are similar as well, we will omit them.

We begin with the uniform convergence version. The contractions $\{K^M\}$ of $B_{\mathbb{Z}}$ converge, pointwise in (u, v), to the contraction K. Observe that because $B_{\mathbb{Z}}$ is given the $E_{\mathbb{Z}}$ sup norm topology the uniformity hypothesis is needed to get $\{K^M(u, v)\}$ to approach $K(u, v)$ in norm. In particular,

$$\lim_{M\to\infty}\{K^M(u^*, v^*)\} = K(u^*, v^*) = (u^*, v^*).$$

Hence, given $\delta > 0$ we can choose M large enough so that $\|K^M(u^*, v^*) - (u^*, v^*)\| < \delta$. Because $L(K^M) < \lambda$ for all M, inequality (10.2) implies that for M large enough $\|(u^{*^M}, v^{*^M}) - (u^*, v^*)\| < \delta/(1-\lambda)$. Thus, $\{(u^{*^M}, v^{*^M})\}$ converges to (u^*, v^*) in $B_{\mathbb{Z}}$.

In the absence of uniformity we can apply exactly the same argument provided that we use the metric d_{λ_1} with the product topology on $B_{\mathbb{Z}}$ instead of the norm topology. That is, we get pointwise convergence of $\{K^M\}$ to K and so with M large we conclude that

$$d_{\lambda_1}((u^{*^M}, v^{*^M}), (u^*, v^*)) < \delta/(1 - (\lambda/\lambda_1)). \quad \square$$

A related application of the metric (10.28) is

17. Lemma. *For every* $\varepsilon_1 > 0$ *there exists a whole number* N *such that if* $\{(u_n, v_n) : |n| \leq N\}$ *is an orbit sequence on* $[-N, N]$, *i.e.,* (10.21) *holds for* $-N \leq n < N$ *then*

$$\|(u_0, v_0) - (u_0^*, v_0^*)\| < \varepsilon_1 .$$

Proof. With λ_1 fixed between λ and 1 let N be a whole number such that

$$\frac{2(\lambda_1)^N \varepsilon}{1 - (\lambda/\lambda_1)} < \varepsilon_1 .$$

Then if $\{(u_n, v_n)\}$ is an orbit sequence on $[-N, N]$ define $(u_i, v_i) = (0, 0)$ in B_i for $|i| > N$ and so obtain an element (u, v) of $B_{\mathbb{Z}}$. Because (u, v) is an orbit sequence on $[-N, N]$, $d_{\lambda_1}(K(u, v), (u, v)) < 2(\lambda_1)^N \varepsilon$. As K is attracting with Lipschitz constant at most λ/λ_1 with respect to d_{λ_1} the estimate (10.2) implies

$$\varepsilon_1 > d_{\lambda_1}((u, v), (u^*, v^*)) \geq \|(u_0, v_0) - (u_0^*, v_0^*)\| . \quad \square$$

Proof of Theorem 13. Identify E with the product $E^+ \times E^-$. We denote by $T^{\pm}$ the restrictions of T to the invariant subspaces $E^{\pm}$ and by $g^{\pm}$ the $E^{\pm}$ coordinate projections of the map g. In applying Theorem 14 we will let $E_n^{\pm} = E^{\pm}$ and $E_n = E$ for all n, and fix ε larger than $\|g(0, 0)\|/(1 - \lambda_g)$.

To prove parts (a) and (b) we assume that the maps are independent of n as well with $F_n(x, y) = T^+(x) + g^+(x, y)$, $R_n(y) = T^-(y)$, and $S_n(x, y) = g^-(x, y)$. Thus, $(F_n, G_n) = f$ for all n. Our results follow by applying the extra bit of structure given by this time independence. If z is a sequence of elements of E we define the shifted sequence $s(z)$ by $s(z)_n = z_{n+1}$, with the appropriately shifted index set. Observe that z is an orbit sequence if and only if $s(z)$ is because the maps are independent of n.

In particular, by part (a) of Theorem 14 there is a unique bounded bi-infinite orbit sequence $\{e_n^*\}$. As its shift is also a bounded bi-infinite orbit sequence, uniqueness implies $e_n^* = e_{n+1}^*$ for all n. Thus, $e_n^* = e$ for some $e \in E$ independent of n. Because the sequence is an orbit sequence $f(e) = e$; e is a fixed point. With λ given by λ_g and ε arbitrarily close above $\|g(0, 0)\|/(1 - \lambda_g)$ we obtain (10.14) because e lies in the ε ball.

Now apply part (b) of Theorem 14 with ε approaching ∞. By the orbit sequence characterizations the stable and unstable manifolds for different ε's fit together to define global manifolds which are the graphs of Lipschitz maps defined on all of E^+ and E^-, respectively. Now the orbit characterizations and shift invariance show that the W_n^+'s are the same for all n. Similarly for W^-. The invariance results then follow from (10.23) and the Lipschitz estimates (10.15) and (10.16) from (10.22) and (10.24).

To prove the Shadowing Lemma, part (c), we begin with a bi-infinite sequence in E $\{z_n = (z_n^+, z_n^-)\}$ such that $\|z_{n+1} - f(z_n)\| \leq \delta$ for all $n \in \mathbb{Z}$. We will construct a unique bounded sequence $\{b_n = (b_n^+, b_n^-)\}$ so that $\{w_n = z_n + b_n\}$ is a bi-infinite orbit sequence for f. Because we want $w_{n+1} = f(w_n)$ we jettison time independence and define instead:

$$
\begin{aligned}
&F(u_n, v_n) = T^+(u_n + z_n^+) + g^+(u_n + z_n^+, v_n + z_n^-) - z_{n+1}^+, \\
(10.29) \qquad &R_n(v_n) = T^-(v_n), \\
&S_n(u_n, v_n) = g^-(u_n + z_n^+, v_n + z_n^-) + T^-(z_n^-) - z_{n+1}^-.
\end{aligned}
$$

Observe that F_n and S_n differ from the previous definition by translations in the domain and range. So the Lipschitz constants are unchanged. On the other hand, $(F_n, G_n)(0, 0) = f(z_n) - z_{n+1}$. Hence, Theorem 14 applies with $\lambda = \lambda_g$ and ε any number larger than $\delta/(1 - \lambda_g)$. The unique bounded bi-infinite orbit sequence for this system is the required $\{b_n\}$.

For the Conjugacy Theorem, part (d), we first apply Proposition 8 to observe that $f: E \to E$ is a homeomorphism. Hence, the map f^n is defined for negative n just as T^n is. Given $z \in E$ define $\{z_n = T^n z\}$ and $\{\tilde{z}_n = f^n(z)\}$ to be the bi-infinite orbit sequences with respect to T and f respectively and with common value z when $n = 0$. Observe that

$$
\begin{aligned}
f(z_n) &= T(z_n) + g(z_n) = z_{n+1} + g(z_n); \\
\tilde{z}_{n+1} &= f(\tilde{z}_n) = T(\tilde{z}_n) + g(\tilde{z}_n).
\end{aligned}
$$

Hence, for all $n \in \mathbb{Z}$ and all $z \in E$:

$$
\|z_{n+1} - f(z_n)\| \leq \|g\|_0, \qquad \|\tilde{z}_{n+1} - T(\tilde{z}_n)\| \leq \|g\|_0.
$$

Thus, the T-orbit sequence $\{z_n\}$ is a $\|g\|_0$ chain with respect to f and so differs by a bounded sequence from a unique f orbit sequence. Denote the $n = 0$ value of this sequence $h(z)$. Thus,

$$
(10.30) \qquad \|T^n(z) - f^n(h(z))\| \leq \|g\|_0/(1 - \lambda_g).
$$

Similarly we define $\tilde{h}(z)$ so that the T orbit of $\tilde{h}(z)$ shadows the f orbit of z

$$
(10.31) \qquad \|T^n(\tilde{h}(z)) - f^n(z)\| \leq \|g\|_0/(1 - \lambda_g).
$$

Substituting $\tilde{h}(z)$ for z in (10.30) we see by (10.31) that the difference between $f^n(z)$ and $f^n(h(\tilde{h}(z)))$ is bounded in norm. So by the uniqueness part of the shadowing Lemma, $z = h(\tilde{h}(z))$. Similarly, $\tilde{h}(h(z)) = z$ and so $\tilde{h}$ is the inverse map to h.

The conjugacy equation comes from a similar argument. By (10.30) $\{f^n(f(h(z)))=f^{n+1}(h(z))\}$ differs from $\{T^{n+1}(z)=T^n(T(z))\}$ by a bounded sequence. Substituting $T(z)$ for z in (10.30) we see that $\{T^{n+1}(z) = T^n(T(z))\}$ differs from $\{f^n(h(T(z)))\}$ by a bounded sequence. Uniqueness in the Shadowing Lemma implies $f(h(z)) = h(T(z))$.

Finally, continuity of h and $\tilde{h}$ follow from Addendum 16 and the proof of the Shadowing Lemma because varying z continuously induces a continuous variation in the data $\{F_n\}$ and $\{G_n\}$ by (10.29). □

Supplementary exercises

18. Let $f: X \to X$ be a continuous map with X compact. The set of ε-fixed points is $\{x: d(x, f(x)) \leq \varepsilon\} = |\overline{V}_\varepsilon \circ f|$. Prove that for every $\varepsilon > 0$ there exists $\delta > 0$ such that $|\overline{V}_\delta \circ f| \subset V_\varepsilon(|f|)$, i.e., every δ-fixed point is ε close to a true fixed point. The Banach Fixed Point Theorem provides an explicit estimate for δ when f is contracting. Prove there exists $\delta_1 > 0$ such that $\overline{V}_{\delta_1}(|f|) \subset |V_\varepsilon \circ f|$. If f is a Lipschitz map, show that $\delta_1 = \varepsilon/(1 - L(f))$ will work.

19. Given $\varepsilon > 0$ let $r: E \to \overline{V}_\varepsilon(0)$ be the retraction of Lemma 10. If E is a Hilbert space prove that $L(r) = 1$. (Reduce to the two dimensional case which is isometric to the Euclidean Plane. Use plane geometry, e.g. the exterior angle of the base angle of an isosceles triangle is obtuse.) On the other hand if E is $\mathbb{R}^2$ equipped with the norm $\|(x, y)\|_1 = |x| + |y|$ then $L(r) > 1$ (use $\varepsilon = 1$ and the vectors $(1, 0)$ and $(1, 1/2)$).

20. Given a B-space E let E^∞ be the set of bi-infinite sequences in E, $E^\infty \equiv \prod\{E_n : n \in \mathbb{Z}\}$ with each E_n a copy of E. With the product topology E^∞ is a complete metric space. On E^∞ let B be the linear subspace of uniformly bounded sequences, i.e., $B = \{b \in E^S: \sup_n \|b_n\| < \infty\}$. Let $\equiv_B$ denote the equivalence relation on E^∞ of congruence $\bmod B$ so that $z \equiv_B \tilde{z}$ when $\{\|z_n - \tilde{z}_n\|\}$ is bounded.

    (a) When $f: E \to E$ is a homeomorphism define $\mathscr{B}_f \equiv \{z \in E^\infty : \{\|z_{n+1} - f(z_n)\|\}$ is bounded in $n\}$. Define $\mathrm{o}_f: E \to E^\infty$ by $\mathrm{o}_f(x)_n = f^n(x)$. Prove that o_f is continuous (remember: the product topology on E^∞) with image in $\mathscr{B}_f$. Furthermore, $e_0: E^\infty \to E$ defined by $e_0(z) = z_0$ provides a continuous left inverse for o_f. Observe that if f satisfies the hypothesis of Theorem 13, the Shadowing Lemma says that each $\equiv_B$ congruence class in $\mathscr{B}_f$ contains a unique member of the form $\mathrm{o}_f(z)$. Furthermore, the left inverse map $r_f: \mathscr{B}_f \to E$ associating to b the unique x such that $b \equiv_B \mathrm{o}_f(x)$ is continuous. Where $s: \mathscr{B}_f \to \mathscr{B}_f$ is the restriction of the shift map, $s(z)_n = z_{n+1}$, prove that $s \circ \mathrm{o}_f = \mathrm{o}_f \circ f$ and $f \circ r_f = r_f \circ s$.

(b) Now suppose $f_1: E \to E$ is a homeomorphism with

$$\sup_{x\in E} \|f(x) - f_1(x)\| < \infty.$$

Prove that $\mathscr{B}_f = \mathscr{B}_{f_1}$ and $h: E \to E$ defined by $r_f \circ o_{f_1}$ is the unique map such that for all $x \in E$

$$\sup_n \|f_1^n(z) - f^n(h(x))\| < \infty.$$

Prove h is continuous. Also, prove $h \circ f_1 = f \circ h$ (use the shift map s on $\mathscr{B}_f$). h is a semiconjugacy from f_1 to f.

(c) When f_1, as well as f, satisfies the conditions of Theorem 13 show that h is a conjugacy, proving again the Conjugacy Theorem part (d) of Theorem 13.

21. (a) For X_1, X_2 compact metric spaces and $0 \leq \lambda < \infty$ prove that $C_\lambda(X_1; X_2) = \{f: X_1 \to X_2 \text{ with } L(f) \leq \lambda\}$ is a compact subset of $C(X_1; X_2)$ with respect to the metric topology (cf. (7.14)), and that on $C_\lambda(X_1; X_2)$ this topology agrees with that induced from the product $X_2^{X_1}$, i.e., a sequence $\{f_n\}$ in C_λ converges uniformly when it converges pointwise (for compactness compare exercise 8.16(b)).

Prove that $c: X_2 \to C_0(X_1; X_2)$ associating to each point the constant map to that point defines an isometric bijection.

Prove that composition of maps restricts to define a map Comp: $C_{\lambda_1}(X_1; X_2) \times C_{\lambda_2}(X_2; X_3) \to C_{\lambda_1\lambda_2}(X_1; X_3)$. Furthermore, Comp has Lipschitz constant ≤ 1 in the first variable and $\leq \lambda_1$ in the second.

Prove that the adjoint association defines an isometric bijection $C_{\lambda_1}(X_1; C_{\lambda_2}(X_2; X_3)) \simeq C_{\lambda_2}(X_2; C_{\lambda_1}(X_1; X_3))$. The adjoint association relates a map in either set to the corresponding map $X_1 \times X_2 \to X_3$ with Lipschitz constant λ_1 in the first variable and λ_2 in the second (see exercise 7.36(d)).

Prove that the induced maps on subsets and measures define isometric inclusions

$$*: C_\lambda(X_1; X_2) \to C_\lambda(C(X_1); C(X_2)),$$

and

$$*: C_\lambda(X_1; X_2) \to C_\lambda(P(X_1); P(X_2)),$$

where, in the latter case, the Hutchinson metric is used on the spaces of measures (see exercises 7.36(b) and 8.16(b)).

Define the natural map

$$\mathrm{v}: C(C_\lambda(X_1; X_2)) \to C_\lambda(X_1; C(X_2))$$

as follows: for A a closed subset of $C_\lambda(X_1; X_2)$ $\mathrm{v}(A)(x) = \{f(x): f \in A\} \subset X_2$. Prove that v is well defined and has Lipschitz constant equal to 1. (*Hint*: Let C_λ denote $C_\lambda(X_1; X_2)$. The identity map in $C_1(C_\lambda; C_\lambda(X_1; X_2))$ is adjoint to the evaluation map ev in $C_\lambda(X_1; C_1(C_\lambda; X_2))$. Composition with $*$ in $C_1(C_1(C_\lambda; X_2); C_1(C(C_\lambda); C(X_2)))$

yields a map in $C_\lambda(X_1; C_1(C(C_\lambda); C(X_2)))$. The result is adjoint to v in $C_1(C(C_\lambda); C_\lambda(X_1; C(X_2))))$.

Similarly, define

$$a: P(C_\lambda(X_1; X_2)) \to C_\lambda(X_1; P(X_2)).$$

With μ a measure on $C_\lambda = C_\lambda(X_1; X_2)$, and $x \in X_1$, $u \in C(X_2; \mathbb{R})$, let

$$\langle u, a(\mu)(x_1)\rangle = \int_{C_\lambda} u(f(x_1))\mu(df).$$

In the case when $X_1 = X_2$ and $\lambda < 1$ define

$$e: C_\lambda(X; X) \to X$$

by associating to f its unique fixed point. Prove that e has Lipschitz constant $\leq (1-\lambda)^{-1}$ (cf. Theorem 1).

Observe that with $\lambda < 1$, $C_\lambda(X; X)$ is a semigroup. For any sequence $\{f_n\}$ in $C_\lambda(X; X)$ $\mathrm{Comp}_N\{f_n\} = f_1 \circ f_2 \circ \cdots \circ f_N$ lies in $C_{\lambda^N}(X; X)$. Let $C_\lambda(X; X)^\infty$ be the space of sequences with the product topology. Prove this topology is given by the metric $d_\lambda(\{f_n\}, \{g_n\}) = \sup\{\lambda^n d(f_n, g_n): n = 0, 1, \ldots\}$. Prove that

$$\mathrm{Comp}_N: C_\lambda(X; X)^\infty \to C_\lambda(X; X)$$

is continuous and converges uniformly on $N \to \infty$ to a map

$$\mathrm{Comp}_\infty: C_\lambda(X; X)^\infty \to X,$$

where we identify X with $C_0(X; X)$. For any sequence $\{f_n\} \in C_\lambda(X; X)^\infty$ the point $\mathrm{Comp}_\infty\{f_n\}$ is the unique point in the decreasing intersection $\bigcap_{N=0}^\infty f_1 \circ \cdots \circ f_N(X)$. Equivalently, it is the $\lim_{N\to\infty}\{e(f_1 \circ \cdots \circ f_N)\}$.

(b) Let $0 < \lambda < 1$ and let A be a fixed closed subset of $C_\lambda(X; X)$. Define $F_A: X \to C(X)$ by $F_A(x) = \{f(x): f \in A\}$. Observe that the map F_A has Lipschitz constant $\leq \lambda$ (F_A is $\mathrm{v}(A)$ of (a)). In particular, regarded as a relation on X, F_A is lsc as well as usc (closed). Prove that the Proposition 1.1 image map $F_A: C(X) \to C(X)$, defined by $F_A(B) = \bigcup\{f(B): f \in A\}$, is Lipschitz with $L(F_A) \leq \lambda$ (F_A is the composition of $F_{A*}: C(X) \to C(C(X))$ with $\mathrm{v}: C(C(X)) \to C(X)$ see Exercise 7.36).

F_A on $C(X)$ has a unique fixed point, which we will denote $e(A) \in C(X)$, characterized as the unique, nonempty, closed subset B of X satisfying $B = \bigcup\{f(B): f \in A\}$. Prove that if $\widetilde{B} \neq \emptyset$ is an A +invariant closed subset, i.e., $f(\widetilde{B}) \subset \widetilde{B}$ for all f in A, then $\{F_A^n(\widetilde{B})\}$ is a decreasing sequence with intersection $e(A)$. In particular, $B \subset \widetilde{B}$. On the other hand, if $\widetilde{B} \neq \emptyset$ satisfies $\widetilde{B} \subset F_A(\widetilde{B})$ (e.g. if $\widetilde{B}$ consists of fixed points of some elements of A) then $\{F_A^n(\widetilde{B})\}$ is an increasing sequence and $e(A) = \bigcup\{F_A^n(\widetilde{B})\}$.

With $\mathrm{Comp}_\infty: C_\lambda(X; X)^\infty \to X$ defined in the previous exercise, prove that $e(A)$ is the image of Comp_∞ applied to $A^\infty \subset C_\lambda(X; X)^\infty$.

(*Hint*: $e(A) = \bigcap\{F_A^n(X)\}$.) Prove also that if $[A]$ is the smallest closed subsemigroup of $C_\lambda(X; X)$ containing A then $e(A)$ is the image of $[A]$ under the fixed point map $e: C_\lambda(X; X) \to X$. (Observe that the fixed point of $f_1 \circ \cdots \circ f_n$ is Comp_∞ applied to the periodic sequence $\{f_j\}$ with $f_j = f_i$ for $j = i + kN$.)

Suppose A acts on A^∞, i.e., $A \times A^\infty \to A^\infty$, by $(f, \{f_1, f_2, \ldots\}) \to \{f, f_1, f_2, \ldots\}$. Give A^∞ the metric d_λ of Part (a). Then by this action A is identified as a subset of $C_\lambda(A^\infty, A^\infty)$ and $A^\infty = e(A)$. (*Hint*: $A^\infty = \bigcup\{f(A^\infty): f \in A\}$.) $\text{Comp}_\infty: A^\infty \to X$ maps f on A^∞ to f on X for all f in A and so maps F_A on $C(A^\infty)$ to F_A on $C(X)$ which provides another proof that the fixed point $e(A)$ in X is the image under Comp_∞ of the fixed point $e(A)$ in A^∞.

Finally, if M is a measure on $C_\lambda(X; X)$ with support A then define $F_M: X \to P(X)$ and the associated $F_M: P(X) \to P(X)$ with $L(F_M) \le \lambda$ with respect to the Hutchinson metric on $P(X)$. Prove that the fixed point measure is the image under Comp_∞ of the Bernoulli measure on A^∞ with M on each factor.

REMARK. The most common applications of these results occur when A is a finite subset of $C_\lambda(X; X)$, i.e., a finite set of contractions (see Barnsley (1988)).

22. Let f be a closed relation on X, or equivalently, $f: X \to C(X)$ a usc map. Let f also stand for the image map $C(X) \to C(X)$.

(a) Prove that $\delta(f(y), X) < \varepsilon$ if and only if for all $x \in X$, $y \in f^{-1}(V_\varepsilon(x))$. Conclude that $\delta(f(A), X) < \varepsilon$ for all $A \in C(X) - \{\varnothing\}$ if and only if $f^{-1}(V_\varepsilon(x)) = X$ for all $x \in X$.

(b) Prove that X is a uniformly attractive fixed point for $f: C(X) - \{\varnothing\} \to C(X) - \{\varnothing\}$, i.e., for every $\varepsilon > 0$ there exists N so that for all $A \in C(X) - \{\varnothing\}$ and all $n \ge N$ $\delta(f^n(A), X) < \varepsilon$, if and only for every $\varepsilon > 0$ there exists N so that for all $x \in X$ and all $n \ge N$ $f^{-n}(V_\varepsilon(x)) = X$. In turn, prove these equivalent to: for every open set $U \ne \varnothing$ in X, there exists N so that for all $n \ge N$ $f^{-n}(U) = X$. (*Hint*: let $\{x_1, \ldots, x_k\}$ be an $\varepsilon/2$ net for X and choose $\{N_1, \ldots, N_k\}$ so that $n \ge N$ implies $f^{-h}(V_{\varepsilon/2}(x_i)) = X$. Observe that every $V_\varepsilon(x)$ contains some $V_{\varepsilon/2}(x_i)$. Let $N = \max(N_1, \ldots, N_k)$).

(c) Prove that $f(x) \ne \varnothing$ for all $x \in X$, i.e., $\text{Dom}(f) = X$ if and only if $f^{-1}(X) = X$. So if $\text{Dom}(f) = X$ and $B \subset X$ such that $f^{-N}(B) = X$, then $f^{-n}(B) = X$ for all $n > N$. If $f^{-1}(X)$ is a proper subset of X then $f^{-n}(B) \subset f^{-1}(X)$ for all $n \ge 1$ and all $B \subset X$. Conclude that X is a uniformly attractive fixed point for $f: C(X) - \{\varnothing\} \to C(X) - \{\varnothing\}$ if and only if for every open set $U \ne \varnothing$ in X there exists N such that $f^{-N}(U) = X$.

Chapter 11. Hyperbolic Sets and Axiom A Homeomorphisms

In this concluding chapter, we describe the concept of hyperbolic invariant set, generalizing the idea of hyperbolic fixed point from the previous chapter. A diffeomorphism of a compact smooth (at least C^1) manifold is said to satisfy Axiom A when its chain recurrent set is hyperbolic. The concept and the beautiful analysis of its consequences were introduced by Smale in a classic Bulletin article in 1967. Much of modern dynamical systems theory has developed from the work of Smale and his colleagues. In particular, as mentioned in the Preface, it inspired the reorientation of topological dynamics presented in this book. We will begin by using the resulting hindsight to describe a topological condition generalizing hyperbolicity and then show how the usual differentiable version implies ours. So for now we will return to our original context of a compact metric space X and study a homeomorphism f on X.

We first study in some detail a property briefly introduced in Theorem 3.6 (b)(3).

1. Definition. *Let $f: X \to X$ be a homeomorphism, and let K be a closed invariant subset of X. K is called* isolated *if it satisfies the following equivalent conditions*:

(a) *K has a neighborhood U such that every invariant subset of U lies in K.*

(b) *There exists U such that*

$$K \subset \operatorname{Int} U, \qquad K = \bigcap_{n=-\infty}^{+\infty} \{f^n U\}. \tag{11.1}$$

(c) *There exists $\varepsilon > 0$ such that*

$$d(f^n x, K) \leq \varepsilon \text{ for all integers } n \Rightarrow x \in K, \tag{11.2}$$

and so the entire orbit of x lies in K.

Furthermore, if $\varepsilon > 0$ is as in (c) *then*:

$$d(f^n x, K) \leq \varepsilon \quad \text{for all nonnegative integers } n \Rightarrow \omega f(x) \subset K. \tag{11.3}$$

Proof. Since $\bigcap_{n=-\infty}^{+\infty}\{f^n U\}$ is the maximum invariant subset of U, (a) and (b) are clearly equivalent. (c) is equivalent to (a) with $U = \overline{V}_\varepsilon(K)$. Finally, the hypothesis of (11.3) implies $\omega f(x) \subset \overline{V}_\varepsilon(K)$ and since $\omega f(x)$ is invariant, (a) implies $\omega f(x) \subset K$.

Remark. A neighborhood U satisfying the conditions of (a) and (b) ($\overline{V}_\varepsilon(K)$ in (c)) is called an *isolating neighborhood* for K. □

2. **Exercise.** (a) *Let K_1, K_2 be compact invariant sets for a homeomorphism $f\colon X \to X$. If K_1 and K_2 are isolated, then $K_1 \cap K_2$ is isolated. If K_1 and K_2 are disjoint, then $K_1 \cup K_2$ is isolated if and only if both K_1 and K_2 are isolated.* (*Hint*: *For the union result let U_1, U_2 be neighborhoods of K_1 and K_2 respectively such that $U_1 \cup f(U_1)$ and $U_2 \cup f(U_2)$ are disjoint. If an orbit remains in $U_1 \cup U_2$, then it lies entirely in U_1 or in U_2.*)

(b) *Let f be the time-one map for the flow* (*in polar coordinates*):

$$\frac{dr}{dt} = 0, \qquad \frac{d\theta}{dt} = r(\sin^2\theta + (1-r)^2).$$

Show that $K_1 = \{(r, \theta): r = 1 \text{ and } \pi \geq \theta \geq 0\}$ and $K_2 = \{(r, \theta): r = 1 \text{ and } 2\pi \geq \theta \geq \pi\}$ are isolated invariant sets but $K_1 \cup K_2$ is not isolated. □

3. **Proposition.** *Let f be a homeomorphism of a compact metric space X.*

(a) *Any attractor or repellor for f is an isolated f invariant set.*

(b) *A basic set for f is an isolated f invariant set if and only if it is an open and closed subset of the chain recurrent set $|\mathscr{C}f|$.*

(c) *All the basic sets for f are isolated if and only if there are only finitely many basic sets. In that case the chain recurrent set $|\mathscr{C}f|$ is isolated.*

Proof. If A is an attractor then there exists an inward neighborhood U such that $\omega f[U] = A$, i.e., $A = \bigcap_{n=0}^{\infty}\{f^n U\}$. A fortiori, A satisfies (11.1). For a repellor an analogous outward neighborhood exists.

If B is a basic set then $B \subset |\mathscr{C}f|$ and B is $\mathscr{C}f$ semi-invariant, i.e., $B = \mathscr{C}f(B) \cap (\mathscr{C}f)^{-1}(B)$. Because $\mathscr{C}f(B)$ is $\mathscr{C}f +$ invariant the inward neighborhoods of $\mathscr{C}f(B)$ form a base for its neighborhood system. Hence, if G is any open set containing B we can find an inward set U^+ and an outward set U^- such that

$$B \subset \operatorname{Int}(U^+ \cap U^-) \subset U^+ \cap U^- \subset G.$$

Since any basic set which meets an inward (or outward) set is contained in it

$$B_1 \cap U^+ \cap U^- \neq \varnothing \quad \Rightarrow B_1 \subset U^+ \cap U^-.$$

Hence, if $U^+ \cap U^-$ is a neighborhood of B isolating it, then $U^+ \cap U^-$ meets no other basic set. Thus, $B = \operatorname{Int}(U^+ \cap U^-) \cap |\mathscr{C}f|$.

On the other hand, if B is open in $|\mathscr{C}f|$ we can choose such a neighborhood $U^+ \cap U^-$ which meets no points of $|\mathscr{C}f| - B$. If the orbit of x

lies entirely in $U^+ \cap U^-$, then $\omega f(x) \cup \alpha f(x) \subset U^+ \cap U^- \cap |\mathscr{C} f| = B$. If $x_- \in \alpha f(x)$ and $x_+ \in \omega f(x)$, then $x \mathscr{C} f x_-$ and $x_+ \mathscr{C} f x$. But x_- and x_+ are $\mathscr{C} f$ equivalent because they both lie in B. So x is in this equivalence class as well. Thus, x lies in B as does its entire orbit.

(c) If the basic sets are isolated, then by (b) they form a disjoint open cover of the compact space $|\mathscr{C} f|$ and so there are only finitely many of them. Conversely, the complement of each basic set in $|\mathscr{C} f|$ is the union of the remaining sets so if there are only finitely many then each is open as well as closed in $|\mathscr{C} f|$. So they are isolated by (b). By the above exercise their finite union $|\mathscr{C} f|$ is then isolated as well.

REMARK. There do exist examples where $|\mathscr{C} f|$ is isolated but consists of infinitely many basic sets. See Exercise 5.18. □

4. DEFINITION. (a) *A homeomorphism* $f: X \to X$ *is called* expansive *when the diagonal* 1_X *is isolated in* $X \times X$. *That is, there exists a positive constant* $\gamma > 0$, *called an* expansivity constant, *that satisfies the following equivalent conditions*:

$$\bigcap_{n=-\infty}^{\infty} \{(f \times f)^n(\overline{V}_\gamma)\} = 1_X, \tag{11.4}$$

or

$$d(f^n x_1, f^n x_2) \leq \gamma \quad \text{for all integers } n \Rightarrow x_1 = x_2. \tag{11.5}$$

(b) *A closed invariant subset* K *is called an* expansive subset *for* f *when there exists a closed neighborhood* U *of* K *and the restriction* f_{K_0} *is expansive where* $K_0 = \bigcap_{n=-\infty}^{+\infty} \{f^n(U)\}$ *is the maximum invariant set in* U. *That is, there exists a positive constant* $\gamma > 0$, *called the* expansivity constant for K in X, *which satisfies the following equivalent conditions*:

$$\bigcap_{n=-\infty}^{+\infty} \{(f \times f)^n(\overline{V}_\gamma \cap (\overline{V}_\gamma(K) \times \overline{V}_\gamma(K)))\} \subset 1_x \tag{11.6}$$

or

$$d(f^n x_1, K) \leq \gamma,\ d(f^n x_2, K) \leq \gamma, \text{ and } d(f^n x_1, f^n x_2) \leq \gamma \text{ for all integers } n \Rightarrow x_1 = x_2. \tag{11.7}$$

PROOF OF EQUIVALENCE. We can replace U by a smaller neighborhood if need be and so can use $U = \overline{V}_\gamma(K)$. With K_0 the associated maximum invariant set, the condition $d(f^n x, K) \leq \gamma$ for all integers n is equivalent to $x \in K_0$. So (11.7) is just (11.5) for K_0.

REMARK. Observe that f expansive is the special case where X is an expansive subset. On the other hand, that K be an expansive subset is clearly a stronger condition, in general, than that the restriction f_K be expansive. □

5. **Exercise.** (a) *Assume that* K_1 *is a closed invariant subset of* K *and that* K *is expansive. Prove that* K_1 *is expansive. For* K_1, K_2 *disjoint expansive subsets, prove that* $K_1 \cup K_2$ *is expansive.*

(b) *Assume that* K *is an isolated invariant subset. Prove that* K *is expansive when the restriction* $f_K: K \to K$ *is expansive. Prove that* K *is isolated and expansive exactly when* 1_K *is an isolated subset of* $X \times X$.

(c) *Let* $K = \{e\}$ *be a fixed point for* f. *Observe that the restriction* $f_e: e \to e$ *is expansive, trivially. Prove that* e *is an expansive subset if and only if it is an isolated invariant subset.* □

We now apply (11.3), or more precisely its proof. But, first, we need

6. **Lemma.** *For* $x_1, x_2, \in X$, $\omega(f \times f)(x_1, x_2) \subset 1_X$ *if and only if*

$$\lim_{n\to\infty} d(f^n x_1, f^n x_2) = 0.$$

Proof. The sequence $\{(f \times f)^n(x_1, x_2)\}$ is eventually in V_ε for every $\varepsilon > 0$ if and only if

$$\omega(f \times f)(x_1, x_2) \subset \bigcap_{\varepsilon>0}\{V_\varepsilon\} = 1_X. \quad \square$$

7. **Proposition.** *Assume* $f: X \to X$ *is a homeomorphism and* K *is an expansive subset with expansivity constant* $\gamma > 0$ *in* X.

(a) *For* $x_1, x_2 \in X$:

$$d(f^n x_1, K) \le \gamma,\ d(f^n x_2, K) \le \gamma, \text{ and } d(f^n x_1, f^n x_2) \le \gamma$$
$$\text{for all nonnegative integers } n \Rightarrow \lim_{n\to\infty} d(f^n x_1, f^n x_2) = 0.$$

(b) *For every* $\varepsilon > 0$ *there exists* N *so that for* $x_1, x_2 \in X$:

$$d(f^n x_1, K) \le \gamma,\ d(f^n x_2, K) \le \gamma, \text{ and } d(f^n x_1, f^n x_2) \le \gamma$$
$$\text{for all integers } n \text{ with } |n| \le N \Rightarrow d(x_1, x_2) < \varepsilon.$$

Proof. (a) The assumption of (a) implies that $\omega(f \times f)(x_1, x_2)$ is an invariant set entirely contained in $\overline{V}_\gamma \cap (\overline{V}_\gamma(K) \times \overline{V}_\gamma(K))$ and so by (11.6) it is contained in the diagonal 1_X. The conclusion follows from Lemma 6.

(b) $\{\bigcap_{n=-N}^{N}\{(f\times f)^n(\overline{V}_\gamma \cap (\overline{V}_\gamma(K) \times \overline{V}_\gamma(K)))\}\}$ is a decreasing sequence of closed sets whose intersection, by (11.6), lies in 1_X. So the sets are eventually contained in V_ε. □

In turning to hyperbolicity properties we will be repeatedly comparing sequences of points $\{x_n\}$ and $\{y_n\}$ in X indexed by the same set I. These can be finite sequences indexed by $I = \{0, \dots, N\}$, infinite sequences with $I = \{0, 1, 2, \dots\}$, or bi-infinite sequences with $I = \{\dots, -1, 0, 1, 2, \dots\}$. In any of these cases we will say $\{x_n\}$ *γ-shadows* $\{y_n\}$ when

$$d(x_n, y_n) \le \gamma \quad \text{for all } n \text{ in } I. \tag{11.8}$$

For example, the expansivity condition (11.5) says that the bi-infinite orbit of x_1, $\{f^n x_1\}$, γ-shadows that of x_2 only when $x_1 = x_2$.

Using this language we extract the key property for hyperbolicity.

8. Definition. *Let f be a homeomorphism on X. We say that a subset A of X satisfies the* Shadowing Property *if for every $\varepsilon > 0$ there exists $\delta > 0$ such that every δ chain in A can be ε-shadowed by a 0 chain in X. This says that for $\{x_0, x_1, \dots, x_N\}$ in A,*

$$(11.9)\qquad \begin{aligned} &d(x_n, f(x_{n-1})) \le \delta, && n = 1, \dots, N, \\ &\Rightarrow d(x_n, f^n(x)) \le \varepsilon, && n = 0, 1, \dots, N \end{aligned}$$

for some $x \in X$. So the 0 to N piece of the orbit of x ε-shadows the original chain $\{x_n\}$. Observe that x need not be in A.

Had we allowed δ to depend on N as well as on ε then (11.9) would always be true. But be warned: the independence of the chain length converts this condition from a simple consequence of uniform continuity to a rather severe demand.

9. Exercise. *Assume that X itself satisfies the Shadowing Property. Prove that the relations $\mathcal{N}f$ and $\mathcal{C}f$ are equal. In particular, every chain recurrent point is nonwandering.* □

The most important sequences for us will be the bi-infinite ones, for which we introduce a special terminology. A *δ orbit* is a bi-infinite sequence $\{x_n : n \in \mathbb{Z}\}$ which is a δ-chain, i.e.,

$$(11.10)\qquad d(x_n, f(x_{n-1})) \le \delta \quad \text{for all integers } n.$$

So a 0-orbit or an *orbit*, satisfying (11.10) with $\delta = 0$, is $\{f^n(x) : n \in \mathbb{Z}\}$ with $x = x_0$. We will also call it *the orbit of x*.

A bi-infinite sequence $\{x_n : n \in \mathbb{Z}\}$ is called *δ-close* to a subset K if it lies in $\overline{V}_\delta(K)$, i.e.,

$$(11.11)\qquad d(x_n, K) \le \delta \quad \text{for all integers } n.$$

As above a sequence $\{x_n : n \in \mathbb{Z}\}$ ε-shadows an orbit $\{f^n(x) : n \in \mathbb{Z}\}$ (and vice-versa) when

$$(11.12)\qquad d(x_n, f^n(x)) \le \varepsilon \quad \text{for all integers } n.$$

10. Proposition. *Assume that K is an f invariant subset of X. K satisfies the Shadowing Property if and only if for every $\varepsilon > 0$ there exists $\delta > 0$ such that every δ orbit, δ close to K is ε-shadowed for some orbit in X, i.e.,* (11.10) *and* (11.11) *imply* (11.12) *for some $x \in X$.*

Proof. If $\{x_0, \dots, x_N\}$ is a δ-chain in K, we can extend it to a δ orbit in K by defining $x_{N+i} = f^i(x_N)$ for $i = 1, 2, \dots$ and $x_n = f^n(x_0)$ for $n = -1, -2, \dots$. Using this extension we easily prove the Shadowing Property.

For the other direction begin with a $\delta = \delta_0$ which works for ε replaced by $\varepsilon/2$ in (11.9). We can assume $\delta_0 \le \varepsilon/2$. Now choose $\delta \le \delta_0/3$ which is a $\delta_0/3$ modulus of uniform continuity for f.

Assume $\{x_n\}$ is a δ-orbit, δ close to K. By (11.11) we can choose $\tilde{x}_n$ in K with $d(\tilde{x}_n, x_n) \leq \delta$. Thus, $\{\tilde{x}_n\}$ δ-shadows $\{x_n\}$. Furthermore, $\{\tilde{x}_n\}$ is a δ_0-orbit because

$$d(\tilde{x}_n, f(\tilde{x}_{n-1})) \leq d(\tilde{x}_n, x_n) + d(x_n f(x_{n-1})) + d(f(x_{n-1}), f(\tilde{x}_{n-1}))$$
$$\leq \delta + \delta + \delta_0/3 \leq \delta_0.$$

By the (11.9) choice of δ_0 we can choose for each positive integer N a point y_N so that the δ_0-chain in K $\{\tilde{x}_{-N}, \dots, \tilde{x}_0, \tilde{x}_1, \dots, \tilde{x}_N\}$ $(\varepsilon/2)$-shadows $\{f^{-N}(y_N), \dots, (y_N), f(y_N), \dots, f^n(y_N)\}$. Here $f^{-N}(y_N)$ is playing the role of x in (11.9). By passing to a subsequence we can assume the sequence $\{y_N\}$ converges to a point x in X. For every fixed n the inequality $d(\tilde{x}_n, f^n(y_N)) \leq \varepsilon/2$ holds for N large enough and so is preserved in the limit with x replacing y_N. So the orbit of x $(\varepsilon/2)$-shadows the δ_0-orbit $\{\tilde{x}_n\}$ in K which δ-shadows the original $\{x_n\}$. As $\delta \leq \varepsilon/2$ the triangle inequality implies the orbit of x ε-shadows $\{x_n\}$, i.e., (11.12) holds. □

If a closed, invariant subset K is expansive and satisfies the Shadowing Property we call it *topologically hyperbolic.*

11. PROPOSITION. *Assume that K is a topologically hyperbolic subset for the homeomorphism $f: X \to X$. There exist constants (which we will call* hyperbolicity constants $\gamma_0 > \gamma_1 > 0$ *such that with $U = \overline{V}_{\gamma_1}(K)$ we have:*

(a) *Every γ_1-orbit in U is γ_0-shadowed by a unique orbit in X (which is contained, of course, in $\overline{V}_{\gamma_0}(U) \subset \overline{V}_{\gamma_0+\gamma_1}(K)$). For every positive $\varepsilon \leq \gamma_0$ there exists a positive $\delta \leq \gamma_1$ such that every δ orbit, δ close to K is ε-shadowed by a unique orbit in X (which is contained in $\overline{V}_{\delta+\varepsilon}(K)$).*

(b) *For every $\varepsilon > 0$ there exists a positive integer N such that for $\{x_n\}$ and $\{\tilde{x}_n\}$ γ_1-orbits in U which are γ_0-shadowed by the orbits of x and $\tilde{x}$, respectively,*

(11.13) $\quad d(x_n, \tilde{x}_n) \leq \gamma_0$ *for all integers n with* $|n| \leq N \Rightarrow d(x, \tilde{x}) < \varepsilon$.

(c) *If $\{x_n\}$ is a γ_1-orbit in U which is γ_0-shadowed by the orbit of x, then*

(11.14) $$\lim_{n\to\infty} d(x_n, K) = 0 \quad \textit{and}$$
$$\lim_{n\to\infty} d(x_n, f(x_{n-1})) = 0 \Rightarrow \lim_{n\to\infty} d(x_n, f^n(x)) = 0,$$

with analogous results as $n \to -\infty$.

(d) *Let $\{x_n\}$ be a γ_1-orbit in U, γ_0-shadowed by the orbit of x. If $\{x_n\}$ is periodic with period N (i.e., $x_{n+N} = x_n$ for all integers n) then x is a periodic point with period N (i.e., $f^n(x) = x$). More generally,*

(11.15) $\quad d(x_n, x_{n+N}) + d(x_n, f^n(x)) \leq \gamma_0$ *for all integers n* $\Rightarrow f^n(x) = x$.

Proof. First choose γ_0 so that $3\gamma_0$ is an expansivity constant for K in X. Next choose a positive $\gamma_1 < \gamma_0$ so that (11.10) and (11.11) with $\delta = \gamma_1$ imply (11.12) with $\varepsilon = \gamma_0$. Any γ_1-orbit in U is γ_1 close to K and so is γ_0-shadowed by an orbit $\{f^n(x_1)\}$. Suppose it is also γ_0-shadowed by $\{f^n(x_2)\}$. Then the orbits of x_1 and x_2 are $\gamma_0+\gamma_1$ close to K and they $2\gamma_0$-shadow each other. Because $3\gamma_0$ is an expansivity constant (11.7) with $\gamma = 2\gamma_0$ implies $x_1 = x_2$. Uniqueness follows, completing the proof of part (a).

(b) Choose N as in Proposition 7(b) with $\gamma = 3\gamma_0$. The orbits of x and $\tilde{x}$ lie in $\overline{V}_{\gamma_0}(U) \subset \overline{V}_\gamma(K)$ and by the triangle inequality $d(f^n(x), f^n(\tilde{x})) \leq 3\gamma_0 = \gamma$ when $|n| \leq N$. So $d(x, \tilde{x}) < \varepsilon$ follows from Proposition 7(b).

(c) Given a positive $\varepsilon \leq \gamma_0$ we will produce a positive integer N_1 so that $n \geq N_1$ implies $d(x_n, f^n(x)) \leq 2\varepsilon$.

To do this first choose $\delta \leq \gamma_1$ to satisfy the condition of (a) for ε. Next choose $\delta_0 \leq \delta/2$ a $\delta/2$ modulus of uniform continuity for f. Finally, choose N so that $n \geq N$ implies $d(x_n, K) < \delta_0$ and $d(x_n, f(x_{n-1})) < \delta_0$.

We now construct a replacement $\{\tilde{x}_n\}$ for $\{x_n\}$ which agrees with $\{x_n\}$ beyond N but whose initial piece is a true orbit. To do this pick $\tilde{x}_N \in K$ with $d(x_N, \tilde{x}_N) < \delta_0$. Now let $\tilde{x}_n = f^{-i}\tilde{x}_N$ for $n < N$ and $i = N - n$. So

$$d(\tilde{x}_n, K) = d(\tilde{x}_n, f(\tilde{x}_{n-1})) = 0 \quad \text{for } n \leq N,$$

$$\begin{aligned} d(\tilde{x}_{N+1}, f(\tilde{x}_N)) &= d(x_{N+1}, f(\tilde{x}_N)) \leq d(x_{N+1}, f(x_N)) + d(f(x_N), f(\tilde{x}_N)) \\ &\leq \delta_0 + \delta/2 \leq \delta, \end{aligned}$$

and

$$\begin{aligned} d(\tilde{x}_n, K) &= d(x_n, K) \leq \delta, \\ d(\tilde{x}_{n+1}, f(\tilde{x}_n)) &= d(x_{n+1}, f(x_n)) \leq \delta \quad \text{for } n > N. \end{aligned}$$

Thus, $\{\tilde{x}_n\}$ is a δ-orbit, δ close to K which is thus ε-shadowed by a unique orbit $\{f^n(\tilde{x})\}$. Observe that

$$d(f^n(x), f^n(\tilde{x})) \leq d(f^n(x), x_n) + d(x_n, \tilde{x}_n) + d(f^n(\tilde{x}), \tilde{x}_n)$$

which, for $n > N$, is at most $\gamma_0 + 0 + \varepsilon \leq 2\gamma_0$. Now Proposition 7(a) applies with $\gamma = 2\gamma_0$ and $(x_1, x_2) = (f^{N+1}(\tilde{x}), f^{N+1}(x))$. This yields $\lim_{n\to\infty} d(f^n(x), f^n(\tilde{x})) = 0$. Choose $N_1 > N$ so that $d(f^n(x), f^n(\tilde{x})) < \varepsilon$ once $n \geq N_1$. Hence, $n \geq N_1$ implies

$$\begin{aligned} d(x_n, f^n(x)) &\leq d(x_n, \tilde{x}_n) + d(\tilde{x}_n, f^n(\tilde{x})) + d(f^n(\tilde{x}), f^n(x)) \\ &\leq 0 + \varepsilon + \varepsilon = 2\varepsilon. \end{aligned}$$

(d) The assumption of (11.15) implies $d(x_m, f^m(f^{-N}(x))) \leq \gamma_0$ for all integers m (let $m = n+N$). So the orbits of x and $f^{-N}(x)$ both γ_0-shadow the γ_1-orbit $\{x_n\}$. By uniqueness $x = f^{-N}(x)$.

In particular, if $\{x_n\}$ has period N and orbit of x γ_0-shadows $\{x_n\}$ then $d(x_n, x_{n+N}) = 0$ and $d(x_n, f^n(x)) \le \gamma_0$. The result follows from (11.15). □

12. EXERCISE. (a) *Assume K_1 is a closed invariant subset of K and that K is a topologically hyperbolic subset for f. Prove that K_1 is topologically hyperbolic. For K_1, K_2 disjoint topologically hyperbolic subsets, prove that $K_1 \cup K_2$ is topologically hyperbolic.*

(b) *Assume K is a topologically hyperbolic subset. Prove that K is an isolated invariant subset if and only if γ_0 can be chosen so that the orbits, i.e., the 0 orbits, all lie in K. Prove that K topologically hyperbolic for f and K isolated implies K is topologically hyperbolic for the restriction $f_K : K \to K$. Conversely, prove that K topologically hyperbolic for f_K and K an expansive subset for f together imply K is topologically hyperbolic for f and isolated.*

(c) *Let K be a closed f invariant subset of X, and let N be a nonzero integer. Prove that K is isolated/expansive/topologically hyperbolic for f if and only if the corresponding property holds for f^n. (Hint: If U is an isolating neighborhood* rel f *then $\bigcap_{i=0}^{N-1}\{f^i(U)\}$ is an isolating neighborhood* rel f^n. *For the other two properties use uniform continuity of $f, f^2, \dots, f^{N-1}$.)*

(d) *Let K be a periodic orbit of period N for f. Prove K is topologically hyperbolic if and only if it is an isolated invariant set. Show that this holds if and only if one (and hence every) point of K is a topologically hyperbolic fixed point for f^n.* □

To describe our main hyperbolicity result, we use a weakening of the isolation condition. We call K *isolated* rel Per f if K admits a neighborhood U so that any periodic orbit which lies entirely in U is in fact contained in K, i.e., $|\mathcal{O}(f_U)|$ (or equivalently its closure $\mathrm{Per}(F_U)$) is contained in K. Notice that there may be periodic points in $U - K$ but the orbits of such points leave U. Thus, it need not be true that $U \cap \mathrm{Per}(f) \subset K$. Clearly, any isolated invariant set is isolated rel Per f.

13. THEOREM. *Let K be a topologically hyperbolic subset for a homeomorphism f of a compact metric space X. Assume K is isolated* rel Per f. *Denote, as usual, by f_K the restriction of f to a homeomorphism of K. The chain recurrent set $\mathcal{C}(f_K)$ consists of finitely many basic sets. Each such basic set B is an isolated invariant set for f on which f is topologically transitive. In fact, given $\varepsilon > 0$ there exists a periodic point p of B whose orbit is ε-dense in B. In particular, the periodic points in K are dense in the chain recurrent set $|\mathcal{C}(f_K)|$. Finally, the chain recurrent set $|\mathcal{C}(f_K)|$ itself is an isolated invariant set for f.*

PROOF. Begin by choosing hyperbolicity constants $\gamma_0 > \gamma_1 > 0$ as in Proposition 11 and so that with $U = \overline{V}_{2\gamma_0}(K)$, $\mathrm{Per}(f_U) \subset K$.

With B a basic set for f_K, assume the orbit of x lies in $\overline{V}_{\gamma_1}(B)$. Fix y

a point of B. We will prove that $x \in K$ and y is $\mathscr{C}(f_K)$ equivalent to x. It then follows that $x \in B$ and so $\overline{V}_{\gamma_1}(B)$ is an isolating neighborhood for B.

Given a positive $\varepsilon < \gamma_1$ choose a positive integer N so that (11.13) holds. Because the entire orbit of x lies in $\overline{V}_{\gamma_1}(B)$ we can choose y_{-1}, y_1 in B so that

$$d(f(y_{-1}), f^{-N}(x)) \leq \gamma_1, \qquad d(y_1, f^{N+1}(x)) \leq \gamma_1.$$

Next recall that a basic set is always chain transitive (cf. Corollary 4.13), i.e., points of B are not merely $\mathscr{C}(f_K)$ equivalent but $\mathscr{C}(f_B)$ equivalent as well. So any two points of B can be joined by a γ_1 chain in B.

We now construct a periodic γ_1 orbit in $\overline{V}_{\gamma_1}(B)$

$$x_n = f^n(x), \qquad -N \leq n \leq N,$$

$\{y_1 = x_{N+1}, \dots, x_{N+k_1} = f^{-N}(y)\}$ is a γ_1-chain in B joining y_1 to $f^{-N}(y)$.

$$x_n = f^{n-2N-k_1}(y), \qquad N + k_1 \leq n \leq 3N + k_1,$$

$\{f^n(y) = x_{3N+k_1}, \dots, x_{3N+k_1+k_2} = y_{-1}\}$ is a γ_1-chain in B joining $f^n(y)$ to y_{-1}.

$$x_{3N+k_1+k_2+1} = f^{-N}(x) = x_{-N}.$$

That is, we begin with a piece of the x orbit of length $2N$ centered on x. Then make a γ_1 jump into B at y_1. Use a γ_1-chain in B to go to $f^{-N}(y)$. Then use a $2N$ long piece of the y orbit. Finally, use a γ_1-chain in B to go to y_{-1} which allows a γ_1 jump back to $f^{-N}(x)$. Repeating the cycle we obtain a γ_1-orbit in $V_{\gamma_1}(B)$ of period $4N + k_1 + k_2$.

By Proposition 11(a) the γ_1 orbit $\{x_n\}$ is γ_0-shadowed by the orbit of a unique point z in X and by part (d) z is a periodic point. As the entire periodic orbit of z lies in $\overline{V}_{\gamma_0+\gamma_1}(B) \subset U$ and $\operatorname{Per}(f_U) \subset K$, we see that z lies in K.

Now apply (11.13) first to $\{x_n\}$ and $\{\tilde{x}_n = f^n(x)\}$ and then to the shifted chain $\{x_{n+2N+k_1}\}$ and $\{\tilde{x}_n = f^n(y)\}$. We obtain

$$d(z, x) < \varepsilon, \qquad d(f^{2N+k_1}(z), y) < \varepsilon.$$

In other words, given an arbitrary positive ε we can find a periodic point z whose orbit passes ε close to x and ε close to y.

As x is the limit of such points z in K when $\varepsilon \to 0$, we see that x lies in K. Then the z orbits provide chains in K joining x to y and y to x. So x and y are $\mathscr{C}(f_K)$ equivalent.

This completes the proof that B is isolated. The rest is easy.

Given a positive $\varepsilon \le \gamma_1$, choose a positive $\delta \le \gamma_1$ so that every δ-orbit in K is $(\varepsilon/2)$-shadowed by a unique orbit in X. Now choose a finite sequence in B $\{x_1, \dots, x_k\}$ which is $(\varepsilon/2)$-dense in B. Construct a periodic δ-orbit in B by joining x_1 to x_2, x_2 to $x_3, \dots, x_{k-1}$ to x_k and finally x_k to x_1 by δ-chains. There exists, by Proposition 11 (d) again, a periodic point z whose orbit $(\varepsilon/2)$-shadows our constructed δ-orbit. In particular, the orbit of z lies in the isolating neighborhood $\overline{V}_{\gamma_1}(B)$ and so $z \in B$. As the orbit of z passes within $\varepsilon/2$ of the points $x_1, \dots, x_k$ it follows the z orbit is an ε-dense subset of B. In particular, if $y_1, y_2 \in B$ we can make an ε jump from y_1 onto the orbit of z and then after traveling along to z orbit make an ε hop off to y_2. Thus, $y_2 \mathscr{N}(f_B) y_1$ and so the restriction f_B is a topologically transitive homeomorphism. Obviously, the periodic points are dense in B.

Finally, applying part (c) of Proposition 3 to f_K we see that there are only finitely many basic sets. As each is isolated in X their union $|\mathscr{C}(f_K)|$ is isolated in X by Exercise 2(a). □

14. Corollary. *Let K be a topologically hyperbolic subset for a homeomorphism f of a compact metric space X. Assume that K is isolated* rel Per f *and that every point of K is chain recurrent in K, i.e., $K = |\mathscr{C}(f_K)|$. Then*

(a) *K is an isolated invariant set in X.*

(b) *The periodic points of K are dense in K.*

(c) *K is the disjoint union of a finite number of compact f invariant sets on each of which f is topologically transitive.*

Proof. Since $K = |\mathscr{C}(f_K)|$ these all follow from the theorem. □

15. Corollary. *Let f be a homeomorphism of a compact metric space X. Let K be the closure of one of the following f invariant sets:*

$|\mathscr{O}f| =$ *periodic points of f,*
$m[f] =$ *union of the minimal sets for f,*
$|\omega f| \cap |\alpha f| =$ *recurrent points of f,*
$\omega f(X) = \omega$-*limit points of f,*
$\alpha f(X) = \alpha$-*limit points of f,*
$\alpha f \cup \omega f(X) =$ *limit points of f,*
$|\mathscr{C} f| =$ *chain recurrent points of f.*

If K is a hyperbolic invariant subset of X, then

(a) *K is an isolated invariant set,*

(b) *$K = \overline{|\mathscr{O}f|}$,*

(c) *K is the disjoint union of a finite number of closed f invariant sets on each of which f is topologically transitive, namely the basic sets of the restriction of f to K.*

PROOF. In each case all periodic points lie in K so K is clearly isolated rel Per f. Next, in each case we claim that K consists of the closure of a (possibly infinite) union of chain transitive subsets $\{K_\alpha\}$. Since $|\mathscr{C}(f_{K_\alpha})| = K_\alpha$ it follows that $\bigcup\{K_\alpha\} \subset |\mathscr{C}(f_K)| \subset K = \overline{\bigcup\{K_\alpha\}}$. Hence, the closed set $|\mathscr{C}(f_K)|$ equals K. To prove the claim, note that a periodic orbit, a minimal set, and the closure of the orbit of a recurrent point are each topologically transitive subsets. For any x, the limit point sets $\omega f(x)$ and $\alpha f(x)$ are chain transitive sets as are the basic sets in $|\mathscr{C} f|$.

The conclusions then follow from Corollary 14.

REMARK. Notice that the argument does *not* apply to $K = |\mathscr{N} f| = |\Omega f|$, the set of nonwandering points. In general, it need not be true that $|\mathscr{C}(f_K)| = K$ for this set. □

We turn now to perturbation properties associated with topological hyperbolicity. As explained in Chapter 7, results for topological perturbations are rather crude.

16. PROPOSITION. *Let K be an isolated topologically hyperbolic invariant subset for $f: X \to X$. Assume that $\gamma_0 > \gamma_1 > 0$ are hyperbolicity constants as in Proposition 11 and that $\overline{V}_{2\gamma_0}(K)$ is an isolating neighborhood for K.*

If $g: X \to X$ is a homeomorphism satisfying $d(f, g) \leq \gamma_1$ (i.e., $d(f(x), g(x)) \leq \gamma_1$ for all $x \in X$) and $K_g = \bigcap_{n=-\infty}^{\infty}\{g^n(\overline{V}_{\gamma_1}(K))\}$ is the maximum g invariant subset of $\overline{V}_{\gamma_1}(K)$, then there is a unique map $k_g: K_g \to K$ satisfying

$$(11.16) \qquad f \circ k_g = k_g \circ g \quad \text{and} \quad d(x, k_g(x)) \leq \gamma_0 \quad \text{for all } x \text{ in } K_g.$$

Furthermore, the function k_g is continuous. Thus, k_g provides a semiconjugacy of the restriction $g_{K_g}: K_g \to K_g$ into the restriction $f_K: K \to K$.

PROOF. If $x \in K_g$ then the g orbit of x defined by $x_n = g^n(x)$ for all integers n (g is a homeomorphism) lies entirely in $\overline{V}_{\gamma_1}(K)$ and $d(x_n, f(x_{n-1})) = d(g(x_{n-1}), f(x_{n-1})) \leq \gamma_1$ implies that $\{x_n\}$ is a γ_1-chain with respect to f. So there is a unique point, which we will denote by $k_g(x)$ whose f orbit γ_0-shadows $\{x_n\}$, i.e.,

$$(11.17) \qquad d(f^n(k_g(x)), g^n(x)) \leq \gamma_0 \quad \text{for all integers } n,$$

and this condition uniquely characterizes the point $k_g(x)$. From (11.17) the f orbit of $k_g(x)$ lies in the isolating neighborhood $\overline{V}_{\gamma_0+\gamma_1}(K) \subset \overline{V}_{2\gamma_0}(K)$. Consequently, $k_g(x) \in K$. Now (11.17) says that the f orbit of $f(k_g(x))$ γ_0-shadows the g orbit of $g(x)$. But this condition characterizes the point $k_g(g(x))$. So $f(k_g(x)) = k_g(g(x))$. This equation, together with (11.17) for $n = 0$, implies (11.16).

Conversely, the semiconjugacy condition $f \circ k_g = k_g \circ g$ implies that k_g maps the g orbit of x to the f orbit of $k_g(x)$, and so $k_g(g^n(x)) = f^n(k_g(x))$. We then obtain (11.17) from (11.16) with x replaced by $g^n(x)$.

But we have seen that (11.17) characterizes the point $k_g(x)$ and so the map k_g is uniquely defined by (11.16).

Now given $\varepsilon > 0$, choose N so that (11.13) is satisfied. Next choose $\delta > 0$ so that $d(x, \tilde{x}) < \delta$ implies $d(g^n(x), g^n(\tilde{x})) \leq \gamma_0$ for all integers n with $|n| \leq N$, i.e., δ is a γ_0 modulus of uniform continuity for $\{g^n : |n| \leq N\}$. Thus, if $d(x, \tilde{x}) < \delta$ in K_g then $\{g^n(x)\}$ and $\{g^n(\tilde{x})\}$ are γ_1 orbits rel f in $\overline{V}_{\gamma_1}(K)$ which are γ_0-shadowed by the f orbits of $k_g(x)$ and $k_g(\tilde{x})$ respectively. Applying (11.13) we see that $d(k_g(x), k_g(\tilde{x})) < \varepsilon$. This proves continuity of the map k_g. □

To get that the map k_g is bijective we have to assume hyperbolicity conditions for g and furthermore impose some uniformity upon these conditions.

17. Addendum. *Assume that g further satisfies the conditions*: (1) $\overline{V}_{2\gamma_0}(K)$ *is an isolating neighborhood with respect to g for K_g and* (2) *every γ_1-orbit for g in $\overline{V}_{\gamma_1}(K)$ is γ_0-shadowed by a unique g orbit in X. Then $k_g : K_g \to K$ is a bijection and so is a homeomorphism providing a conjugacy between g_{K_g} and f_K.*

Proof. We reverse the above argument. Given $y \in K$, the f orbit of y is a γ_1 orbit rel g and so, by (2), is γ_0-shadowed by the g orbit of a unique point $x \in X$. By assumption (1), $x \in K_g$. Thus, for every $y \in K$ there is a unique $x \in K_g$ such that

$$d(f^n(y), g^n(x)) \leq \gamma_0 \quad \text{for all integers } n.$$

By the (11.17) characterizations of $k_g(x)$ we see that $y = k_g(x)$. This means, for every $y \in K$ there exists a unique $x \in K_g$ such that $k_g(x) = y$, i.e., k_g is a bijection. Since k_g is continuous and K_g is compact, it follows that k_g is a homeomorphism. □

Assumption (2) is a strong demand and we will obtain it later by resorting to C^1 perturbations. But assumption (1) holds provided g is close enough to f in the topological sense.

18. Exercise. (a) *Let f be a closed relation on X and A and U be closed subsets. Prove that*

$$\bigcap_{n=-\infty}^{+\infty} \{f^n(A)\} \subset \operatorname{Int} U \Rightarrow \bigcap_{n=-N}^{N} \{(\overline{V}_\varepsilon \circ f \circ \overline{V}_\varepsilon)^n(A)\} \subset \operatorname{Int} U$$

for some whole number N and some $\varepsilon > 0$. (Hint: First replace the infinite intersection by a finite one, choosing N, and then apply Lemma 3.2.) In particular, prove that for some $\varepsilon > 0$, $\overline{V}_{2\gamma_0}(K)$ an isolating neighborhood for K and $d(f, g) \leq \varepsilon$ imply $\bigcap_{n=-\infty}^{\infty} \{g^n(\overline{V}_{2\gamma_0}(K))\} \subset V_{\gamma_1}(K)$ and so $\overline{V}_{2\gamma_0}(K)$ is an isolating neighborhood with respect to g for K_g.

(b) *Let* g_ε *be the time-one map for the flow on* $[0, 1]$ *satisfying*:

$$\frac{dx}{dt} = \sin^2(2\pi x) + \varepsilon \sin^2(\pi x).$$

Let f *denote* g_0. *Prove that for* $\varepsilon \geq 0$ *this is a positive interval example with closed set of equilibria* $\{0, 1\}$ $(\varepsilon > 0)$ *and* $\{0, 1/2, 1\}$ $(\varepsilon = 0)$. *Let* $K = \{1/2\}$ *and* $U = [1/3, 2/3]$. *Prove that* $K_g = \emptyset$ *for* $\varepsilon > 0$ *while for* $\varepsilon < 0$, K_g *is the interval*

$$\left[\frac{1}{\pi}\cos^{-1}(|\varepsilon|^{1/2}/2), \frac{1}{\pi}\cos^{-1} -(|\varepsilon|^{1/2}/2)\right]$$

containing $1/2$. *Observe that* $\{1/2\}$ *is a topologically hyperbolic subset for* f. □

19. Theorem. *We will call a homeomorphism* f *of a compact metric space* X *an* Axiom A homeomorphism *if the chain recurrent set* $|\mathscr{C}f|$ *is topologically hyperbolic. For such a homeomorphism there are finitely many basic sets on each of which* f *is topologically transitive. Furthermore,* $|\mathscr{C}f|$ *is an isolated invariant set in which the set of periodic points* $|\mathscr{O}f|$ *is dense.*

For $\varepsilon > 0$, *there exists* $\delta > 0$ *such that for each homeomorphism* g *of* X *with* $d(f, g) \leq \delta$, $|\mathscr{C}g| \subset V_\varepsilon(|\mathscr{C}f|)$ *and there is a unique map* $k_g\colon |\mathscr{C}g| \to |\mathscr{C}f|$ *such that for all* x *in* $|\mathscr{C}g|$

$$f(k_g(x)) = k_g(g(x)) \quad \text{and} \quad d(x, k_g(x)) \leq \varepsilon.$$

Furthermore, the semiconjugacy map k_g *is continuous.*

Proof. The first paragraph is a restatement of Corollary 15 with $K = |\mathscr{C}f|$. Then choose $\gamma_0 > \gamma_1 > 0$ hyperbolicity constants so that $\overline{V}_{2\gamma_0}(|\mathscr{C}f|)$ is an isolating neighborhood. By Exercise 18(a) we can choose δ small enough that $\overline{V}_{2\gamma_0}(|\mathscr{C}f|)$ is an isolating neighborhood for $K_g = \bigcap_{n=-\infty}^{+\infty}\{g^n(\overline{V}_{\gamma_1}(|\mathscr{C}f|))\}$. By Theorem 7.23, the map associating $|\mathscr{C}g|$ to g is upper semicontinuous and so for $\delta > 0$ small enough $|\mathscr{C}g| \subset V_{\gamma_1}(|\mathscr{C}f|)$. Consequently, $|\mathscr{C}g| \subset K_g$. The unique semiconjugacy map k_g is constructed in Proposition 17. □

From the next result we see that the homeomorphisms g_ε of Exercise 18 (b) are all Axiom A and so yield examples where $|\mathscr{C}g|$ is a proper subset of K_g.

20. Proposition. *The chain recurrent set of a homeomorphism* f *is finite if and only if* f *is Axiom* A *and* $|\mathscr{C}f|$ *consists entirely of periodic points, i.e,* $|\mathscr{C}f| = |\mathscr{O}f|$.

Proof. If $|\mathscr{C}f|$ is finite then each chain recurrent point is periodic. Furthermore, the periodic orbits are isolated. For if U is a closed neighborhood of a periodic orbit B disjoint from the rest of $|\mathscr{C}f|$ and the orbit of x

lies entirely in U then $\alpha f(x)$ and $\omega f(x)$ are subsets of $U \cap |\mathscr{C}f|$ and so equal B. It follows that x is $\mathscr{C}f$ equivalent to the points of B and so also lies in $U \cap |\mathscr{C}f| = B$. So by Exercise 12(d) and (a), $|\mathscr{C}f|$ is topologically hyperbolic.

Conversely, for an Axiom A homeomorphism $|\mathscr{C}f|$ consists of finitely many basic sets B. f is topologically transitive on B and so B contains a dense orbit. $|\mathscr{C}f| = |\mathscr{O}f|$ implies this dense orbit is periodic and so is closed and equal to B. Thus, $|\mathscr{C}f|$ consists of finitely many periodic orbits. □

The original version of the Axiom A condition did not use the chain recurrent set.

21. PROPOSITION. *Let f be a homeomorphism of a compact metric space X and assume that $l[f]$, the closure of the set of limit points, is a topologically hyperbolic invariant set. Then $l[f]$ is an isolated invariant set in which the periodic points are dense. $l[f]$ is the disjoint union of a finite number of closed invariants sets on which f is topologically transitive, namely the basic sets of $f_{l[f]}$. The collection $\mathscr{F}$ of these $f_{l[f]}$ basic sets forms a decomposition for f. f is a Axiom A homeomorphism if and only if $l[f]$ is topologically hyperbolic and the decomposition $\mathscr{F}$ satisfies the no-cycle condition (cf. Exercise* 5.16(j)).

PROOF. $l[f]$ is the closure of $\omega f(X) \cup \alpha f(X)$ and so we can again apply Corollary 15. Now if $\mathscr{F}$ satisfies the no-cycle condition then by Exercise 5.16(j) $\mathscr{F} = \mathscr{F}_b$ and

$$l[f] = \bigcup \mathscr{F} = \bigcup \mathscr{F}_b = |\mathscr{F}_b| \supset |\mathscr{C}f| \supset l[f].$$

So $|\mathscr{C}f| = l[f]$ and $|\mathscr{C}f|$ is topologically hyperbolic.

On the other hand, if $|\mathscr{C}f|$ is topologically hyperbolic then because the periodic points are dense in $|\mathscr{C}f|$, we have

$$\overline{|\mathscr{O}f|} \subset l[f] \subset |\mathscr{C}f| = \overline{|\mathscr{O}f|}$$

and so $l[f] = |\mathscr{C}f|$. The decomposition $\mathscr{F}$ then consists of the basic sets for f. As these sets are $\mathscr{C}f$ semi-invariant, Exercise 5.16(j) implies the no-cycle condition. □

22. EXERCISE. (a) *Applying Proposition* 5.17, *prove that every basic set for f is isolated if and only if f admits a fine decomposition. On the other hand, for $\mathscr{F}$ any decomposition for f, prove that $|\mathscr{F}|$ is an isolated invariant set. In particular, if $\mathscr{F}$ is a decomposition with $|\mathscr{F}| = |\mathscr{C}f|$, then $|\mathscr{C}f|$ is an isolated invariant set. Thus, the example of Exercise* 17(b) *yields a homeomorphism with infinitely many basic sets but with $|\mathscr{C}f|$ isolated. With $K = \{0\} \cup \{n^{-1} : n = 1, 2, \dots\}$ and f an associated interval example, show that $|\mathscr{C}f| = K$ is not an isolated invariant set.*

(b) *Prove that a homeomorphism f is Axiom A if and only if the following two conditions hold:*

(i) *The set of nonwandering points* $|\mathcal{N}f| = |\Omega f|$ *is a topologically hyperbolic invariant set in which the periodic points are dense (i.e., we* assume $|\mathcal{O}f| = |\mathcal{N}f|$).

(ii) *The decomposition* $\mathcal{F}$ *consisting of the basic sets of the restriction* $f_{|\mathcal{N}f|}$ *satisfies the no-cycle condition.*

□

23. **Theorem.** *We will call a homeomorphism* f *of a compact metric space* X *an* Anosov homeomorphism *if the entire space* X *is topologically hyperbolic. An Anosov homeomorphism is Axiom A as well.*

For $\varepsilon > 0$, *there exists a* $\delta > 0$ *such that for each homeomorphism* g *of* X *with* $d(f, g) \le \delta$, *there is a unique map* $k_g: X \to X$ *with* $f \circ k_g = k_g \circ g$ *and* $d(1_X, k_g) \le \varepsilon$. *The semiconjugacy maps* k_g *are continuous and satisfy* $k_g(|\mathcal{C}g|) \subset |\mathcal{C}f|$, *and the association* $g \mapsto k_g$ *is continuous regarded as a mapping of a neighborhood of* f *in* $\mathrm{Cis}(X; X)$ *into* $C(X; X)$.

Proof. As $|\mathcal{C}f|$ is an f invariant subset of X, it is topologically hyperbolic when X is. So f is Axiom A. Choose δ as in Theorem 19 and then shrink δ if necessary so that k_g of Proposition 16 can be constructed on all of X.

Now given $\varepsilon_1 > 0$ choose N to satisfy Proposition 11(b) for ε_1. Then choose $\delta_1 > 0$ so that $d(g, \tilde{g}) < \delta_1$ and $d(g^{-1}, \tilde{g}^{-1}) < \delta_1$ imply $d(g^n, \tilde{g}^n) \le \gamma_1$ for all n with $|n| \le N$. The f orbits of $k_g(x)$ and $k_{\tilde{g}}(x)$ γ_0-follow $\{g^n(x)\}$ and $\{\tilde{g}^n(x)\}$ respectively and so $d(k_g(x), k_g(\tilde{x})) < \varepsilon_1$. That is, $d(g, \tilde{g}) < \delta_1$ and $d(g^{-1}, \tilde{g}^{-1}) < \delta_1$ imply $d(k_g, k_{\tilde{g}}) \le \varepsilon_1$.

Remark. If X is a manifold without boundary then for ε small enough the maps $k_g: X \to X$ are surjective. This is because a map sufficiently close to 1_X is homotopic to 1_X and so has degree equal to 1 (e.g. induces the identity of the homology group $H_n(X; \mathbb{Z}_2)$ when dimension $X = n$), while a map with image a proper subset of X has degree zero. □

Observe that Exercise 12(b) says that if K is a topologically hyperbolic and isolated invariant subset of X then the restriction $f_K: K \to K$ is an Anosov homeomorphism. Conversely, if $f_K: K \to K$ is Anosov and K is an expansive subset then K is topologically hyperbolic and isolated.

Now, as an alternative approach to the differential theory, we will begin the exposition anew and apply the results of the previous chapter to study fixed points.

We will suppose now that X is a smooth Banach manifold. A chart at a point x of X is an open set U of X, containing x, together with a coordinate mapping h of U onto an open subset of a Banach space E. That is, h is a homeomorphism of U onto $h(U)$ and is compatible with the differential structure on X (see Lang (1972) Chapter II). By translating and restricting if necessary we can assume that $h(x) = 0$ and that the image

$h(U)$ is an open ball in E centered at 0. We say that such a chart is *based at* x.

Let $f: X \to X$ be a C^1 diffeomorphism and let x be a fixed point for f. We say that x is a *nondegenerate* or *hyperbolic* fixed point if the linearization at x, the tangent map $T_x f: T_x X \to T_x X$, is a nondegenerate or hyperbolic linear map. This statement is interpreted by using a chart $h: U \to E$ about x. hfh^{-1} is a C^1 map defined on $U_0 = U \cap f^{-1}(U)$, an open subset of U which still contains the fixed point x. From the following commutative diagram

$$\begin{array}{ccc} U_0 & \xrightarrow{h} & h(U_0) \\ {\scriptstyle f}\downarrow & & \downarrow{\scriptstyle hfh^{-1}} \\ U & \xrightarrow[h]{} & E \\ \cap & & \\ X & & \end{array} \tag{11.18}$$

we see that the restriction of f to a small enough neighborhood of x is conjugate via h, to the map hfh^{-1} of the open set $h(U_0)$ into E. Taking the derivative of the latter map at its fixed point $h(x)$ we get the linear automorphism $D_{h(x)}(hfh^{-1})$ on E which represents $T_x f$.

If $h_1: U_1 \to E_1$ is a different chart about x then because hfh^{-1} and $h_1 f h_1^{-1}$ are both conjugate to f near x they are conjugate to each other by using hh_1^{-1} and its inverse. Differentiating, the chain rule yields the commutative diagram:

$$\begin{array}{ccc} E_1 & \xrightarrow{D_{h_1(x)}(hh_1^{-1})} & E \\ {\scriptstyle D_{h_1(x)}(h_1 f h_1^{-1})}\downarrow & & \downarrow{\scriptstyle D_{h(x)}(hfh^{-1})} \\ E_1 & \xrightarrow[D_{h_1(x)}(hh_1^{-1})]{} & E. \end{array}$$

This says that the vertical maps, the alternative representatives of $T_x f: T_x X \to T_x X$, are similar automorphisms. In particular, they have the same spectrum. So calling x nondegenerate (or hyperbolic) when $D_{h(x)}(hfh^{-1})$ is nondegenerate (resp. hyperbolic) is well defined, that is, it is independent of the choice of chart.

24. THEOREM (Hartman's Theorem). *Let f be a C^1 diffeomorphism of a smooth manifold X, and let x be a fixed point of f.*

If x is a nondegenerate fixed point, then there exists a neighborhood U which contains no other fixed point. If x is hyperbolic, then it is an isolated invariant set, i.e., for suitable choice of U, x is the only point whose entire orbit remains in U. Furthermore, in the hyperbolic case f is locally conjugate *to its linearization* $T_x f\colon T_x X \to T_x X$. *This means there is a homeomorphism* $k\colon U \to T_x X$ *from an open neighborhood of* x *in* X *to an open subset of* $T_x X$ *such that* $k(x) = 0$ *and on* $U_0 \equiv U \cap f^{-1}(U)$ *the following diagram commutes*

(11.19)
$$\begin{array}{ccc} U_0 & \xrightarrow{k} & T_x X \\ {\scriptstyle f}\downarrow & & \downarrow{\scriptstyle T_x f} \\ U & \xrightarrow[k]{} & T_x X. \\ \cap & & \\ X & & \end{array}$$

Finally, the neighborhoods U *can be chosen to work uniformly for all diffeomorphisms* g *sufficiently close to* f *in the* C^1 *sense. That is, we can choose* U *sufficiently small and then choose* $\mathscr{U}$ *a neighborhood of* f *in the* C^1 *topology (details below) so that each* g *in* $\mathscr{U}$ *has a unique fixed point* x_g *in* U *and* x_g *is nondegenerate for* g *(or hyperbolic if* x *is). In the hyperbolic case* x_g *is the only point whose* g *orbit remains in* U. *Furthermore, there is a unique homeomorphism defined on* U *mapping into* $T_x X$ *with* $k_g(x_g) = 0$ *and yielding a local conjugacy of* g *with the linearization of* f, $T_x f\colon T_x X \to T_x X$.

PROOF. We begin with an open chart $h\colon U \to E$ based at x and choose $\varepsilon > 0$ so that the closed ball $\overline{V}_\varepsilon(0)$ is contained in $h(U_0)$, where $U_0 = U \cap f^{-1}(U)$. Denote $h^{-1}(V_\varepsilon(0))$ by U_ε.

As in (11.18) hfh^{-1} is defined on $\overline{V}_\varepsilon(0)$ and we can write this map as

$$hfh^{-1}(y) = T(y) + w(y),$$

where T is the derivative $D_0(hfh^{-1})$ which represents the linearization $T_x f$ via the chart h.

Thus, the error term w is a C^1 map on $\overline{V}_\varepsilon(0)$ and satisfies $D_0 w = 0$. So given any $\delta > 0$ we can choose ε small enough that $\|D_v w\| < \delta$ for v in $\overline{V}_\varepsilon(0)$. $hfh^{-1}(0) = 0$ and the Mean Value Theorem imply that w is a Lipschitz map satisfying

$$w(0) = 0, \qquad L(w) \le \delta \quad (\text{on } \overline{V}_\varepsilon(0)). \tag{11.20}$$

Now recall the retraction $r\colon E \to \overline{V}_\varepsilon(0)$ defined by $r(y) = \varepsilon y/\|y\|$ when $\|y\| \ge \varepsilon$. Use it to extend the definition of w to all of E by $\tilde{w}(y) = w(r(y))$. Then define $\tilde{f}\colon E \to E$ by

$$\tilde{f}(y) = T(y) + \tilde{w}(y).$$

By Lemma 10.10 $L(r) \le 2$ and so from (11.20)

$$\tilde{w}(0) = 0, \qquad L(\tilde{w}) \le 2\delta \quad (\text{on } E).$$

Thus, we have constructed a map $\tilde{f}$ which is a small Lipschitz perturbation of the linear map T and so we can apply to it the results of the previous chapter. On the other hand, $\tilde{f}$ agrees with hfh^{-1} on $V_\varepsilon(0)$ and so is locally conjugate to f.

To say that x is nondegenerate is to say that $1 - T$ is an isomorphism of E. If δ is chosen so that $2\delta < \|(1-T)^{-1}\|^{-1}$, then by Proposition 10.9, 0 is the unique fixed point for $\tilde{f}$ on E. By the conjugacy x is the only fixed point for f in U_ε.

In the hyperbolic case we can choose the norm on E adapted to T so that

$$\lambda = \max(\|T|E^+\|, \|(T|E^-)^{-1}\|) < 1.$$

Choose δ small enough that $\lambda + 2\delta < 1$. Theorem 10.13 applies to $\tilde{f}$ and says that 0 has the only bounded $\tilde{f}$ orbit. A fortiori, x has the only orbit for f which remains in U_ε. If we further assume that $\|(T|E^+)^{-1}\|2\delta < 1$ then we obtain a homeomorphism $\tilde{k}\colon E \to E$ providing a conjugacy of $\tilde{f}$ with T. The composition $\tilde{k} \circ h$ gives the local conjugacy of f with T.

For the perturbation results, given $\delta > 0$ and the resulting $\varepsilon > 0$ and also given a positive $\delta_1 < \varepsilon$, we define the neighborhood $\mathscr{U}$ of f to consist of all diffeomorphisms g such that (a) $\overline{V}_\varepsilon(0) \subset h(U \cap g^{-1}(U))$, (b) $\|h(g(x))\| < \delta_1$, and (c) for all $y \in V_\varepsilon(0)$ $\|D_y[hfh^{-1} - hgh^{-1}]\| < \delta$. Conditions (a) and (b) are C^0 conditions of closeness of g to f while (c) says that the derivatives are close on a fixed neighborhood of x. By (a) hgh^{-1} is defined on $\overline{V}_\varepsilon(0)$ and we can define first w_g on $\overline{V}_\varepsilon(0)$ to be $hgh^{-1} - T$ and then $\tilde{w}_g$ on E by $\tilde{w}_g(y) = w_g(r(y))$. So we get

$$\tilde{g}(y) = T(y) + \tilde{w}_g(y), \qquad \|\tilde{w}_g(0)\| = \|h(g)(x))\| \le \delta_1, \quad L(\tilde{w}_g) < 4\delta,$$

and $\tilde{g}$ is conjugate to g on U_ε via h. The above arguments for $\tilde{f}$ work just the same for $\tilde{g}$ (2δ in the conditions on δ must be replaced by 4δ) provided that δ_1 is restricted by

$$\frac{\|(1-T)^{-1}\|}{1 - \|(1-T)^{-1}\|4\delta}\delta_1 < \varepsilon$$

in the nondegenerate case (c.f. (10.10)) to obtain the fixed point x_g in U_ε. In the hyperbolic case we require $\delta_1 < (1 - (\lambda + 4\delta))\varepsilon$. □

Hartman's Theorem says two things about a hyperbolic fixed point x. First, is its local structural stability. If g is a perturbation C^1 close enough to f then g has a hyperbolic fixed point which is close to x, and depends continuously on g. Furthermore, g near its fixed point is locally conjugate to f near x. Secondly, we have a description of this local behavior: f near x is locally conjugate to its linearization T near 0.

In particular, we observe that a hyperbolic fixed point for a diffeomorphism is a topologically hyperbolic invariant set. Exercise 18(b) provides an example of a topologically hyperbolic fixed point which is degenerate. On the other hand if f is the rotation of the unit disc in the plane through an irrational multiple of π then the origin 0 is a nondegenerate fixed point, isolated rel Per f but not a topologically hyperbolic invariant set because of the invariant circles arbitrarily close to 0.

We now extend the concept of hyperbolicity from a fixed point to a compact f invariant subset. For a fixed point x hyperbolicity is a condition on the tangent map at x, $T_xf\colon T_xX \to T_xX$. So for an invariant set K we consider all of the maps T_xf $(x \in K)$ at once, with the added complication that as $f(x)$ is not usually equal to x, the domain and range of $T_xf\colon T_xX \to T_{f(x)}X$ are not the same space. Even if the two spaces are identified by some isomorphism so that iterating the single linear map T_xf becomes possible, such iterations no longer are usefully related to iterations of the original function f. Instead, if we begin with a vector v based at x, which we think of as an infinitesimal variation at x, then the induced variations along the orbit $f(x)$, $f^2(x)$, $f^3(x)$, ... are obtained by first applying T_xf, then $T_{f(x)}f$, $T_{f^2(x)}f$, etc. which are usually different maps.

Returning now to the fixed point situation, hyperbolicity of the linear automorphism T_xf was first introduced in spectral terms (no eigenvalues on the unit circle) then we derived the splitting of T_xX into stable and unstable subspaces $E^{\pm}$. While a spectral definition is possible for subsets, it applies to an associated linear map on a space of sections of the tangent bundle restricted to K (see Hirsch and Pugh (1970)). Instead of considering such section maps we will use the splitting into stable and unstable subspaces to build our definition of hyperbolicity for an invariant set.

25. Definition. *Let f be a C^1 diffeomorphism of a smooth manifold X and let K be a compact, f invariant subset of X. K is called a hyperbolic invariant set if the restriction to K of the tangent bundle of X admits an f invariant splitting: $T_xX = E_x^+ \oplus E_x^-$ (for all $x \in K$) with $T_xf(E_x^{\pm}) = E_{f(x)}^{\pm}$, such that for some constants $C > 0$ and $0 < \lambda < 1$*

$$\|T_xf^n|E_x^+\| \le C\lambda^n, \qquad \|T_xf^{-n}|E_x^-\| \le C\lambda^n \tag{11.21}$$

for all $x \in X$ and all whole numbers n.

Notice that the operator norms in (11.21) require a choice of norms at each tangent space T_xX. Such a continuously varying choice of norms is called a Finsler. The norms associated with a Riemannian metric are a Finsler. A particular Finsler on the tangent bundle is called *adapted to* K if it provides the max norm on T_xX regarded as the product of E_x^+ and E_x^- and if the constant C in (11.21) can be chosen to be 1. In that case (11.21) is equivalent to

$$\|T_xf\,|E_x^+\| \le \lambda, \qquad \|T_xf^{-1}|E_x^-\| \le \lambda \quad \text{(for all } x \in K). \tag{11.22}$$

By a proof analogous to that of Lemma 10.12 one can show that Finslers adapted to K always exist.

For a hyperbolic fixed point we used the infinitesimal information about $T_x f$ to study f near x. We did this by writing f in local coordinates as a Lipschitz perturbation of $T_x f$. The next proposition does the analogous job for a hyperbolic set.

26. Proposition. *Let f be a C^1 diffeomorphism of a smooth manifold X and let K be a compact, f invariant subset of X. K is a hyperbolic invariant subset if and only if there exist constants $0 < \lambda < 1$, $\ell > 0$, and $e > 0$ and for each x in $V_\ell(K)$ a chart $h_x\colon U_x \to E_x$ based at x with $E_x = E_x^+ \times E_x^-$ a product with the max norm, such that the following conditions hold*:

If $x, y \in V_\ell(K)$ and $d(y, f(x)) < e$ then $h_x(U_x \cap f^{-1}(U_y))$ contains the ball $\overline{V}_\ell(0)$ in E_x, which we will denote $B_x = B_x^+ \times B_x^-$. Writing $(F_{(x,y)}, G_{(x,y)})$ for the restriction $h_y \circ f \circ h_x^{-1}\colon B_x \to E_y^+ \times E_y^-$, there exist linear isomorphisms $T_{(x,y)}\colon E_x^+ \to E_y^+$ and $R_{(x,y)}\colon E_x^- \to E_y^-$ so that for $(u, v) \in B_x = B_x^+ \times B_x^-$:

$$F_{(x,y)}(u, v) = T_{(x,y)}(u) + S^+_{(x,y)}(u, v),$$
$$G_{(x,y)}(u, v) = R_{(x,y)}(v) + S^-_{(x,y)}(u, v) \tag{11.23}$$

with S^+ and S^- C^1 (and hence Lipschitz) maps satisfying

$$\|T_{(x,y)}\| + L(S^+_{(x,y)}) < \lambda, \qquad \|R^{-1}_{(x,y)}\| + L(S^-_{(x,y)}) < \lambda, \tag{11.24}$$

and

$$\|(S^+_{(x,y)}, S^-_{(x,y)})(0, 0)\| < (1 - \lambda)\ell. \quad \square \tag{11.25}$$

Let us sketch how the conditions of Proposition 26 are obtained from those of Definition 25. One begins with a Finsler adapted to K and uses it to define charts based at x for all x in K with h_x mapping to $T_x X$ which is expressed as the product $E_x^+ \times E_x^-$ by the hyperbolic splitting. When $y = f(x)$, $h_y f h_x^{-1}$ preserves the origin and we write its derivative at $(0, 0)$ as $(T, R)_{(x,y)}$. By Tf invariance of the splitting, T and R map E_x^+ to E_y^+ and E_x^- to E_y^- respectively. So (11.23) with the estimate (11.24) follows from (11.22) just as in the fixed point case provided ℓ is small enough. The same estimates continue to hold if y is merely close to $f(x)$ because for y_1 close to y_2 in K, $h_{y_1} h_{y_2}^{-1}$ differs from a factor preserving linear isomorphism of norm near one by a C^1 small term. For x in $\overline{V}_\ell(K)$ choose $\tilde{x}$ in K ℓ close to x and translate the chart based at $\tilde{x}$ to get one based at x. Again by shrinking ℓ and e if necessary (11.23) and (11.24) hold.

While the gross outline is clear enough there are obviously a number of subtleties which we will omit. Nor have we discussed the converse at all. For (11.25) notice that $h_y(y)$ is $(0, 0)$ in $E_y^+ \times E_y^-$ while for $f(x)$ close to y

$$\begin{aligned}(S^+_{(x,y)}(0, 0), S^-_{(x,y)}(0, 0)) &= (F_{(x,y)}(0, 0), G_{(x,y)}(0, 0)) \\ &= h_y f h_x^{-1}(0, 0) = h_y(f(x)).\end{aligned}$$

Thus, (11.25) is a matter of comparing the metric distance $d(f(x), y)$ with the norm distance between the images under h_y. As we will repeatedly make this kind of comparison, we will state the result we need (see, e.g., the Metric Estimate of Akin (1978) Theorem III 2.2):

27. ADDENDUM. *In Proposition 26 the charts* $\{h_x : U_x \to E\}$ *can be chosen so that the families* $\{h_x\}$ *and* $\{h_x^{-1}\}$ $(x \in V_\ell(K))$ *are uniformly equicontinuous, that is, for every* $\varepsilon > 0$ *there exists an* ε *modulus of uniform continuity* $\delta > 0$ *such that*

$$(11.26)\qquad \begin{aligned} &x_1, x_2 \in U_x \quad \text{and} \quad d(x_1, x_2) < \delta \Rightarrow \|h_x(x_1) - h_x(x_2)\| < \varepsilon, \\ &v_1, v_2 \in h_x(U_x) \quad \text{and} \quad \|v_1 - v_2\| < \delta \Rightarrow d(h_x^{-1}(v_1), h_x^{-1}(v_2)) < \varepsilon. \end{aligned}$$

In particular, $x, y \in V_\ell(K)$ *and* $d(f(x), y) \le \min(e, \delta)$ *imply*

$$(11.27)\qquad \|(S^+_{(x,y)}, S^-_{(x,y)})(0, 0)\| < \varepsilon.$$

In addition, there exists $\delta^* > 0$ *such that* δ^* *is an* ℓ *modulus of uniform continuity and*

$$(11.28)\qquad x \in V_\ell(K) \quad \text{and} \quad d(x, x_1) \le \delta^* \Rightarrow x_1 \in U_x.$$

28. EXERCISE. (a) *Let* K *be the orbit of a periodic point* x *of period* N *for* f. *Prove that* K *is hyperbolic invariant subset for* f *(in which case we call* x *a hyperbolic periodic point) if and only if* x *is a hyperbolic fixed point for* f^n. *(Hint: Start with the* $T_x f^n$ *invariant splitting of* $T_x X$ *and move it around* K *by using* $T_x f$, $T_{f(x)} f$, *etc.)*

(b) *Let* $K_1 \subset K$ *be compact* f *invariant sets. Prove that* K_1 *is hyperbolic if* K *is. In particular, if* K *is a hyperbolic invariant set then any periodic point in* K *is hyperbolic.* □

The transition from an infinitesimal definition to a local description, as provided by Proposition 26, is a crucial technical step in differential topology. Analogous, but much deeper, results are the Inverse Function Theorem and the Frobenius Integrability Theorem identifying certain distributions with foliations. The reader should observe some useful properties hidden in the foliage of details. First, if we replace f by a sufficiently C^1 close diffeomorphism g we can use the same charts, the same linear maps (T, R) and even the same constants ℓ, e, and λ. Only the family of C^1 functions (S^+, S^-) are replaced by a uniformly C^1 close family, and the f invariant set K is replaced by K_g the maximal g invariant subset of $V_{\ell/2}(K)$. Secondly, the local description invites comparison with Conley's Theorem of the previous chapter. In fact, it is by applying that theorem that we make the transition back to what we earlier called topological hyperbolicity.

29. THEOREM (The Shadowing Lemma). *Let K be a compact hyperbolic invariant set for a C^1 diffeomorphism f of a smooth manifold X. There exists a neighborhood U of K, hyperbolicity constants $\gamma_0 > \gamma_1 > 0$, and a neighborhood $\mathscr{U}$ of f in the C^1 topology so that for $g \in \mathscr{U}$, every γ_1-orbit for g in U is γ_0-shadowed by a unique g orbit in X. Furthermore, for every positive $\varepsilon \leq \gamma_0$ there exists a positive $\delta \leq \gamma_1$ so that every δ-orbit for g in U is ε-shadowed by a unique g orbit in X.*

PROOF. As described above, we use Proposition 26 to apply Conley's Hyperbolicity Theorem 10.14. With λ, ℓ, and e as in the proposition the conditions $h_x(U_x \cap g^{-1}(U_y)) \supset \overline{V}_\ell(0)$ and (11.24), (11.25) with (S^+, S^-) defined via g are conditions open under C^1 perturbation. We use these conditions to define $\mathscr{U}$. While we will conduct the proof using f the same proof applies for any such g.

First, we define the constants γ_0 and γ_1 by using choices from Addendum 27:

(a) Choose γ_0 with $0 < \gamma_0 < e$, and so that $x \in U$ and $d(x, y) \leq \gamma_0 \Rightarrow y \in U_x$ and with $w = h_x(y)$

$$\|w\| = \|h_x(y) - h_x(x)\| \leq \ell.$$

(b) Choose ℓ_0 with $0 < \ell_0 < \ell$ and so that $w \in E_x$ and $\|w\| \leq \ell_0 \Rightarrow$ with $y = h_x^{-1}(w)$ (11.29)

$$d(x, y) = d(h_x^{-1}(0), h_x^{-1}(w)) \leq \gamma_0.$$

(c) Choose γ_1 with $0 < \gamma_1 < \gamma_0$ and so that $x \in U$ and $d(f(x), y) \leq \gamma_1 \Rightarrow \|(S^+_{(x,y)}, S^-_{(x,y)})(0, 0)\| \leq \ell_0(1 - \lambda)$.

Now if $\{x_n\}$ is an e orbit in U then we use the charts based at $\{x_n\}$ to replace the sequence of pieces of f (mapping points near x_n to those near x_{n+1}) by a sequence of maps on origin neighborhoods of different B spaces. Thus, we let $E_n = E_{x_n}$ and $(F_n, G_n) = (F_{(x_n, x_{n+1})}, G_{(x_n, x_{n+1})}) = h_{x_{n+1}} \circ f \circ h_{x_n}^{-1}$. Conditions (10.19) and (10.20) (with ε replaced by ℓ) follow from (11.24) and (11.25). If $w_n \in E_n$ and $w_{n+1} \in E_{n+1}$ with $\|w_n\|, \|w_{n+1}\| < \ell$, then we can define $y_n = h_{x_n}^{-1}(w_n)$ and $y_{n+1} = h_{x_{n+1}}^{-1}(w_{n+1})$. Notice that

$$f(y_n) = h_{x_{n+1}}^{-1}(h_{x_{n+1}} f h_{x_n}^{-1})(w_n) = h_{x_{n+1}}^{-1}((F_n, G_n)(w_n)).$$

So we obtain that

$$y_{n+1} = f(y_n) \Leftrightarrow w_{n+1} = (F_n, G_n)(w_n). \tag{11.30}$$

Conversely, if $\{y_n\}$ is a sequence in X which γ_0-shadows $\{x_n\}$ then by (11.29)(a), $y_n \in U_{x_n}$ and $w_n = h_{x_n}(y_n)$ satisfies $\|w_n\| \leq \ell$. So we can apply (11.30).

We now assume that $\{x_n\}$ is a γ_1 orbit in U. Conclusion (a) of Conley's Theorem says that there is a unique orbit sequence $\{w_n\}$ with $\|w_n^*\| \leq \ell$. Furthermore, by (11.29)(c) we have $\|w_n^*\| \leq \ell_0$ and so by (11.29)(b) $y_n^* = h_{x_n}^{-1}(w_n^*)$ satisfies $d(x_n, y_n^*) \leq \gamma_0$. By (11.30) $\{y_n^*\}$ is an orbit γ_0-shadowing $\{x_n\}$. Conversely, if $\{y_n\}$ is an orbit γ_0-shadowing $\{x_n\}$, then by (11.30) and (11.29)(a) $\{w_n = h_{x_n}(y_n)\}$ is an orbit sequence with $\|w_n\| \leq \ell$. By uniqueness in Conley's Theorem $w_n = w_n^*$ and so $y_n = y_n^*$. Thus, $\{x_n\}$ is γ_0-shadowed by a unique orbit in X.

If a positive $\varepsilon \leq \gamma_0$ is chosen, then in (11.29)(b) we define ℓ_ε (rather then ℓ_0) using ε instead of γ_0. Then in (11.29)(c) we define δ (rather than γ_1) using ℓ_ε instead of ℓ_0. By the proof above any δ orbit in U is ε followed by a unique orbit in X. □

30. Corollary. *Let K be a closed hyperbolic invariant set for a C^1 diffeomorphism f of a compact smooth manifold X. K is a topologically hyperbolic invariant subset for the homeomorphism f. There exists a closed neighborhood U of K and a neighborhood $\mathcal{U}$ of f in the C^1 topology so that for every g in $\mathcal{U}$ the maximal invariant set $K_g = \bigcap_{n=-\infty}^{+\infty}\{g^n(U)\}$ is a hyperbolic invariant set for the diffeomorphism g. If, furthermore, K is an isolated f invariant subset with isolating neighborhood U, then for every $\varepsilon > 0$ the neighborhood $\mathcal{U}_\varepsilon$ can be chosen so that for $g \in \mathcal{U}_\varepsilon$ K_g is an isolated g invariant subset with isolating neighborhood U and there is a unique map $k_g : K_g \to K$ with $f \circ k_g = k_g \circ g$ and $d(x, k_g(x)) \leq \varepsilon$ for all x in K_g. The map k_g is a homeomorphism providing a conjugacy between f_K and g_{K_g}. Furthermore, $\delta(K, K_g) \leq \varepsilon$ (Hausdorff metric δ on closed subsets, cf. Chapter 7).*

Proof. By uniqueness in the Shadowing Lemma, K is an expansive subset of X and by Proposition 10 it satisfies the Shadowing Property. As mentioned before Theorem 29, the conditions of Proposition 26 remain true when f is varied by a C^1 perturbation. (By the Inverse Function Theorem and the homotopy argument alluded to after Theorem 23, a map g sufficiently close in the C^1 sense to a diffeomorphism of a compact manifold is still a diffeomorphism.) Consequently, for g in some neighborhood $\mathcal{U}$, each K_g is a hyperbolic g invariant subset. If U is an isolating neighborhood for K with respect to f then by Exercise 18(a) we can choose $\mathcal{U}$ so that U is an isolating neighborhood for K_g. The existence and uniqueness of k_g follow from Proposition 16, after replacing γ_0 by ε, and $\mathcal{U}$ by $\mathcal{U}_\varepsilon$. That k_g is a homeomorphism follows from Addendum 17. Then because

$k_g: K_g \to K$ and $k_g^{-1}: K \to K_g$ are onto and $d(x, k_g(x)) \leq \varepsilon$ for $x \in K_g$, it follows that $\delta(K, K_g) \leq \varepsilon$. □

31. THEOREM. *A C^1 diffeomorphism f of a compact smooth manifold X is called an* Axiom A diffeomorphism *if the chain recurrent set $|\mathscr{C} f|$ is a hyperbolic subset. f is then an Axiom* A *homeomorphism and so $|\mathscr{C} f|$ consists of finitely many basic sets on each of which f is topologically transitive. Furthermore, $|\mathscr{C} f|$ is an isolated invariant set in which the set of periodic points $|\mathscr{O} f|$ is dense.*

For $\varepsilon > 0$, there is a C^1 neighborhood $\mathscr{U}_\varepsilon$ of f such that each g in $\mathscr{U}_\varepsilon$ is an Axiom A *diffeomorphism and there is a unique map $k_g: |\mathscr{C} g| \to |\mathscr{C} f|$ such that for all $x \in |\mathscr{C} g|$*

$$f(k_g(x)) = k_g(g(x)), \qquad d(x, k_g(x)) \leq \varepsilon.$$

Furthermore, $k_g: |\mathscr{C} g| \to |\mathscr{C} f|$ is a homeomorphism and so defines a conjugacy of the restrictions.

PROOF. The results follow from Corollary 30 and Theorem 19. We need only note that if U is an isolating neighborhood for $|\mathscr{C} f|$ we can choose $\mathscr{U}_\varepsilon$ so that $|\mathscr{C} g| \subset U$ and so $|\mathscr{C} g| \subset K_g$. But by Corollary 30, $k_g: K_g \to |\mathscr{C} f|$ defines a conjugacy of the restriction. Hence, g on K_g consists of a finite number of topologically transitive pieces. Thus,

$$K_g = |\mathscr{C}(g_{K_g})| \subset |\mathscr{C} g| \subset K_g$$

implies $K_g = |\mathscr{C} g|$.

REMARK. The analogues of Proposition 21 and Exercise 22(b) are true for diffeomorphisms with hyperbolic replacing topologically hyperbolic. These yield equivalent descriptions of Axiom A diffeomorphisms. □

32. COROLLARY. *Let f be a C^1 diffeomorphism of a compact smooth manifold X. Call f a* simple diffeomorphism *if the following equivalent conditions hold*:

(a) *The chain recurrent set is hyperbolic and consists entirely of periodic points (i.e., f satisfies Axiom* A *and $|\mathscr{C} f| = |\mathscr{O} f|$).*

(b) *The chain recurrent set is finite and every periodic point is hyperbolic.*

If f is a simple diffeomorphism then there is a neighborhood $\mathscr{U}$ of f in the C^1 topology consisting of simple diffeomorphisms.

PROOF. The equivalence of (a) and (b) follows from Proposition 20 and Exercise 28(b). If f is simple and g is Axiom A with g on $|\mathscr{C} g|$ conjugate to f on $|\mathscr{C} f|$, then $|\mathscr{C} g|$ is finite and so g is simple. The existence of the neighborhood $\mathscr{U}$ then follows from Theorem 31. □

For simple diffeomorphisms we can do better than the bijection of finite sets $k_g: |\mathscr{C} g| \to |\mathscr{C} f|$. By applying Hartman's Theorem (Theorem 24) to each periodic orbit separately we can extend k_g to a local conjugacy on neighborhoods of the chain recurrent sets.

At the opposite extreme from simple diffeomorphisms are Anosov diffeomorphisms.

33. PROPOSITION. *Let f be a C^1 diffeomorphism of a compact smooth manifold X. f is called an* Anosov diffeomorphism *if the entire space X is a hyperbolic invariant set. If f is an Anosov diffeomorphism there exists a neighborhood $\mathcal{U}$ of f in the C^1 topology so that each g in $\mathcal{U}$ is an Anosov diffeomorphism and there exists a homeomorphism $k_g: X \to X$ providing a conjugacy between g and f.*

PROOF. If $K = X$ then K is isolated. Corollary 30 applies and $U = K_g = X$ as well. □

Simple diffeomorphisms are happily rather common. If L is an C^2 real valued function on a compact manifold X, then after a slight perturbation it can be assumed that L has only finitely many critical points, i.e., points at which the derivative $d_x L: T_x X \to \mathbb{R}$ is zero. If ξ is the gradient vector field of L with respect to a Riemannian metric on X, then L is a vector field Lyapunov function for ξ and $|\mathscr{C} f|$ is the finite set $|L|$, where f is the time-one map of the flow for ξ. By a further perturbation, if necessary, we can assume that L is a *Morse function* which means that: (a) L distinguishes the points of $|L|$ (i.e., L is injective on $|L|$), and (b) each point of $|L|$ is a hyperbolic fixed point for f. The construction of such Morse functions, the introductory example for transversality theory, yields a rich supply of simple diffeomorphisms.

Anosov diffeomorphisms on the other hand, are rather hard to imagine. Thom's beautiful torus map examples came as something of a shock.

Begin with $\tilde{f}$ an $n \times n$ matrix with integer entries and determinant $= \pm 1$, i.e., $\tilde{f} \in \mathrm{GL}(n, \mathbb{Z})$. The associated linear map, also denoted $\tilde{f}$, on $\mathbb{R}^n$ preserves the lattice $\mathbb{Z}^n$ of points with integer coordinates, as does its inverse. So $\tilde{f}$ induces a diffeomorphism f on the n dimensional torus, the quotient of $\mathbb{R}^n$ by the discrete subgroup $\mathbb{Z}^n$: $X = \mathbb{R}^n/\mathbb{Z}^n$. The projection map $\Pi: \mathbb{R}^n \to X$ and the natural identification of $T_v\mathbb{R}^n$ with $\mathbb{R}^n$ for each $v \in \mathbb{R}^n$, together allow us to regard $\mathbb{R}^n$ as the tangent space $T_x X$ for any $x \in X$. Because $\tilde{f}$ is linear the tangent map of f satisfies

$$T_x f = \tilde{f}: \mathbb{R}^n \to \mathbb{R}^n \quad \text{for all } x \in X. \tag{11.31}$$

Now assume that the linear map $\tilde{f}$ is hyperbolic, i.e., no eigenvalues lie on the unit circle. Since the product of the eigenvalues is the determinant, which has absolute value 1, the eigenvalues are neither all inside nor all outside the unit circle. So the stable and unstable subspaces $E^{\pm}$ for $\tilde{f}$ are proper subspaces for $\mathbb{R}^n$. By the identification of $T_x X$ with $\mathbb{R}^n$ we can regard

$\mathbb{R}^n = E^+ \oplus E^-$ as a splitting of the tangent bundle of X and by (11.31) it is an invariant splitting. So f is an Anosov diffeomorphism.

34. EXERCISE. *If k is any positive integer, show that $(1/k)\mathbb{Z}^n$ is an $\tilde{f}$ invariant subset of $\mathbb{R}^n$ and $\Pi((1/k)\mathbb{Z}^n)$ is a finite subset of X for each k. Conclude that the points of $\mathbb{R}^n$ with rational coordinates map under Π to periodic points for f. If m is a positive integer and $z \in \mathbb{Z}^n$ prove that the equation $\tilde{f}^m(v) = v + z$ has a unique solution v in $\mathbb{R}^n$ and that the coordinates of v are rational. (Invert the matrix $1 - f^m$.) Conclude that $|\mathscr{O}f|$ is the image under Π of Q^n, the points of $\mathbb{R}^n$ with rational coordinates.*
□

Now we will restrict to the case $n = 2$ and a particular example,

$$f_0 = \begin{pmatrix} 1 & 2 \\ 1 & 1 \end{pmatrix}. \tag{11.32}$$

The image of the origin in $\mathbb{R}^2$ is a fixed point $e \in X$ for f, and the stable and unstable manifolds for e, $W^{\pm}(e)$ are the images, under Π, of the subspaces $E^{\pm}$. Each of these subspaces is a line in $\mathbb{R}^2$ with irrational slope and so the images in X are each dense in X. They are, in fact, typical examples of an orbit of the irrational flow on the torus described in Chapter 9. Thus, the projection $\Pi\colon E^{\pm} \to W^{\pm}(e)$ are examples of continuous, injective immersions which are not homeomorphisms onto their image. As an isolated invariant set, e is the only point whose entire f orbit remains close to e, but there is a countable infinity of points in $W^+(e) \cap W^-(e)$, the so-called homoclinic points associated with e. In fact, for each $z \in \mathbb{Z}^2$ there is a unique point where the lines E^+ and $z + E^-$ intersect. The images of all of these points forms the intersection $W^+(e) \cap W^-(e)$.

35. EXERCISE. (a) *With $p_1 \neq e$ in $W^+(e) \cap W^-(e)$, let K_1 be the closure of the orbit of p_1, i.e., $K_1 = \{e\} \cup \{\dots f^{-1}(p_1), p, f(p_1), \dots\}$. Show that K_1 is a hyperbolic invariant set for f with $|\mathscr{C}(f_{K_1})| = K_1$ but $|\mathscr{O}(f_{K_1})| = \{e\}$. Show directly that K_1 is not isolated* rel Per f.

(b) *Compute e_1, the second fixed point for f (with f_0 given by* (11.32)). *With $p_2 \in W^+(e) \cap W^-(e_1)$ let K_2 be the closure of the orbit of p_2, i.e., $K_2 = \{e_1, e\} \cup \{\dots, f^{-1}(p_2), p_2, f(p_2), \dots\}$. Show that K_2 is a hyperbolic invariant set for f with $|\mathscr{C}(f_{K_2})| = |\mathscr{O}(f_{K_2})| = \{e, e_1\}$. Prove that K_2 is isolated.* □

Supplementary exercises

36. Let x be a fixed point for a C^1 diffeomorphism f on X; let T be a linear isomorphism of a B space E and assume $k\colon U \to E$ is a homeomorphism of a neighborhood of x onto a neighborhood of 0 providing a local conjugacy of f with T, i.e., $T \circ k = k \circ f$ on $U_0 = U \cap f^{-1}(U)$.

If k is differentiable with $T_x k$ an isomorphism prove that $T_x k$ provides a similarity (i.e., a linear conjugacy) between the linear maps $T_x f$ and T. In particular, $T_x f$ and T have the same spectrum. This shows that the homeomorphism k_g in Hartman's Theorem usually cannot be taken to be C^1. Even when $T = T_x f$ so that the similarity problem does not arise, it may not be possible via a smooth change of coordinates to replace f near x by its linearization. For Sternberg's Linearization Theorem describing when such smooth coordinates exist, see Nelson (1969) §3.

37. Let X be the space of bi-infinite sequences on some finite set A, $X = A^{\pm\infty}$ and $s: X \to X$ be the shift homeomorphism defined by (9.19). Prove that s is an Anosov homeomorphism. A finite sequence $a_0, a_1, \dots, a_{n-1}$ in A^n is called a *word* (of length n) in the alphabet A. A subset K of X is called a *subshift of finite type* provided there is a subset W of A^n so that

$$x \in K \Leftrightarrow x_i x_{i+1} \cdots x_{i+n-1} \in W \quad \text{for all } i \in Z.$$

Prove that a subshift of finite type is a closed s-invariant subset of X. Prove that a closed s-invariant subset is an isolated invariant set if and only if it is a subshift of finite type. For example, with $A = \{0, 1\}$ let $W = \{00, 01, 11\}$. Prove that the associated subshift consists of the orbit of x^* with $x_i^* = 0$ $(i < 0)$ and $= 1$ $(i \geq 0)$ together with the two fixed points $\cdots 000 \cdots$ and $\cdots 111 \cdots$ (While this is a purely topological example, it can be embedded as the hyperbolic subset of a diffeomorphism of the sphere. This is Smale's "horseshoe" example, see Shub (1987) Chapter 4).

38. Prove the assertions in the following discussion of the stable/unstable foliations for an Anosov homeomorphism.

(a) Let K be a closed, expansive, isolated, invariant subset for a homeomorphism f of a compact metric space X. Assume $\gamma > 0$ is an expansivity constant with $\overline{V}_\gamma(K)$ an isolating neighborhood for K. So $\overline{V}_\gamma \cap [\overline{V}_\gamma(K) \times \overline{V}_\gamma(K)]$ is an isolating neighborhood for 1_K in $X \times X$. Now define

$$\begin{aligned} W_\gamma^+ &= \bigcap_{n=0}^{\infty} \{(f \times f)^{-n}(\overline{V}_\gamma \cap [\overline{V}_\gamma(K) \times \overline{V}_\gamma(K)])\} \\ &= \{(x_1, x_2): d(f^n(x_1), K) \leq \gamma,\ d(f^n(x_2), K) \leq \gamma, \\ &\qquad \text{and } d(f^n(x_1), f^n(x_2)) \leq \gamma \text{ for all integers } n \geq 0\}. \end{aligned} \tag{11.33}$$

$$W_\gamma^- = \bigcap_{n=0}^{\infty} \{(f \times f)^n(\overline{V}_\gamma \cap [\overline{V}_\gamma(K) \times \overline{V}_\gamma(K)])\}.$$

$(x_1, x_2) \in W_\gamma^+$ implies $\omega(f \times f)(x_1, x_2) \subset 1_K$ and so $f^n(x_1) \to K$, $f^n(x_2) \to K$, and $d(f^n(x_1), f^n(x_2)) \to 0$.

W_γ^+ is $f\times f+$ invariant and for every $\varepsilon > 0$

$$(f\times f)^N(W_\gamma^+) \subset W_\varepsilon^+ \quad \text{for some integer } N>0. \tag{11.34}$$

(Observe that $\bigcap_{N=0}^{\infty}\{(f\times f)^N W_\gamma^+\} = 1_K$ and it suffices to show that $(f\times f)^N(W_\gamma^+) \subset \overline{V}_\varepsilon \cap [\overline{V}_\varepsilon(K)\times \overline{V}_\varepsilon(K)]$ by $f\times f+$ invariance).

Define:

$$\begin{aligned} W^+ &= \bigcup_{N=0}^{\infty}\{(f\times f)^{-N}(W_\gamma^+)\} \\ &= \{(x_1, x_2)\colon \lim_{n\to\infty} d(f^n(x_1), K) = \lim_{n\to\infty} d(f^n(x_2), K) \\ &= \lim_{n\to\infty} d(f^n(x_1), f^n(x_2)) = 0\}. \end{aligned} \tag{11.35}$$

$$W^- = \bigcup_{N=0}^{\infty}\{(f\times f)^N(W_\gamma^-)\}.$$

Observe that

$$\begin{aligned} W^+\circ W^+ &= W^+, \qquad (W^+)^{-1} = W^+, \\ W^-\circ W^- &= W^-, \qquad (W^-)^{-1} = W^-. \end{aligned} \tag{11.36}$$

Now assume in addition that K is topologically hyperbolic. It follows that

$$W^+(K) = \{x\colon \omega f(x)\subset K\}.$$

That is, $\omega f(x)\subset K$ if and only if there exists $x_1\in K$ such that

$$\lim_{n\to\infty} d(f^n(x), f^n(x_1)) = 0.$$

(b) Now let $f\colon X\times X$ be an expansive homeomorphism with expansivity constant $\gamma>0$. With $K=X$ in (a), let $W_N^+ = (f\times f)^{-N}(W_\gamma^+)$, $W_N^- = (f\times f)^N(W_\gamma^-)$ for all integers N. $\{W_N^+\}$ and $\{W_N^-\}$ are increasing bi-infinite sequences of closed, symmetric subsets with intersections 1_X and unions the equivalence relations W^+ and W^- respectively. On $X\times X$ define

$$\begin{aligned} d^+(x_1, x_2) &= \sup_{n=0}^{\infty}\{d(f^n(x_1), f^n(x_2))\}, \\ d^-(x_1, x_2) &= \sup_{n=0}^{\infty}\{d(f^{-n}(x_1), f^{-n}(x_2))\}. \end{aligned} \tag{11.37}$$

These are metrics on the set X with

$$\begin{aligned} W_\varepsilon^+ &= \{(x_1, x_2)\colon d^+(x_1, x_2)\le \varepsilon\}, \\ W_\varepsilon^- &= \{(x_1, x_2)\colon d^-(x_1, x_2)\le \varepsilon\}. \end{aligned} \tag{11.38}$$

Let $X^{\pm}$ denote the set X with the topologies associated with $d^{\pm}$ respectively.

$$(11.39) \qquad \begin{aligned} d^{+} &\geq d, & d^{-} &\geq d, \\ d^{+} &\geq (f \times f)^{*} d^{+}, & d^{-} &\geq (f^{-1} \times f^{-1})^{*} d^{-}. \end{aligned}$$

$f \colon X^{+} \to X^{+}$ is a homeomorphism which is distance nonincreasing. $f \colon X^{-} \to X^{-}$ is a homeomorphism with distance nondecreasing. The "identity" maps $X^{+} \to X$ and $X^{-} \to X$ are continuous. On each $W_{N}^{+}(x)$ the d^{+} and d topologies agree and are compact. (Similarly, for $W_{N}^{-}(x)$.) *Hint*: use (11.34) and uniform continuity of $f \times f, \dots, (f \times f)^{N-1}$.

The equivalence relation W^{+} is open and closed with respect to the d^{+} topology. Hint: $W_{\gamma}^{+} \circ W^{+} \circ W_{\gamma}^{+} = W^{+}$. Consequently the equivalence classes $W^{+}(x)$ are open and closed, σ-compact, locally compact subsets of X^{+}. We call X^{+} the *stable foliation* with the W^{+} equivalence classes being the *leaves of the foliation*. Similar results for W^{-} on X^{-} define the *unstable foliation*.

Because $(x, x_1) \in W^{+}$ if and only if $(f(x), f(x_1)) \in W^{+}$ it follows that $f \colon W^{+}(x) \to W^{+}(f(x))$ is a bijection for each x. The bi-infinite sequence of leaves $\{W^{+}(f^{n}(x)) \colon n \in \mathbb{Z}\}$ usually consist of pairwise distinct terms and so the orbit of x is a discrete set in X^{+}. The exception is the following list of equivalent conditions:

$$(11.40) \qquad \begin{array}{ll} (1) & f^{n}(W^{+}(x)) = W^{+}(x), \\ (2) & (x, f^{n}(x)) \in W^{+}, \\ (3) & \omega f(x) \text{ consists of the orbit of a periodic} \\ & \text{point } p \text{ with } f^{n}(p) = p. \\ (4) & W^{+}(x) \text{ contains a (necessarily unique)} \\ & \text{periodic point } p \text{ and } f^{n}(p) = p. \end{array}$$

(Observe (1) and (2) say that $d(f^{n}(x), f^{n+N}(x)) \to 0$ and so $p = \lim f^{n_i}(x)$ implies $f^{n}(p) = p$. A closed invariant subset on which $f^{n} = 1_X$ is necessarily finite by expansivity and a finite set is indecomposable if and only if it is a single orbit. Also, on a finite set W^{+} is just the diagonal and so $(p_1, p_2) \in W^{+}$ with p_1, p_2 periodic implies $p_1 = p_2$.)

By (11.34), 1_X is an attractor for the map $f \times f$ on W_{γ}^{+}. The dual repellor is empty. By using Exercise 3.24 one can construct $v \colon W^{+} \to [0, \infty)$ with $v^{-1}(0) = 1_X$; with $v(f(x_1), f(x_2)) = (1/2)v(x_1, x_2)$, with v continuous on W_{γ}^{+} (and so on each W_{N}^{+}); with $v^{-1}[0, K]$ compact, and so contained in some W_{N}^{+}, for every finite K.

The bundle topology on W^{+} is the weak topology generated by the increasing sequence of compacta $\{W_{N}^{+}\}$. Equivalently, it is the topology so that the map $V: W^+ \to X \times X \times [0, \infty)$ defined by $(x_1, x_2) \to (x_1, x_2, v(x_1, x_2))$ is a homeomorphism. The image of V is closed and so W^+ with the bundle topology is a locally compact, σ-compact metrizable space. The metric d^+ regarded as a map of W^+ to $[0, \infty)$ is continuous when the bundle topology is used on W^+.

We regard W^+ as the total space of a bundle by mapping $\pi_1: W^+ \to X$ with $\pi_1(x_1, x_2) = x_1$. The fiber $\pi_1^{-1}(x) = \{x\} \times W^+(x)$ receives from the bundle topology the topology induced by d^+, i.e., as in $W^+(x) \subset X^+$. On the other hand, the diagonal 1_X $(= v^{-1}(0))$ receives the original topology of X under the identification $x \to (x, x)$. Notice that for each $\varepsilon > 0$ W_ε^+ is a neighborhood of 1_X in the bundle topology.

By applying these results to f^{-1} we obtain the analogues for W^-.

(c) Now let $f: X \to X$ be an Anosov homeomorphism such that 2γ is an expansivity constant for f and $\gamma_0 > \gamma_1 > 0$ are hyperbolicity constants so that, in addition, $\gamma_0 < \gamma/2$ is a γ modulus of uniform continuity for f.

LEMMA. *$d(x_1, x_2) \le \gamma_1$ implies there is a unique point z in $W_\gamma^+(x_2) \cap W_\gamma^-(x_1)$.*

(*Hint*: Let $w_n(x_1, x_2) = f^n(x_1)$ for $n < 0$ and $f^n(x_2)$ for $n \ge 0$. Observe that $\{w_n(x_1, x_2)\}$ is an ε orbit if and only if $d(x_1, x_2) \le \varepsilon$. Observe that $z \in W_\gamma^+(x_2) \cap W_\gamma^-(x_1)$ implies the orbit of z γ-follows $\{w_n(x_1, x_2)\}$ and conversely if the orbit of z γ_0-follows $\{w_n(x_1, x_2)\}$ then $z \in W_\gamma^+(x_2) \cap W_\gamma^-(x_1)$.)

Define the map for $x \in X$ by

$$\overline{V}_{\gamma_1}(x) \to W_\gamma^+(x) \times W_\gamma^-(x), \qquad x_1 \mapsto (z_1, z_2) \tag{11.41}$$

with

$$\{z_1\} = W_\gamma^+(x) \cap W_\gamma^-(x_1), \qquad \{z_2\} = W_\gamma^-(x) \cap W_\gamma^+(x_1).$$

This is a well defined, injective and continuous map whose inverse is defined on a subset by $(\gamma_2 \equiv (1/2)\gamma_1)$

$$\begin{gathered} W_{\gamma_2}^+(x) \times W_{\gamma_2}^-(x) \to \overline{V}_\gamma(x), \qquad (z_1, z_2) \mapsto x_1, \\ \{x_1\} = W_\gamma^-(z_1) \cap W_\gamma^+(z_2). \end{gathered} \tag{11.42}$$

These maps describe the so-called "local product structure" for an isolated topological hyperbolic subset. Observe that the results imply $\overline{V}_{\gamma_1} \subset W_\gamma^+ \circ W_\gamma^-$. Also, we can regard $\overline{V}_{\gamma_1} \to X$ as the pullback of $W_\gamma^+ \to X$ and $W_\gamma^- \to X$ (projection on the first coordinate in all three cases).

39. (a) Let $\{f_i: X_i \to X_i\}$ be a sequence of homeomorphisms. If each of the f_i's satisfies the Shadowing Property on its domain, then so does the

product homeomorphism $f \equiv \prod_i f_i \colon \prod_i X_i \to \prod_i X_i$. If each of the f_i's is expansive then the product f is expansive when the sequence is finite but never when the sequence is infinite. (In the latter case we exclude the trivial situation where all but finitely many of the factors are singletons). In particular, the identity map on the Cantor set of sequences satisfies the Shadowing Property, but is not expansive.

(b) Assume a homeomorphism $f\colon X \to X$ satisfies the Shadowing Property on its domain. Prove that f is topologically transitive (or topologically mixing) if it is chain transitive (resp. chain mixing). (*Hint*: compare Exercise 9.)

40. For any compact metric space X prove that the shift homeomorphism $s\colon X^{\pm\infty} \to X^{\pm\infty}$ (cf. (9.19)) satisfies the Shadowing Property (use a diagonal argument) but is not expansive unless X is finite. Conclude that every homeomorphism $f\colon X \to X$ can be embedded in one which satisfies the Shadowing Property.

Historical Remarks

The relations of the tower in Chapter 1 are essentially Joseph Auslander's "higher prolongations" (cf. Auslander (1964)), but our view of relations as subsets of $X \times X$, rather than as maps from X to 2^X, makes the dynamic viewpoint easier. These results, as well as those on invariant sets and Lyapunov functions in Chapters 1 and 2, are in large part reformulations of parts of Bhatia and Szego (1970), and the earlier version (1967). The idea of chain recurrence and its relationship with attractors comes from Conley (see Conley (1978)). The equivalences for attractor are an assemblage of folklore. Some results are from Conley (1978) and several key proofs are from Smale (1970).

The concepts in Chapter 4 of topological transitivity and minimal set are central to topological dynamics (see, e.g. Auslander (1988)). In Chapter 5 the concept of decomposition is used as a substitute for filtration (see Shub and Smale (1972) and Shub (1987)).

The work in Chapter 7 on topological robustness is an expansion of Takens' ideas on tolerance stability in Takens (1971) and (1975).

The invariant measure results of Chapter 8 are a reworking of results from Nemytskii and Stepanov (1960) with additional results taken from Denker, Grillenberger, and Sigmund (1970).

Conley's time dependent version of the Stable Manifold Theorem in Chapter 10 and its utility for the hyperbolicity results come from unpublished notes by Fenichel (1975) but the proof is a version of Irwin's (see, e.g. Irwin (1980)).

The final chapter presents Smale's Axiom A results from Smale (1967) revised to highlight the purely topological parts. Compare, especially, Shub (1987) and Newhouse (1980).

References

E. Akin, *The metric theory of Banach manifolds*, Lecture Notes in Math., vol. 662, Springer-Verlag, Berlin and New York, 1978.

A. A. Andronov, A. A. Vitt, and S. E. Khaikin, *Theory of Oscillations* (reprint of 1966 translation of 1937 second Russian edition), Dover, New York, 1987.

V. I. Arnold, *Ordinary differential equations*, MIT Press, Cambridge, MA, 1973

J. Auslander, *Generalized recurrence in dynamical systems*, Contributions to Differential Equations, vol. 3, Wiley, New York, (1964), pp. 55–74.

——, *Minimal flows and their extensions*, North-Holland, Amsterdam, 1988.

M. Barnsley, *Fractals everywhere*, Academic Press, San Diego, CA, 1988.

N. P. Bhatia and G. P. Szegő, *Stability theory of dynamical systems*, Springer-Verlag, Berlin and New York, 1970.

——, *Dynamical systems: stability theory and applications*, Lecture Notes in Math., vol. 35, Springer-Verlag, Berlin and New York, 1967.

P. Billingsley, *Ergodic theory and information*, Wiley, New York, 1965.

R. P. Bowen, *Equilibrium states and the ergodic theory of Anosov Diffeomorphisms*, Lecture Notes in Math., vol. 470, Springer-Verlag, Berlin and New York, 1975.

C. Conley, *Some abstract properties of the set of invariant sets of a flow*, Illinois J. Math. **16** (1972), 663–668.

——, *Isolated invariant sets and the Morse index*, CBMS Regional Conf. Ser. in Math., vol. 38, Amer. Math. Soc., Providence, RI, 1978.

M. Denker, C. Grillenberger, and K. Sigmund, *Ergodic theory on compact spaces*, Lecture Notes in Math., vol. 527, Springer-Verlag, Berlin and New York, 1970.

N. Fenichel, *Hyperbolicity conditions for dynamical systems*, preprint, 1975.

J. E. Franke and J. F. Selgrade, *Hyperbolicity and chain recurrence*, J. Differential Equations **26** (1977), 27–36.

H. Furstenberg, *Disjointness in Ergodic theory, minimal sets and a problem in Diophantine approximation*, Math. Systems Theory **1** (1967), 1–49.

——, *Recurrence in ergodic theory and combinatorial number theory*, Princeton Univ. Press, Princeton, NJ, 1981.

P. Halmos, *Measure theory*, Van Nostrand, Princeton, NJ, 1950.

P. Hartman, *Ordinary differential equations*, Wiley, New York, 1964.

M. Hirsch and C. C. Pugh, *Stable Manifolds and Hyperbolic Sets*, Global Analysis (Shing-Shen Chern and Stephen Smale, eds.), Proc. Sympos. Pure Math. vol. XIV, Amer. Math. Soc., Providence, RI, 1970, 133–163.

M. Hirsch and S. Smale, *Differential equations dynamical systems, and linear algebra*, Academic Press, San Diego, CA, 1974.

R. B. Holmes, *A Formula for the Spectral Radius of an Operator*, MAA Monthly **75** (1968), 163–166.

P. Huber, *Robust statistics*, Wiley, New York, 1981.

J. Hutchinson, *Fractals and Self-Similarity*, Indiana Univ. Math. J. **30** (1981), 713–747.

M. C. Irwin, *Smooth dynamical systems*, Academic Press, San Diego, CA, 1980.

J. Kelley, *General topology*, Van Nostrand, Princeton, NJ, 1955.

J. C. Lagarias, *The* $3X + 1$ *Problem and Its Generalizations*, MAA Monthly, **92** (1985), 3–21.

S. Lang, *Differential manifolds*, Addison-Wesley, Reading, MA, 1972.

R. McGehee, *Attractors for Closed Relations on Compact Hausdorff Spaces*, Indiana Univ. Math. J. (to appear).

R. Mañé, *Ergodic theory and differentiable dynamics*, Springer-Verlag, Berlin and New York, 1983.

E. Nelson, *Topics in Dynamics*: I. *Flows*, Princeton Univ. Press, Princeton, NJ, 1969.

V. V. Nemytskii and V. V. Stepanov, *Qualitative theory of differential equations*, Princeton Univ. Press. Princeton, NJ, 1960.

S. Newhouse, *Lectures on dynamical systems*, Dynamical Systems. C.I.M.E. Lectures, Bressanone, Italy, June, 1978, Birkhauser, Boston, 1980, 1–115.

Z. Nitecki, *Differentiable dynamics*, MIT Press, Cambridge, MA, 1971.

Z. Nitecki and M. Shub, *Filtrations, decompositions and explosions*, Amer. J. Math. **107** (1975), 1029.

J. Palis, C. Pugh, M. Shub, and D. Sullivan, *Genericity theorems in topological dynamics*, Dynamical Systems-Warwick, 1974, Lecture Notes in Math., vol. 468, Springer-Verlag, Berlin and New York, 1975, 234–240.

M. Shub and S. Smale, *Beyond hyperbolicity*, Ann. of Math. **96** (1972), 587–591.

M. Shub, *Global stability of dynamical systems*, Springer-Verlag, Berlin and New York, 1987.

S. Smale, *Differentiable dynamical systems*, Bull. Amer. Math. Soc. **73** (1967), 747–817.

——, *The Ω-stability theorem*, Global Analysis, Shing-Shen Chern and Stephen Smale, eds.), Proc. Sympos. Pure Math, vol. XIV, Amer. Math. Soc., Providence, RI, 1970, 289–297.

——, *The mathematics of time*, Springer-Verlag, Berlin and New York, 1980.

F. Takens, *On Zeeman's tolerance stability conjecture*, Manifolds-Amsterdam 1970, Lecture Notes in Math., vol. 197, Springer-Verlag, Berlin and New York, 1971, 209–219.

——, *Tolerance Stability*, Dynamical systems-Warwick 1974, Lecture Notes in Math. vol. 468, Springer-Verlag, Berlin and New York, 1975, 293–304.

K. Yoshida, *Functional analysis*, Springer-Verlag, Berlin and New York, 1965.

Subject Index

ISBN 0-8218-3800-8

9 780821 838006